INNOVATION IN DESIGN, COMMUNICATION AND ENGINEERING

Smart Science, Design and Technology

ISSN: 2640-5504
eISSN: 2640-5512

Book series editors

Stephen D. Prior
Faculty of Engineering and Physical Sciences, University of Southampton, Southampton, UK

Siu-Tsen Shen
Department of Multimedia Design, National Formosa University, Taiwan, R.O.C.

Volume 3

PROCEEDINGS OF THE 8TH ASIAN CONFERENCE ON INNOVATION, COMMUNICATION AND ENGINEERING (ACICE 2019), OCTOBER 25–30, 2019, ZHENGZHOU, CHINA

Innovation in Design, Communication and Engineering

Editors

Artde Donald Kin-Tak Lam
Fujian University of Technology, P.R. China

Stephen D. Prior
University of Southampton, Southampton, UK

Siu-Tsen Shen
National Formosa University, Huwei District, Taiwan

Sheng-Joue Young
National Formosa University, Huwei District, Taiwan

Liang-Wen Ji
National Formosa University, Huwei District, Taiwan

Routledge
Taylor & Francis Group

LONDON AND NEW YORK

Routledge is an imprint of the Taylor & Francis Group, an informa business

© 2020 Taylor & Francis Group, London, UK

Typeset by Integra Software Services Pvt. Ltd., Pondicherry, India

Library of Congress Cataloging-in-Publication Data

Applied for

Published by: CRC Press/Balkema
　　　　　　Schipholweg 107C, 2316XC Leiden, The Netherlands
　　　　　　e-mail: Pub.NL@taylorandfrancis.com
　　　　　　www.routledge.com – www.taylorandfrancis.com

ISBN: 978-0-367-17777-5 (Hbk)
ISBN: 978-0-429-05766-3 (eBook)
DOI: 10.1201/9780429057663
https://doi.org/10.1201/9780429057663

Innovation in Design, Communication and Engineering – Lam et al. (eds)
© 2020 Taylor & Francis Group, London, ISBN 978-0-367-17777-5

Table of contents

Green technology & Architecture engineering

Xuanzang theory

Preface

We have great pleasure in presenting this conference proceeding for technology applications in engineering science and mechanics, consisting of the selected articles of the Asian Conference on Innovation Communication and Engineering (ACICE 2019), organized by Taiwanese Institute of Knowledge Innovation, from the 25th to the 30th of October 2019, in Zhengzhou, China.

The ACICE 2019 conference was a forum that brought together Asian users, manufacturers, designers, and researchers involved in the structures or structural components manufactured using smart science. The Asian forum provided an opportunity for exchange of the researches and insights from scientists and scholars thereby promoting research, development and use of computational science and materials. The conference theme for ACICE 2019 was "Computational Science & Engineering" and tried to explore the important role of innovation in the development of the technology applications including articles dealing with design, research, and development studies, experimental investigations, theoretical analysis and fabrication techniques relevant to the application of technology in various assemblies, ranging from individual to components to complete structure were presented at the conference. The major themes on technology included Material Science & Engineering, Communication Science & Engineering, Computer Science & Engineering, Electrical & Electronic Engineering, Mechanical & Automation Engineering, Architecture Engineering, IOT Technology, and Innovation Design. About 150 participants representing 8 countries came together for the 2019 conference and made ACICE 2019 a highly successful event. We would like to thank all those who directly or indirectly contributed to the organization of the conference.

The articles presented at the ACICE 2019 conference has been brought out as a special issue in various journals. In this conference proceeding we have some selected articles from various themes. A committee consisting of experts from leading academic institutions, laboratories, and industrial research centres was formed to shortlist and review the articles. The articles in this conference proceeding have been peer reviewed to the standards. We are extremely happy to bring out this conference proceeding and dedicate it to all those who have made their best efforts to contribute to this publication.

Professor Siu-Tsen Shen & Dr. Stephen D. Prior

Editorial board

Hao-Ying Lu, Associate Professor
Department of Electronic Engineering, National Quemoy University, Taiwan, R.O.C.
hylu@nqu.edu.tw
ORCID: 0000-0002-1574-7751

Material science & Engineering

Innovation in Design, Communication and Engineering – Lam et al. (eds)
© 2020 Taylor & Francis Group, London, ISBN 978-0-367-17777-5

An experimental study on wear properties of the drive elements with deep cryogenic treatment under the extreme contact pressure

Y.P. Chang*
Department of Green Energy Technology Research Center, Kun Shan University, Tainan, Taiwan

L.M. Chu
Interdisciplinary Program of Green and Information Technology, National Taitung University, Taiwan

C.T. Liu & H.Y. Wang
Department of Green Energy Technology Research Center, Kun Shan University, Tainan, Taiwan

ABSTRACT: The wear of the transmission components is the main cause of mechanical damage. There will be deformation and loss between the ball and the track after a long time under extreme pressure conditions. Moreover, the temperature rise after the friction action will have a great impact on the wear level. These will cause work efficiency and operation life reduction. Therefore, it is very important to establish the key technology of wear resistant for the drive elements under high temperature and extreme contact pressures. To decrease the wear, it is necessary to improve the surface engineering. Moreover, heat treatment has been used widely for the purpose of wear resistance and low friction in the industry. Hence, it is necessary to further establish the key technology of wear resistance under high temperature and extreme pressure conditions by the novel heat treatments. Based on the above statements, the effects of special induction hardening with deep cryogenic treatment on the drive elements for the serious conditions will be investigated in this paper. The proper parameters for wear resistance under high temperature and extreme contact pressures could be developed. These results will be very beneficial to the development of the machine tool and vehicle industry.

Keywords: wear, deep cryogenic treatment, Extreme contact pressure, drive elements

1   INTRODUCTION

Due to the need to replace traditional high-energy-consuming hydraulic equipment, the demand for special high-load transmission components has increased greatly. Under such extreme pressure conditions, the conventional design alone will not be able to cope with the actual situation, so it must be applied from extreme pressure. The next performance improvement study begins. Although it is unclear to obtain extreme pressure conditions and the best performance of materials at relatively high temperatures, the application of special induction hardening is extremely high, and the expected mechanical properties can be obtained by changing the material structure (Nasreldin et al. 1993).

After the base material is subjected to basic heat treatment such as quenching and tempering, substantially the overall material properties will be expected to be improved, and then the most important surface hardening engineering must be carried out (Guodong et al. 2008, Thelning, 1975, Clayton et al. 1992, Bennighoff et al. 1945). Since the carburizing process must have sufficient time for the carbon atoms to undergo sufficient thermal diffusion, it is

* Corresponding Author: ypc0318@mail.ksu.edu.tw

relatively time consuming and consumes power. However, induction hardening does not have such problems. Since the induced current is more likely to concentrate on the surface of the workpiece, the surface is rapidly heated and then quenched to achieve surface hardening. The depth of the heating layer varies depending on the frequency of the cycle, generally 0.5 to 8 mm. Induction hardening has the following advantages [5]: (a) the surface of the product can be hardened locally; (b) energy saving; (c) the furnace is not required to protect the gas; (d) continuous operation is not required.

Based on the above literature, the wear resistance of the transmission element under extreme pressure conditions is improved. The most effective method is the improvement of material properties and surface engineering, which is a key technology in the future of the industry. In this paper, the influence of cryogenic treatment on the wear of transmission components under extreme pressure conditions is studied experimentally. The goal is to develop wear-resistant transmission components and processing parameters in accordance with extreme pressure conditions.

2 EXPERIMENTS

2.1 *Experimental apparatus*

The reciprocating friction tester, heating controller and various measurement systems are used to conduct the experiments. The whole set of experimental machine and measurement equipment are shown in the Figure 1 and 2.

The motion of the machine experiment is the crank slider mechanism, the upper test piece (SUJ2 ball) and the lower test piece are respectively fixed by the fixture, and the fixtures are also fixed on the upper and lower V-shaped slides respectively and insulated by insulating spacers. The upper V-shaped sliding seat is connected to the crank for linear reciprocating motion, the moving stroke can be adjusted by the length of the crank, and the lower V-shaped sliding seat is connected with the load meter. The heating piece for heating is designed in the lower test piece fixture, wherein the lower test piece is equipped with a thermocouple and is connected to the heating controller to monitor the heating temperature of the lower test piece throughout the process.

The application of the load uses a lever mechanism to press the lower test piece upward to the upper test piece, and a soft spring and oil damping are attached to the weight hanging end to avoid vibration during the friction process.

Figure 1.   The fretting wear tester with the measuring system.

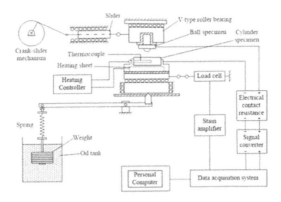

Figure 2.   Schematic diagram of the fretting wear tester.

2.2   Test specimens

The upper piece is made of high carbon chromium alloy steel ball (Φ 6.35mm). The lower piece is made of high carbon chromium alloy steel cylindrical test piece, and the composition is shown in Table 1.

The lower piece uses the different induction heat treatment process that shown in Table 2. After the heat treatment is completed, it is processed into a polygonal column shape conforming to the shape of the heating fixture, and a schematic view of the experimental test piece is shown in Figure 3.

Table 1.   Contents of the high-carbon chromium alloy steel.

C	Cr	Mn	Si	Mo	P	S
0.95~1.10	1.3~1.6	<0.5	0.15~0.35	≤0.08	<0.025	<0.025

Table 2.   Heat treatment process for the lower specimens.

Code	Heat treatment process
B4	Original material → Quenching → tempering → Induction Heat Treatment (normal power)
B6	Original material → Quenching → tempering → Induction Heat Treatment (normal power + Cryogenic treatment)

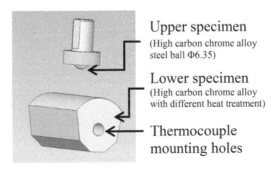

Upper specimen
(High carbon chrome alloy
steel ball Φ6.35)

Lower specimen
(High carbon chrome alloy
with different heat treatment)

Thermocouple
mounting holes

Figure 3.   The size and shape of the test specimens.

2.3 Experimental conditions

The experimental parameters are shown in Table 3. The response time of the overall measurement system is less than 1 ms, and the measurement accuracy is 0.1% of the entire measurement scale.

Before the experiment, the surface of the upper and lower test pieces is cleaned, and the stroke and sliding rate of the reciprocating friction tester are adjusted to the parameter settings required for the experiment; then the upper and lower test pieces are locked and the lower test piece is heated. After the temperature that measured in the lower specimen is reach to the set temperature, the test machine and the program are turned on, and the friction coefficient and the contact resistance are measured and captured simultaneously.

3 RESULTS AND DISCUSSIONS

3.1 Dynamic response of friction coefficient and electrical contact resistance

Figure 4 is a dynamic response diagram of the friction coefficient and electrical contact resistance of high carbon chromium alloy steel B4 at 25 °C. It can be seen from the figure that the friction coefficient fluctuates between 1.2 and 2.5, and the average friction coefficient after

Table 3. The experimental conditions.

Set temperature at lower specimen (°C)	25, 80, 120
Reciprocating speed (cpm)	400
Stroke (mm)	6
Normal load (N)	100
Experimental time (sec)	120
Wear condition	Dry wear

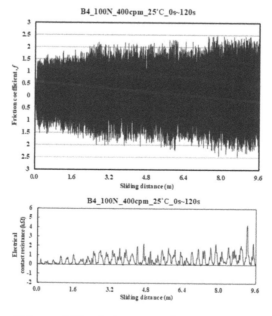

Figure 4. The friction coefficient and Electrical contact resistance of the B4 specimen at set temperature of 25°C.

steady state is about 2.3; the contact resistance fluctuates between 0 and 4.2 kΩ, and it is estimated that the wear particles of the interface are caused by rolling phenomenon. This causes the contact resistance to become large.

It can be seen from Figure 5 that the friction coefficient fluctuates between 0.8 and 2.3, and the average friction coefficient after steady state is about 2.0; the electrical contact resistance fluctuates between 0 and 1.5 kΩ, and it is speculated that the temperature of the lower test piece is softened due to the temperature rise relationship. It is easy to puncture (the true contact area becomes larger - the contact resistance becomes smaller), so that it gradually produces a ploughing phenomenon (reduction in rolling).

Figure 6 shows that the friction coefficient fluctuates between 0.6 and 2.2, and the average friction coefficient after steady state is about 1.7; the contact resistance fluctuates between 0 and 0.9 kΩ, it is speculated that the temperature of the test piece is softened due to the temperature rise relationship. It is easier to puncture (the true contact area becomes larger and the contact resistance becomes smaller), making the ploughing phenomenon more obvious.

Figure 7 shows that the friction coefficient fluctuates between 1.3 and 2.6, and the average friction coefficient after steady state is about 2.2; then the contact resistance fluctuates between 0 and 3.1 kΩ, and it is presumed that the wear particles of the interface are caused by rolling phenomenon. This causes the contact resistance to become large.

It can be seen from Figure 8 that the friction coefficient fluctuates between 1.0 and 2.4, and the average friction coefficient after steady state is about 1.9; the contact resistance fluctuates between 0 and 1.8 kΩ, and it is speculated that the temperature of the lower test piece is softened due to the temperature rise relationship. It is easy to puncture (the true contact area becomes larger - the contact resistance becomes smaller), so that it gradually produces ploughing. (Reduction in rolling).

Figure 9 shows the dynamic response of the friction coefficient and electrical contact resistance of high carbon chromium alloy steel B6 at 120 °C. It can be seen from the figure that the friction coefficient fluctuates between 1.0 and 2.0, and the average friction coefficient after steady state is about 1.5; the contact resistance fluctuates between 0 and 0.7 kΩ, and it is speculated that the temperature of the lower test piece is softened due to the temperature rise relationship. It is easier to puncture (the true contact area becomes larger and the contact resistance becomes smaller), making the ploughing phenomenon more obvious.

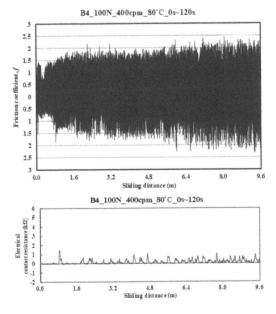

Figure 5.   The friction coefficient and Electrical contact resistance of the B4 specimen at set temperature of 80°C.

7

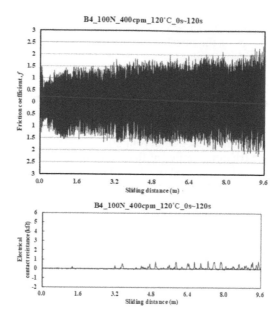

Figure 6. The friction coefficient and Electrical contact resistance of the B4 specimen at set temperature of 120°C.

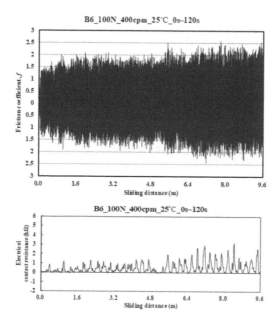

Figure 7. The friction coefficient and Electrical contact resistance of the B6 specimen at set temperature of 25°C.

3.2 *OM observation of wear surface*

Figure 10 shows the optical microscopic observation (X 100) of B4 and B6 after friction test. It clearly shows that the set temperature is increased; the ploughing phenomenon becomes more apparent.

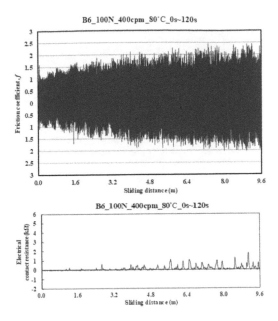

Figure 8. The friction coefficient and Electrical contact resistance of the B6 specimen at set temperature of 80°C.

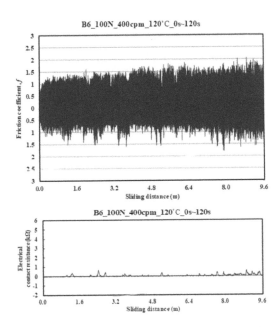

Figure 9. The friction coefficient and Electrical contact resistance of the B6 specimen at set temperature of 120°C.

The combination of the above electrical contact resistance and optical micrographs prove that the contact resistance becomes small, which is due to the temperature rise relationship, which causes the surface peak end to soften and is easier to puncture (the true contact area becomes larger), and thus the more obvious plow Ploughing phenomenon. Therefore, contact resistance and optical micrographs have mutually corresponding results.

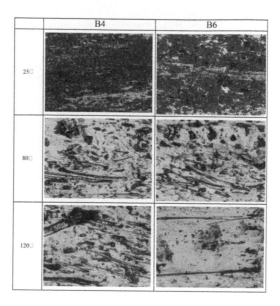

Figure 10.   Optical microscope (X100) of the B4 and B6 at different temperature after friction test.

3.3   *Quantitative analysis of surface wear scars*

Figure 11 shows the surface wear scar analysis of high carbon chromium alloy steels B4 and B6 at different set temperatures. In order to present the accuracy of the experiment, the next test piece was subjected to three reproducibility experiments.

Figure 12 shows the average wear scar depth of high carbon chromium alloy steels B4 and B6 at different temperatures. It can be seen from the figure that as the temperature increases, the average wear scar depth of B4 and B6 will increase and the growth of cracks (the deepest wear marks) will become more serious. Under the same temperature comparison, it is found that the induction heat treatment is deep. The cold B6's anti-wear ability is slightly worse than the B4 which only induces heat treatment.

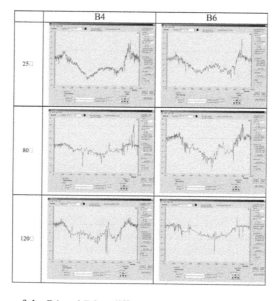

Figure 11.   The wear scar of the B4 and B6 at different set temperature after friction test.

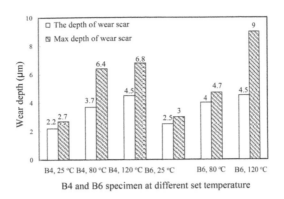

Figure 12. Average wear depth of the specimens of B4 and B6 at different set temperature after friction test.

Based on the above-mentioned spectroscopic resistance, optical micrograph, and wear scar depth, it is proved that the contact resistance becomes small, which is due to the temperature rise relationship, which causes the surface peak end to soften and is easier to puncture (the true contact area becomes larger), and thus The more obvious the ploughing phenomenon, and the more serious the wear and crack growth. Therefore, the contact resistance, the optical micrograph, and the depth of the wear scar have mutually corresponding results.

4 CONCLUSIONS

In this study, the effects of cryogenic treatment on the wear of transmission components under extreme pressure conditions were investigated using a reciprocating friction tester and a measurement system. The results of the comprehensive experiment can be concluded as follows:

(1) The friction coefficient and the electrical contact resistance are decreased with increasing the set temperature of lower specimen.
(2) As the set temperature is lower, the wear particles rolling at the interface causing the contact resistance become larger.
(3) When the set temperature is increased, the soften material at contact interface make the ploughing phenomenon become more obvious. Therefore, the contact resistances become smaller.
(4) The wear resistance is slightly decrease when the specimen with the induction heat treatment and deep cryogenic treatment. It could be the deep cryogenic does not help to improve surface fatigue life.

ACKNOWLEDGEMENTS

The authors would like to express their appreciation to the National Science Council in Taiwan, R. O. C. for their financial support under grant numbers MOST 108-2221-E-168-010 and MOE 107-N-270-EDU- T-142.

REFERENCES

A.M. Nasreldin, M.M. Ghoneim, F.H. Hammad, et al. 1993. *Journal of Material Engineering and Performance, 2(3)*, 413–420.
J. Guodong, & H. Maolin, 2008. *Material & Heat Treatment, 16-0074-02*, 1001–3814.
K.E. Thelning, 1975. *Bofors Handbook*, 246–250.
P. Clayton, & R. Devanathan, 1992. *Wear, 156*, 121–131.
W.E. Bennighoff, & H.B. Osborn, 1945. Trans., *A.S.M. 34*, 310–350.

Oxide whisker growth mechanism of 304 stainless steel coated with hot-dipping pure Al and Si-modified aluminide coating

Te Yuan Chiang*
Baise University, Baise City, China

ABSTRACT: This study applied hot-dipping pure Al and Si-modified Al coating to 304 stainless steel, and subsequently placed the steel in high-temperature corrosive environments with NaCl vapor to study the reaction. Additionally, this study discussed the phases of aluminide coating and growth mechanisms of oxide crystals generated by air and NaCl vapor. Furthermore, it analyzed the microstructure influence of external stress on the Al layer of Al-coated nickel. Subsequently, this study adopted a nickel-plating pretreatment and substrates with silicon content to create an Al layer that prevented oxide whisker growth mechanisms on 304 stainless steel. This study revealed that high temperatures and the NaCl vapor concentration substantially increased the volume diffusion speed of oxides, thereby causing oxide whiskers to rapidly form in accordance with surface dents. Additionally, the experiment discovered that oxide whisker growth was faster and generated longer crystals in a 750°C environment than in an 850°C environment. To prevent the diffusion of oxide whisker growth, the temperature should be maintained beneath 550°C to ensure slow and steady oxide whisker growth as well as chlorination reactions. Finally, oxide whiskers could not grow in 990-vppm NaCl vapor environments.

1 INTRODUCTION

Within numerous types of aluminide-surface processing, hot-dipping aluminide layers are the most suitable for mass-produced workpieces, including steel boards, materials, and pipes, because of the required costs and operation convenience. Studies have proven that hot-dipping Al layers or Si-modified aluminide coating layers can effectively enhance the hot temperature oxidation, vulcanization of steel materials, and heat corrosion property of chlorides. The current industrialized Al hot-dipping process adopts a continuous surface treatment process in which steel materials are dipped into a molten bath comprising pure Al and Al alloy. After dipping at a constant temperature for a certain period of time, the steel material is extracted from the Al bath, thereby completing the surface treatment process. However, long durations of exposure to high temperatures cause Al layers and substrates to constantly diffuse, thereby depleting the Al content within the coating layer. This results in complex phase transformation problems, including excessively thick Al layers and surface dents. With the aim of effectively ameliorating the difficulties faced in aluminizing technology, the current study's experimental stage revealed that the high-temperature oxidation process causes Fe–Al–Si to formulate compounds with other participating elements. Additionally, the researchers conducted structural analysis to enhance the properties of Al hot-dipping layers. Coating steel materials with an aluminide layer enables modification of the chemical, physical, and electrical properties of the alloy surface; furthermore, it enhances the original alloy's oxidation, corrosion, and erosion properties while simultaneously retaining its mechanical properties. Therefore, Al coating is the most

---

*Corresponding author: kelly4921@gmail.com

effective and economical method for protecting alloy substrates and is commonly applied to mechanical equipment that requires anti-erosion, -oxidation, and -abrasion properties.

2 EXPERIMENTAL

2.1 *Al hot-dipping*

This experiment employed a laser-cutting machine to cut 304 stainless steel (304SS) into four $20 \times 8 \times 0.5$-mm test pieces (Figure 1). Before immersion plating, the test pieces were bathed in acetone, alkaline, and acid cleaning solutions. The test pieces were washed and dried with an ultrasonic oscillator to remove surface oil stains. First, the test pieces were coated with a layer of liquid flux to increase the wettability of the pieces and the Al bath. After being dried, the test pieces are soaked in 700°C molten baths comprising 0.5 wt% of Al and pure Al, 5 wt% of Si and Al, 10 wt% of Si and Al, and 2.5 wt% of Si and Al. After the test pieces were extracted from the molten bath, their surfaces were cleansed of foreign substances with a mixture of nitric acid, phosphoric acid, and water (1:1:1). Additionally, the test pieces were washed with substantial amounts of water and dried, thereby completing the preparation process.

The SIGMA D-Sorbitol (98%) doped the Clevios PH 500 PEDOT: PSS was used as solution for preparation of thin films by a spin-coating method. The Sorbitol was added to the PEDOT: PSS directly, and then the doped PEDOT: PSS was stirred for 30 min at room temperature. The mixed solution was doped again by adding different molar concentrations of H_2SO_4. At last, the mixed PEDOT: PSS solution is the so-called the double doped PEDOT: PSS solution. The double doped PEDOT: PSS solution was coated by spinner on 2×2 cm^2 glass substrates and formed the double doped PEDOT: PSS film. The glass substrates were pre-cleaned with acetone, methanol and de-ionized (DI) water in an ultrasonic bath, sequentially. The spin-coating was performed at a rotation rate of 3500 rpm for 20 sec. The double doped PEDOT: PSS film was heated at 150 °C for 20 min on a hotplate in ambient lab conditions.

2.2 *Nickel-plating pretreatment*

Before the test pieces were electroplated, they were subjected to alkaline cleansing and washed with an ultrasonic oscillator. The electroplating solution was a mixture of 45 g/L of NiCl$_2$, 29 g/L of NiSO$_4$, and 45 g/L of H$_3$BO$_3$. The process was conducted at 55°C with the current density set to 0.20 A/cm^2. After the electroplating, the test pieces were cleansed with alcohol and dried, resulting in an approximately 20-μm-thick nickel coating that consisted of 1at% of steel. This is because acid electroplating solutions, pH 4 in this instance, corroded the test pieces, causing the steel dissolve into the electroplating solution. Therefore, the Ni in the solution and dissolved steel simultaneously deposited onto the test piece surfaces.

2.3 *High-temperature erosion from air and NaCl vapor*

In the high-temperature erosion experiment, standard 304SS and hot-dipping aluminide 304SS (304 HDAS) were placed into 750°C and 850°C air and NaCl vapor mixtures (500 and 990 vppm, respectively) for 30 minutes to 100 hours. After erosion, the researchers conducted layer grinding analysis with an X-ray to understand the surface erosion product and in situ phase

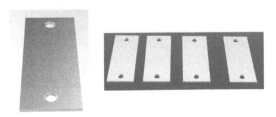

Figure 1.   The test pieces, each measuring $20 \times 8 \times 0.5$ mm.

transformation within the Al layer. Subsequently, they employed an electron probe microanalyzer and a field emission scanning electron microscope with energy dispersive spectroscopy to observe the Al layer's surface topography and cross-section morphology. Through this, they analyzed the composition and element distribution.

2.4 *High-temperature stress experiment*

The 304 HDAS test piece was placed in a 750°C hot-air environment and subjected to various levels of stress. The researchers observed changes in the Al coating of the 304SS under high-temperature stress at different time points. After increasing the temperature in the heating furnace to 750°C and ensuring it had stabilized, the researchers extracted the test pieces at different times. Once the test pieces had cooled to room temperature, the researchers observed the metallography microstructure.

3 RESULTS AND DISCUSSION

This study investigated oxide whisker growth mechanisms in high-temperature erosive environments. During the initial erosion period in 500-vppm NaCl vapor, Al_2O_3 whiskers formed on the surface of the hot-dipped, processed Si-modified aluminide coatings of 304SS. After a long duration of erosion, continuous chlorination and oxidation rendered the Al coating effect useless and generated Fe2O3 crystal flakes that formed an Fe2O3 film. Therefore, Al2O3 whisker growth mechanisms were attributed to the initial stage of high-temperature erosion. During this process,

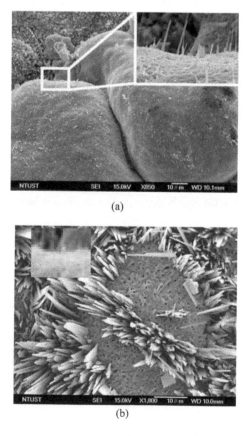

(a)

(b)

Figure 2. Scanning electron microscopy analysis of the morphology and topography of 304 HDAS under high-temperature (750°C) NaCl(g) corrosion: (a) 9 hours, and (b) 72 hours.

halides on the Al coating surface volatilized and oxidized toward the outer sides of the film, thereby causing different Al2O3 oxidization speeds and generating anisotropic Al2O3 whiskers. Figure 2(a) indicates that Fe2O3 whisker growth mechanisms were attributed to the chlorine gas generated by chlorination and oxidization circulation. This circulation replaced oxygen ions located inside the Fe2O3 film with Cl ions, and further created excessive ionized cation vacancies, thereby increasing the ion diffusion speed within the Fe2O3 film and generating lath-shaped Fe2O3 whiskers.

Additionally, the Fe2O3 film grain boundary served as a fast diffusion channel, enabling partial Fe2O3 whiskers to form in circular arrangements at the grain boundary (see 2(b)). Figure 3(a), (b), (c), (d), and (e) display the growth mechanism of oxidized whiskers on the surface coating of 304 HDAS and the weakening of the oxidized coating under high-temperature corrosion. This process was divided as follows:

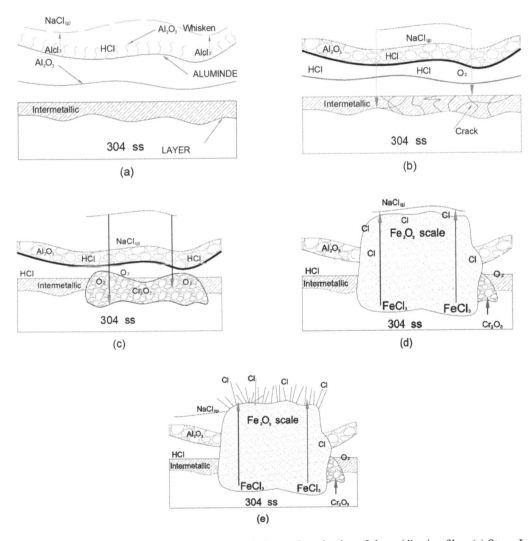

Figure 3. Whisker growth as well as the formulation and weakening of the oxidization film: (a) Stage I: Al2O3 whiskers grew; (b) Stage II: the aluminide layer was rendered useless and Al_2O_3 whiskers vanished; (c) Stage III: Cr2O3 was rendered useless; (d) Stage IV: Fe2O3 stuck out; and (e) Stage V: Fe2O3 whiskers grew.

15

Stage I: Al2O3 whiskers grew; Stage II: the aluminide layer was rendered useless and Al2O3 whiskers vanished; Stage III: chlorination and oxidation corroded the Cr2O3 substrate, rendering it useless; Stage IV: Fe2O3 stuck out from the aluminide coating surface; and Stage V: Fe2O3 whiskers grew. Furthermore, high temperatures and concentrations of NaCl vapor substantially increased in the volume diffusion of oxides, thereby causing oxidized whiskers to grow in accordance with the rapid diffusion. Oxide whiskers grown under 750°C were slenderer and longer than those grown under 850°C. In addition, whiskers were unable to grow under high-concentration (990-vppm) NaCl vapor environments.

4 CONCLUSION

Because a substantial amount of water is required to cool equipment used for high-temperature reactions, high-temperature factories are commonly situated in locations with abundant water resources. Because air circulation causes NaCl to react with factory equipment, the oxidization protection film on the surface of materials is subjected to oxidization and chlorination. Chloride metals exhibit low melting points, low boiling points, and high vapor pressure properties. During corrosion reactions between chlorides and metal alloys, chloride metals are generated, commonly during the liquid phase. This enables a short circuit for diffusion and further accelerates corrosion and oxidation speeds. This study revealed that materials in environments with high temperatures and chlorite concentrations exhibited whisker growth similar to that displayed in Figure 3 (a), (b), (c), (d), and (e). Literature has stated that whisker growth is mainly caused by chloride reaction circulation between air particles and the test piece surface as well as the generation of liquid phase chloride metals. These factors accelerate the transmission speed of matter inside the oxidized film, thereby increasing the oxidation speed and causing oxide whiskers to grow. In environments where NaCl was present, surface coatings deteriorated because of whisker growth and high-temperature corrosion environments. This study revealed that under 750–850°C environments with high chlorite contents, the oxide volume diffusion speed of 304 SS substantially increased because of whisker growth and high NaCl vapor concentrations. The experiment, aimed at preventing or mitigating Al coatings being rendered useless, revealed that the chlorination reaction was slow and stable under 550°C. However, the reaction was unachievable in environments with an NaCl vapor concentration of 990 vppm.

REFERENCES

Y. Y. Chang, C. C. Tsaur and J. C. Rock, Surf. Coat. Technol, 200 (2006) 6588.
C. J. Wang and S. M. Chen, Surf. Coat. Technol, 200 (2006) 6601.
R.W. Richards, R.D. Jones, P.D. Clements and H. Clarke, Int. Mater. Rev.
T. Heumann and S. Dittrich, Z. Metallkunde, 50 (1959) 617.
S. C. Kwon and J. Y. Lee, Can. Metall. Quart, 20 (1981) 351.
W. Deqing, S. Ziyuan and Z. Longjiang, Appl. Surf. Sci, 214 (2003) 304.
M. V. Akdeniz, A. O. Mekhrabov and T. Yilmaz, Scr. Metall, 33 (1994) 1723.
D. I. Lainer and A. K. Kurakin, Fiz. Met. Metalloved, 18 (1964) 145.
R. W. Richards, PhD thesis, University of Wales, Cardiff, 1989.
W.J. Cheng and C. J. Wang, Mater. Char, 61 (2010) 467.
W. J. Cheng, Y. Y. Chang and C. J. Wang, Surf. Coat. Technol, 203 (2008) 401.
H. Xiaoxia, Y. Hua, Z. Yan and P. Fuzhen, Mater. Lett. 58 (2004).
F. Bosselet, D. Pontevichi, M. Sacerdote-Peronnet, and J.C. Viala, Measurement of theIsothermal Section at 1000 K of Al-Fe-Si, J. Phys. IV France, (122) 2004.National Science Council Research Project Outcome Report NSC 98-2221-E-011-038-.Te Yuan Chiang, Ay Su, Int. J. Electrochem. Sci., 10 (2015)9556–9570.

Innovation in Design, Communication and Engineering – Lam et al. (eds)
© 2020 Taylor & Francis Group, London, ISBN 978-0-367-17777-5

The effect of stirring time varieties in stir casting on the tensile strength and hardness of Al-Cu alloys with alumina reinforcement

Ching-Hua Chiu
Graduate School of Technological and Vocational Education, National Yunlin University of Science and Technology, Douliu, Taiwan

Chiung-Hsien Huang
Department of Electro-Optical Engineering, National Formosa University, Hu-Wei, Yunlin, Taiwan

Didik Nurhadi*, Vega Rizki & Putut Murdanto
Department of Mechanical Engineering Education, Universitas Negeri Malang, Malang, Indonesia

ABSTRACT: This research aimed to obtain the tensile strength and hardness of Al-Cu alloy reinforced with alumina in various stirring durations. This research used the experimental method with pre-experimental research design. The data analysis in this research used the descriptive technique. The results showed that the mechanical properties experienced decreases in tensile strength and increases in hardness value with increasing duration. Based on the average tensile strength and Vickers hardness results, the 60-second stirring duration resulted in tensile strength of 181.4 MPa and hardness of 53.03 HV. The 180-second stirring duration resulted in tensile strength of 158.3 MPa and hardness of 57.94 HV. Meanwhile, the 300-second stirring duration resulted in tensile strength and hardness of 139.1 MPa and 73.7 HV. In conclusion, stirring duration in stir casting method affected the tensile strength and hardness of Al-Cu alloy reinforced with alumina.

1 INTRODUCTION

Additional non-metal reinforcements could improve aluminum properties and increase its mechanical properties as desired. Alumina is a non-silicate ceramic material and often acts as a reinforcement in the making of metal composite (Low 2014; Galusek et al., 2015). Generally, alumina has three phases: α, β, and γ alumina. These phases have different properties, thus, have unique applications. Beta alumina (β-Al2O3) has tremendous fireproof property, so it is useful in ceramic applications such as furnaces (Sadik et al., 2014; Beitollahi et al., 2010). Gamma alumina (γ-Al2O3) frequently performs as a catalyst material; for example, in the petroleum-refining process and automotive area (Yajima et al., 2003). Alpha alumina (α-Al2O3) is a hexagonal crystallite with lattice parameters a = 4.7588 and c = 12.9910 nm. Alpha alumina usually acts as a refractory material and reinforcement from the oxide group because of its excellent physical, mechanical, and thermal properties (Branch, 2011).

There are adaptations of processing methods or techniques to produce composite reinforcement metals (e.g., stir casting, metallurgy powder, squeeze casting, and vacuum hot pressing) (Luan et al., 2011). The direct factor in affecting the properties and quality of metal with ceramic reinforcement is whether the distribution of reinforcement particles using stir casting methods is even or not (Pawar et al., 2014; Abolkassem et al., 2018). The stir casting method is the most suitable method to create non-metal reinforcement. The stir casting method has several parameters that affect aluminum's mechanical properties. Stirring time is one of those parameters. Stirring

---

*Corresponding author: didik.nurhadi.ft@um.ac.id

time in the stir casting process, hopefully, plays a significant role in increasing and improving the mechanical properties of Al-Cu metal with alumina reinforcement.

2   METHODS

This research used experimental process with a quantitative approach. The design in this research was pre-experimental design with One-shot Case Study model; where a sample was treated before being analyzed. The data analysis in this research was descriptive analysis from the tensile strength and hardness tests.

Pre-experimental research is pseudo-experimental research because of the many external variables that affect the dependent variables. Research that uses this design uses two variables: independent and dependent, without the controlled variable. The indicator for the independent variables here was the stirring time of 60 seconds, 180 seconds, and 300 seconds. Meanwhile, the signs for the dependent variables were the tensile strength and hardness value.

There were three specimens for each tensile strength test and three points for each hardness test in each sample. Specimens in the pouring process adopted the shape of the existing metal molds. The metal molds in this research were cylinders based on the American Association State Highway and Transportation Official Standard (AASHTO) No. T68 an American National Standard (ASTM A 370). The hardness test pattern took the remaining pieces of the tensile strength test specimens. The test pattern of Vickers hardness test did not require particular particles, unlike the tensile strength. Besides, using leftover pieces left no waste.

This research used descriptive analysis technique to analyze the data. The data analysis in this research covered explanations of the tensile strength and Vickers hardness test results of Al-Cu metal with alumina reinforcement. Then, this research processed the collected data to obtain the average values of each test. The data was communicated in graphs and bar chart in Microsoft Word to facilitate the reading process. Besides graphs and bar charts, there was also a narrative description.

3   RESULTS

There were two types of research in this study and resulted in two types of data: tensile strength and micro Vickers hardness. Below is the explanation on both data.

3.1   *Tensile strength of Al-Cu alloys with alumina reinforcement*

These tests used the Universal Testing Machine with the force of 5,000 kg in standard specimen based on the ASTM A370 tensile test. Table 1 displays the results.

Specimen III with 60 seconds stirring time has the highest tensile strength of 19.9 kgf/mm^2, whereas Specimen II with 300 seconds stirring time has the lowest tensile strength of 12.43 kgf/mm^2. Other than the tensile strength, Table 1 also shows strain value. Specimen III with 60 seconds stirring time has the highest strain of 1.9%, whereas Specimen II with 60 seconds stirring time has the lowest strain value of 0.2%. Ultimate Tensile Strength was directly proportional with the stress. The highest stress is in Specimen III with 60 seconds stirring time in the value of 199 MPa. Meanwhile, the lowest stress is in Specimen II with 300 seconds stirring time in the value of 124.3 MPa.

3.2   *The hardness of Al-Cu alloy with alumina reinforcement*

The conduction of hardness tests used the micro Vickers hardness test in three different points from each specimen's surface. Each location received 200 gr force for 10 seconds. The results were in numbers with the HV unit. Table 2 are the results of Al-Cu hardness test with alumina reinforcement and various stirring time.

Table 1. Tensile strength test results.

Stirring Time	Spe-cimen	L_0 (mm)	L_1 (mm)	A_0 (mm^2)	P_{Max} (kgf)	σ (kgf/mm^2)	UTS (MPa)
60 seconds	I	72.3	72.9	126.73	1,987.5	15.68	156.8
	II	80.4	80.6	120.74	2,275	18.84	188.4
	III	79.9	81.4	117.75	2,343.74	19.9	199
	Average					**18.14**	**181.4**
180 seconds	I	77.9	79.3	122.74	2,212.5	18.02	180.2
	II	78.3	79.1	121.74	1,975	16.22	162.2
	III	79	79.7	120.74	1,600	13.25	132.5
	Average					**15.83**	**158.3**
300 seconds	I	76.4	76.6	123.73	1,862.5	15.05	150.5
	II	76.2	76.8	123.73	1,537.5	12.43	124.3
	III	78.7	79	122.74	1,750	14.26	142.6
	Average					**13.91**	**139.1**

Table 2. Micro Vickers hardness test results.

Stirring Time	Specimen	Test Point	Hardness (HV)	Average per Specimen	Average per Variation
60 seconds	I	1	51.5	54.73	53.03
		2	52		
		3	60.7		
	II	1	51.6	52.83	
		2	49.1		
		3	57.8		
	III	1	57.1	51.53	
		2	45.5		
		3	52		
180 seconds	I	1	51.9	57.13	57.94
		2	60.9		
		3	58.6		
	II	1	48.1	51.3	
		2	48		
		3	57.8		
	III	1	57.6	65.4	
		2	78.1		
		3	60.5		
300 seconds	I	1	44.3	56.6	73.7
		2	51.6		
		3	73.9		
	II	1	74.5	105.7	
		2	90.6		
		3	151.9		
	III	1	67.2	58.8	
		2	51.5		
		3	57.6		

Table 2 showed three variations in each stirring duration – 60 seconds, 180 seconds, and 300 seconds – in which there were three different test points in each specimen. Table 2 presents that the highest hardness is in the 3rd point of Specimen II with 300 seconds in the value of 151.9 HV whereas the lowest hardness is in the 1st point of Specimen I with 300 seconds stirring time in the value of 44.3 HV. The data in Table 2 also presents the average value per specimen. In summary, Specimen II with 300 seconds stirring time had the highest average hardness of 105.7 HV while Specimen II with 180 seconds stirring time had the lowest average hardness with the value of 51.3 HV.

4 DISCUSSION

This Research produced two types of data: tensile strength and micro Vickers hardness. Hence, discussion of two reviews about the obtained data, as below.

4.1 *Tensile strength of Al-Cu Alloys with alumina reinforcement*

There were decreases in the tensile strength from the stirring time of 60 seconds to 300 seconds because of gas-cavity defects in the specimens with 180 seconds and 300 seconds stirring time that turned into fractures after the test. Figure 1 displays the fractures.

Air-cavity defects formed themselves on the inside part of the specimen that had 180 and 300 seconds stirring time. According to Basak et al. (2010), air cavity or porosity was due to the upright shape of the model. Gómez-Díaz et al. (2009) stated that very high temperature causes a decrease in viscosity. A reduction in thickness meant a rise in the dilution level, creating a quick liquid rate and resulting in turbulence. The turbulence trapped gas in the liquid. The lower the viscosity, the higher the turbulence possibility was, and in turn, the higher the chance to catch the gas.

Those theories were in line with this research, and along, with increases in stirring time, became the main reason for decreases in tensile strength. In a previous study, Arifin and Junaidi (2017) declared that in research with various stirring times of 180 seconds, 300 seconds, and 480 seconds, the specimens improved their mechanical properties in terms of the hardness, tensile strength, and impact values. That research proved that the decreases in this research were due to porosity with a long diameter that significantly reduced real tensile strength.

Hartsuijker and Welleman (2017), stated the F - Δℓ diagram is generally useful to measure the force (F) and extension (Δℓ) from a material in a tensile test. The chart is also helpful to

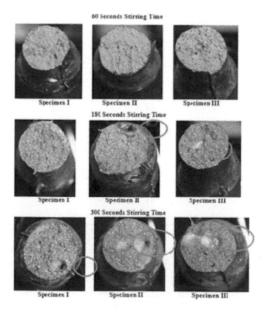

Figure 1. Tensile strength test results with stirring time variations.

obtain an object's dimensions. This theory correlates with the porosity that practically reduces the real diameter of the specimen. Automatically, the sample experienced a fracture in the area with the smallest diameter. In conclusion, porosity affected the tensile strength decrease of the sample with alumina reinforcement in 180 seconds and 300 seconds stirring time because of the decline in the testing diameter, resulting in the unreliable values of tensile strength.

4.2 The hardness of Al-Cu alloy with alumina reinforcement

Hardness values experienced increases from stirring times of 60 seconds to 300 seconds. These results were in line with the research of Tariq et al. (2019). The results showed that the additional stirring time in stir casting method improved the mechanical properties of the aluminium material. Also, an extra level of stirring time generated material wettability that caused the distribution of reinforcement in metal liquid. The better the wettability of a material, the better the mechanical properties.

Several parameters influenced the mechanical properties of metal in the stir casting method, particularly the hardness value. The parameters are stirring speed, stirring time, holding time, mixing temperature, and dimensions and position of the impeller. These parameters improved wettability (Kandpal et al., 2017). This statement is in line Singh et al.'s (2017) declaration that there are several things to mind during the stir casting process. First, stirring has to form a vortex. Number and total of blades determine the pattern of liquid flow. All these support distribution of the reinforcement, perfecting the bonds between materials, and avoiding materials clumping. Second, stirring time and speed are essential to reach wettability between metal liquid and reinforcement.

5 CONCLUSION

The Al-Cu alloy with alumina reinforcement that had 60 seconds stirring time had the highest tensile strength compared to specimens with 180-second and 300-second stirring times. The decreases in the tensile strength were due to porosity in the samples with 180-second and 300-second stirring times. The dissolution of hydrogen in a significant amount during the additional stirring time trapped air cavities in the casting. The correct solution was to decrease the pouring temperature and administrate digitizer as a degassing treatment.

Additional stirring time influenced the hardness of Al-Cu with alumina reinforcement. The longer stirring time generated higher hardness. The Al-Cu with alumina reinforcement and 300-second stirring times had the highest hardness compared to specimens with 60-second and 180-second stirring times. The rise in hardness was due to the wettability process. Wettability was the ability of the metal liquid to wet the reinforcement particle to make the metal and reinforcement homogenous. The evenly distributed reinforcement particle underlies the increased hardness of alumina-strengthened Al-Cu alloys.

REFERENCES

Abolkassem, S. A., Elkady, O. A., Elsayed, A. H., Hussein, W. A., & Yehya, H. M. (2018) *Results in Physics*, 9, 1102–1111.
Arifin, A., & Junaidi. (2017) *FLYWHEEL: J. Tek. Mes. Untirta*, 3, 21–31.
Basak, T., Roy, S., Singh, S. K., & Pop, I. (2010) *Int. J. of Heat & Mass Trans.*, 53, 1819–1840.
Beitollahi, A., Hosseini-Bay, H., & Sarpoolaki, H. (2010) *J. of Mater. Sci.e: Mater in Elec.*, 21, 130–136.
Branch, M. (2011) *Ceramics–Silikáty*, 55, 378–383.
Galusek, D., & Galusková, D. (2015) *Nanomaterials*, 5, 115–143.
Gómez-Díaz, D., Navaza, J. M., & Quintáns-Riveiro, L. C. (2009) *Int. J. of Food Propert.*, 12, 396–404.
Hartsuijker, C., & Welleman, J. W. (2007) *Engineering Mechanics: Stresses, Strains, Displacements* (Vol. 2). Delft: Springer.
Kandpal, B. C., Kumar, J., & Singh, H. (2017) *Mater. Today: Proc.*, 4, 2783–2792.

Low, I. M. (2014) "Contributor contact details." In *Adv. in Ceramic Matrix Composites.* Cambridge Woodhead Publishing.

Luan, B. F., Hansen, N., Godfrey, A., Wu, G. H., & Liu, Q. (2011) *Mater. & Design,* 32, 3810–3817.

Pawar, P. B., & Utpat, A. A. (2014) *Procedia Mater. Sci.,* 6, 1150–1156.

Sadik, C., El Amrani, I.E., & Albizane, A. (2014) *J. of Asian Ceramic Soc.,* 2, 83–96.

Singh, S., Singh, I., & Dvivedi, A. (2017) *Int. J. of Cast Metals Res.,* 30, 356–364.

Tariq, M., Khan, I., Hussain, G. & Farooq, U. (2019) *Int. J. of Lightweight Mater. & Manufac.,* 2, 123–130.

Yajima, Y., Hida, M., Taruta, S., & Kitajima, K. (2003) *J. of the Ceramic Soc. of Japan,* 111, 419–425.

Innovation in Design, Communication and Engineering – Lam et al. (eds)
© 2020 Taylor & Francis Group, London, ISBN 978-0-367-17777-5

Aseismatic and wind pressure design evaluation and optimization design of horizontal water tower

Jui-Chang Lin, Chia-Wei Yang & Yu-De Lin*
Department of Mechanical Design Engineering, National Formosa University, Hu-Wei, Yunlin, Taiwan

ABSTRACT: This article uses the finite element analysis software Abaqus to simulate the resistance of stainless steel horizontal water towers under the influence of earthquakes and typhoon forces. The method of analysis is based on the seismic design of buildings and the design rules of wind resistance. In this experiment, the analysis of the angle frame structure is carried out. The seismic force (wind pressure) is converted into horizontal (or vertical) force using the formula defined by the Building Construction Department's seismic and wind pressure resistance specifications, and static simulation analysis is performed by the finite element method.

The seismic test results show that the structure meets the seismic design specified in the seismic code of the building and can effectively resist the external force of the earthquake. The structure has been tested to withstand a seismic acceleration of 470.4 Gal, which is equivalent to a magnitude 7 earthquake.

The wind resistance test results show that the structure has a high wind resistance. Under the influence of the strongest typhoon "Dip Typhoon" on record, the structure is subjected to a stress of only 156.7 Mpa, which is calculated to be the most sustainable. The wind pressure is 1674.9 kgf/m2.

The results of the optimization analysis show that after reducing the volume by 10%, the maximum external force originally given can be withstood, and safety will not be reduced.

Keywords: finite element, optimization, water tower, earthquakes, typhoon

1 INTRODUCTION

The water tower is a product that Taiwan needs in daily life, and it is closely related to our life. Most of the water towers are installed outdoors, and they are directly impacted when disaster strikes, so their safety must reach a certain level.

For the safety issue, this article uses the Abaqus finite element software for safety analysis of seismic and wind pressure resistance to understand how much strength this product can withstand in the face of two natural disasters, the earthquake and the typhoon [1–4].

2 EXPERIMENTAL EQUIPMENT

This study will use the finite-element software Abaqus to simulate behavior under the influence of natural disasters (earthquakes and typhoons).

---

* Corresponding author: nhit100@gmail.com

The products analyzed in this study are horizontal water towers, and the product entity diagram is shown in Figures 1 and 2. This experiment simulates the strength of this product against the influence of natural forces. Product specifications are shown in Table 1.

The analysis structure is made of 304 stainless steel. The 304 steel grade has a bright, clean surface, is resistant to atmospheric corrosion, and has excellent formability and weldability. It can be widely used in food containers, food processing equipment, building structures, decorative panels, transportation, etc. It is the steel material with the greatest quantity and use. The mechanical properties of the materials are shown in Table 2.

3 EXPERIMENTAL METHODS

The analysis method of this earthquake simulation experiment is to use the seismic design specifications of buildings, and refer to the static analysis method and the calculation equation of the seismic force of the non-building structure. For buildings with regular shape, without the need for dynamic analysis, one can calculate seismic forces according to the provisions of this chapter, and perform structural analysis by the static method.

Figure 1. Horizontal water tower tripod entity diagram.

Figure 2. Horizontal water tower entity diagram.

Table 1. Structural specification.

Storage Tank Capacity	Structure Length	Structure Width		Structure Height
2000 Liter	137 CM	125 CM		63 CM
Storage Tank Diameter	Storage Tank Storage		Water Tower Total Height	
122 CM	193.5 CM		140 CM	

Table 2. Material mechanical properties.

Elastic modulus	Hardness	Yield Strength(YS) N/mm^2	tensile strength(TS) N/mm^2
210E3 N/mm^2	Ann	205 min	520 min
	1/2 H	470 min	780 min
	3/4 H	665 min	930 min
	H	880 min	1130 min

The minimum design equations for the earthquakes of each axis of the structure are respectively: (1) The horizontal total transverse force V is calculated according to the following formula:

$$V = \frac{S_{aD}I}{1.4\alpha_y F_u} W \tag{1}$$

The coefficient 1.4 in the formula (1) is the static indefinite coefficient. In this experiment, the static indefinite coefficient of the water tower is conservatively considered. Therefore, the coefficient is changed to 1.0, and the minimum design level total transverse force V of the experimental earthquake is:

$$V = \frac{S_{aD}I}{\alpha_y F_u} W \tag{2}$$

In this experiment, form reference data to find that the toughness capacity R of this product is 1.2. The toughness capacity R_α of the general work site and the near fault zone is calculated as (3):

$$R_a = 1 + \frac{(R-1)}{1.5} \tag{3}$$

3.1 Wind resistance test method

Wind power is set according to wind resistance design regulations specified by CNS standards. The products of this experiment belong to the open architecture. According to the specifications, at least two walls have openings of 80% or more. The designed wind F is the equivalent static wind used to design the wind. The formula is (4), where A_c is the characteristic area affected by wind, C_f is the wind force coefficient, G is the gust response factor, q is the wind speed pressure, and z_{Ac} is the centroid height of the wind-induced characteristic area.

$$F = q(z_{Ac})GC_f A_c \tag{4}$$

The calculation formula of wind speed pressure q is as follows (5), where K(z) is the wind speed pressure ground condition coefficient, K_{zt} is the topographic coefficient, I is the use coefficient, and V_{10} (C) is the basic design wind speed. This experiment uses typhoon values to simulate. The typhoon wind speed pressure and various numerical classifications are shown in Table 3.

$$q(z) = 0.06K(z)K_{zt}[IV_{10}(C)]^2 \tag{5}$$

3.2 Analysis steps

The analysis step flow is:

(1) Create part: drawn using SolidWorks, imported using Abaqus.
(2) Set material parameters: set stainless steel material parameters for parts.
(3) Establish assembly: load the parts into the combined module for analysis.
(4) Establish analysis steps: analyze the experiment using static analysis.
(5) Set interaction: establish constraints and interactions between parts.
(6) Setting the load: giving the corresponding external force.

Table 3.   Typhoon level and wind speed.

Strength	Wind Level	Wind Speed	Wind Pressure
Mild Typhoon	Level 8	17.2-20.7(m/s)	35-52 kgf/m^2
	Level 9	20.8-24.4(m/s)	52-72 kgf/m^2
	Level 10	24.5-28.4(m/s)	72-97 kgf/m^2
	Level 11	28.5-32.6(m/s)	97-128 kgf/m^2
Moderate Typhoon	Level 12	32.7-36.9(m/s)	128-164 kgf/m^2
	Level 13	37.0-41.4(m/s)	164-206 kgf/m^2
	Level 14	41.5-46.1(m/s)	206-256 kgf/m^2
	Level 15	46.2-50.9(m/s)	256-312 kgf/m^2
Strong Typhoon	Level 16	51.0-56.0(m/s)	312-377 kgf/m^2
	Level 17	56.1-61.2(m/s)	377-499 kgf/m^2
	Level>17	>61.2(m/s)	>499 kgf/m^2

(7)   Create a grid: analyze by finite element analysis.
(8)   Establish an optimization module: set optimization parameters.
(9)   Establish work: create analytical work and conduct analytical work.
(10) Post-processing: Review the analysis results.

4   ANALYSIS AND RESULTS

4.1   *Analysis results of seismic simulation experiments*

In this analysis, seismic forces such as earthquake-free and seismic forces of 0.28W, 0.35W, 0.45W, 0.48W, and 0.5W are analyzed separately to explore how much maximum seismic force this structure can withstand, and X, Y, respectively. Z analysis in three directions, the analysis results are as follows (Tables 4–7).

4.2   *Displacement analysis results*

According to the seismic design code of the building, the total horizontal force of the seismic minimum design level is 0.28W as the design safety standard. The seismic force in the X and Z directions is 5531.9N, and the Y direction is 2766N. The analysis results are as follows (Table 8).

Table 4.   Stress analysis results with a horizontal seismic force of 0.28W.

Seismic Force of 0.28W	Force Direction	Stress
Horizontal Seismic Force (5531N)	X Direction	128Mpa
	Z Direction	170.4Mpa
Vertical Seismic Force (2766N)	+Y Direction	116.4Mpa
	-Y Direction	156.8Mpa

Table 5.   Stress analysis results with a horizontal seismic force of 0.35W.

Seismic Force of 0.35W	Force Direction	Stress
Horizontal Seismic Force (6914.9N)	X Direction	144.3Mpa
	Z Direction	178.7Mpa
Vertical Seismic Force (3457.5N)	+Y Direction	111.5Mpa
	-Y Direction	162Mpa

Table 6. Stress analysis results with a horizontal seismic force of 0.45W.

	Force Direction	Stress
Horizontal Seismic Force (8890.6N)	X Direction	168.4Mpa
	Z Direction	198.4Mpa
Vertical Seismic Force (4445.3N)	+Y Direction	102.1Mpa
	-Y Direction	166.4Mpa

Table 7. Stress analysis results with a horizontal seismic force of 0.48W.

Seismic Force of 0.48W	Force Direction	Stress
Horizontal Seismic Force (9483.3N)	X Direction	177.3Mpa
	Z Direction	204.5Mpa
Vertical Seismic Force (4741.7N)	+Y Direction	101.5Mpa
	-Y Direction	168.6Mpa

Table 8. Displacement analysis results of horizontal seismic force of 0.28W.

Horizontal Seismic Force of 0.28W	Force Direction	Displacement
Horizontal Seismic Force (5531N)	X Direction	1.69 E-3mm
	Z Direction	2.12 E-3mm
Vertical Seismic Force (2766N)	+Y Direction	1.67 E-3mm
	-Y Direction	2.43 E-3mm

The above analysis shows that the force given by the earthquake is not enough to destroy the structure of the water tower. Therefore, the next analysis will continue to enhance the seismic force to explore the maximum seismic force that the structure can withstand. The analysis results are as follows (Tables 9–11).

Table 9. Displacement analysis results of horizontal seismic force of 0.35W.

Horizontal Seismic Force of 0.35W	Force Direction	Displacement
Horizontal Seismic Force (6914.9N)	X Direction	1.9 E-3mm
	Z Direction	2.21 E-3mm
Vertical Seismic Force (3457.5N)	+Y Direction	1.58E-3mm
	-Y Direction	2.53 E-3mm

Table 10. Displacement analysis results of horizontal seismic force of 0.45W.

Horizontal Seismic Force of 0.45W	Force Direction	Displacement
Horizontal Seismic Force (8890.6N)	X Direction	2.21 E-3mm
	Z Direction	2.33 E-3mm
Vertical Seismic Force (4445.3N)	+Y Direction	1.54 E-3mm
	-Y Direction	2.81 E-3mm

Table 11.   Displacement analysis results of horizontal seismic force of 0.48W.

Horizontal Seismic Force of 0.48W	Force Direction	Displacement
Horizontal Seismic Force (9483.3N)	X Direction	2.39 E-3mm
	Z Direction	2.49 E-3mm
Vertical Seismic Force (4741.7N)	+Y Direction	1.47 E-3mm
	-Y Direction	2.86 E-3mm

4.3   Wind pressure simulation experiment analysis results

In this analysis, the three directions of X, Y, and Z are analyzed separately for mild, moderate, and intense typhoons. In addition, the typhoon Hebe in 1996 was simulated. This is the strongest typhoon in Taiwan. The highest wind speed is 260 km/hr, and the 1979 typhoon (also known as Taipei) is the strongest typhoon in the world. The maximum wind speed is 305 km/hr. The analysis results are as follows (Tables 12–14).

4.4   Optimization results

This experiment uses strain energy optimization and volume optimization, and optimization is used to reduce material cost and rigidity.

It can be seen from the above analysis that when the structure is subjected to a thrust of 0.48W, the stress is 204.5 MPa. In this experiment, under the influence of the same external

Table 12.   Level-11 wind stress.

	Force Direction	FORCE	STRESS
Light typhoon (Level 11)	X-Horizontal Force	741 N	103.4 Mpa
	Z-Horizontal Force	725 N	136.5 Mpa
	+Y-Horizontal Force	395 N	106.6 Mpa
	-Y-Vertical Force		136.5 Mpa

Table 13.   Level-15 wind stress.

	Force Direction	FORCE	STRESS
Moderate typhoon (15 levels)	X-Horizontal Force	1806 N	104.1 Mpa
	Z- Horizontal Force	962 N	143 Mpa
	+Y- Horizontal Force	1767 N	103.2 Mpa
	- Y-Vertical Force		140.6 Mpa

Table 14.   Level-15 wind stress.

	Force Direction	FORCE	STRESS
Strong typhoon (Level 17)	X-Horizontal Force	2894 N	113.8 Mpa
	Z- Horizontal Force	2831 N	149.7 Mpa
	+Y- Horizontal Force	1542 N	100 Mpa
	- Y-Vertical Force		144.9 Mpa

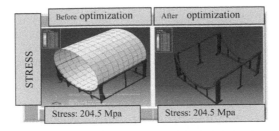

Figure 3.   Results of pre-optimal stress analysis.

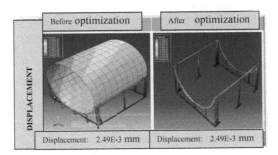

Figure 4.   Results after optimal displacement analysis.

Table 15.   Optimization results.

Seismic Force of 0.48W	Before of optimization	After of optimization
Stress	204.5 Mpa	204.5 Mpa
Displacement	2.49E-3 mm	2.491E-3 mm

force, the volume percentage of the original structure can be reduced to resist the same external force..

The results of the optimization analysis showed that the original volume was retained at 90% while the withstand stress remained unchanged. This shows that the structure can still have an impact on the external force of the antigen under the condition that 10% of the material is saved. The results are shown in Figures 3 and 4 and Table 15.

5   CONCLUSION

It can be seen from the analysis results that the structure receives the external force in the Z direction and the stress on the structure is the largest, indicating that the strength of the structure in the Z direction is weak relative to the other directions,

The seismic capacity of this structure:

In the seismic design code for buildings, the total transverse force V of the earthquake is 0.28W and the vertical seismic force $V_v = 1/2$ V. For this experiment, the horizontal force of 5531.9N and 2766N are applied to the structure. The vertical external force is equivalent to a magnitude 6 earthquake in the earthquake classification table. From the analysis results, this product meets the design safety standards.

After confirming that the structure meets the safety design criteria, we will continue to increase the seismic force to explore how many levels of earthquake the structure can withstand. After conducting multiple experimental analyses of the data, when the seismic coefficient reaches 0.48, the horizontal seismic force is 0.48W, which is equivalent to 9483.3N. At this time, the stress is 204.5 Mpa, which is close to the material 205 Mpa, which shows that the maximum seismic external force of this structure is 0.48W, which is equivalent to 470.4 Gal. According to the earthquake classification table, the intensity is a magnitude 7 earthquake.

The wind resistance of this structure:

In this anti-wind pressure test, the external force corresponding to the wind pressure level is applied according to the formula formulated in the building windproof design specification. The 11th, 15th, and 17th typhoons and the two strongest typhoons that have occurred since the record began were analyzed.

The analysis results show that the wind resistance of this structure is quite high. When the strong typhoon strikes, the structure is far from the material's fall. Under the influence of the strongest typhoon "Dip Typhoon" recorded, the structure is only subjected to a stress of 156.7 Mpa.

The result of this structure optimization:

This experiment uses strain energy optimization and volume optimization, and optimization is used to reduce material cost and rigidity.

When the seismic coefficient is 0.48 (470.4 Gal), it is equivalent to the horizontal thrust of 9483.3N. The simulation results show that the stress of this structure is 204.5 Mpa, which is quite close to the fall, and the volume is reduced at the same intensity, so that the structure is not damaged. Optimize the experiment under the premise.

The results of the optimization analysis show that the original volume is retained at 90% and maximum stress remains unchanged. The results show that the structure can still have an impact on the external force of the antigen under the condition that 10% of the material is saved.

REFERENCES

Fan H., Li, Q. S., Tuan A. Y., Xu, L. (2009) "Seismic analysis of the world's tallest building." *Journal of Constructional Steel Research*, 65(5), 1206–1215.

Wang, Yuanbin., Guo, H., & Zhang, C. (2007) "Seismic response analysis of portal water injection sheet pile." *Journal of Ocean University of China (Ocean and Coastal Sea Research)*, 6(1), 90–94.

Aliberti, D., Cascone, E., & Biondi, G. (2016)"Seismic performance of the San Pietro dam."*ScienceDirect-Procedia Engineering*, 158, 362–367.

Gurkalo, F., Du, Y.G., Poutos, K., & Jimenez-Bescos, C. (2017) "The nonlinear analysis of an innovative slit reinforced concrete watertower in seismic regions." *ScienceDirect-Engineering Structures*, 134, 138–149.

Uematsu, Y., Yamaguchi, T., & Yasunaga, J. (2018) Effects of wind girders on the buckling of open-topped storage tanks underquasi-static wind loading." *ScienceDirect-Thin-Walled Structures*, 124, 1–12.

Communication science & Engineering

Innovation in Design, Communication and Engineering – Lam et al. (eds)
© 2020 Taylor & Francis Group, London, ISBN 978-0-367-17777-5

An efficient hybrid two-dimensional DOA estimation scheme for massive MIMO systems

Ann-Chen Chang*
Department of Information Technology, Ling Tung University, Taichung, Taiwan

ABSTRACT: This article deals with efficient hybrid two-dimensional (2D) direction of arrival (DOA) estimation for massive multi-input multi-output (MIMO) systems based on a large-scale uniform rectangular array (URA). The proposed hybrid scheme first applied the 2D discrete Fourier transform (2D-DFT) to the received data matrix for acquiring coarse initial DOA estimates under single data snapshot. Then, through rearranging the received signal matrix, the fine azimuth and elevation angles can be separately estimated by using the iterative search estimator within a very small region. Furthermore, the estimate angles pairing is given automatically. Meanwhile, it iteratively searches for correct DOA vector by minimizing the cost function, using a first-order Taylor series approximation of the DOA vector with the one initially estimated. Compared with conventional noise subspace–based estimators, it avoids the direct computation of full-dimensional sample autocorrelation matrix and its eigenvalue decomposition. Therefore, it provides lower computation complexity, especially in large-scale URA scenarios. Simulation results verify the efficiency of the proposed hybrid 2D DOA estimate scheme, particularly when the number of snapshots is small.

1 INTRODUCTION

Multiple-input multiple-output (MIMO) has been one of the key technologies to improve the data rate and link reliability of wireless communication systems because it can provide more degrees of freedom. As a promising technology in 5G, massive MIMO has been attracting significant attention owing to its unprecedented potential of high spectral efficiency (Shu et al., 2018). In massive MIMO systems, each base station (BS) is equipped with a large number of antennas with the objective of serving a number of single-antenna mobile stations that simultaneously occupy the same set of time and frequency resources. The performance of massive MIMO systems greatly relies on direction of arrival (DOA) estimation, since accurate DOA knowledge is essential for the BS to perform downlink precoding/beamforming. Therefore, to estimate DOA efficiently and accurately is a vital problem.

DOA estimation has been of interest to the signal processing community for decades. Due to the form factor limitation at the BS, two-dimensional (2D) massive MIMO systems are introduced to fit a large number of antenna elements on the BS in reality. As a 2D array is needed to estimate azimuth and elevation angles of the source signal, the uniform rectangular array (URA) has become a research focus due to the high degree of regularity in its structure. For the 2D DOA estimation problem in URA, some eigenstructure-based estimation methods, such as the 2D multiple signal classification (MUSIC) method (Chen et al., 1993) and the estimation of signal parameters via rotational invariance technique (ESPRIT) (Hu et al., 2014) have been proposed. The 2D-MUSIC needs a 2D spectral peak search and is therefore less practical. Thus, a 2D unitary ESPRIT (U-ESPRIT) (Wang et al., 2015) is developed as an extension of 2D ESPRIT (Hu et al., 2014) with automatic pairing. However, the

*Corresponding author: acchang@teamail.ltu.edu.tw

application of these subspace-based methods involve the calculation of sample autocorrelation matrix and its eigenvalue decomposition (EVD), which bear heavy computational burdens especially for massive URA. A possible alternative to the subspace-based methods for DOA estimation with computational saving is the orthogonal projection (OP) method (Chang et al., 2014). However, the complexity and estimation accuracy of these spectral searching-based estimators strictly depend on the grid size used during the search. It is time consuming and the search grid is ambiguous. A reduced dimension MUSIC (RD-MUSIC) method (Zhang et al., 2010) was proposed for 2D angle estimation, which can reduce the computational cost by replacing the 2D searching with one-dimensional (1D) searching. However, using least squares fitting principle of the RD-MUSIC method to estimate phase angles is very sensitive to the noise effect and the total estimate performance will be declined. By rearranging the received signal matrix to estimate DOA, a separate MUSIC (S-MUSIC) method (Li et al., 2016) requires two 1D spectral searching procedures to reduce the 2D search computation load. Nevertheless, the angle pair matching procedure is still required (Li et al., 2016).

To lessen the above-mentioned problems, this article presents a hybrid 2D DOA estimation scheme for a large-scale URA. First, the proposed hybrid scheme obtains coarse initial DOA estimates through a 2D discrete Fourier transform (DFT) (Fan et al., 2017). Next, through rearranging the received signal matrix, the fine electrical azimuth and elevation angles can be separately estimated by using the iterative search estimator within a very small region, which only performs two 1D angle estimates. Furthermore, the estimate angles pairing is given automatically. Specifically, in conjugation with the 2D-DFT initial DOA estimator and the noise subspace–based iterative fine (NSIF) estimator (Chang, 2015), the resulting hybrid scheme is termed the DFT-NSIF scheme. The basic idea of the NSIF estimator is based on the principle that the output power of an estimator achieves a local maximum if the direction vector coincides with the desired signal. The DFT-NSIF scheme iteratively searches for the correct DOA vector by minimizing the objective function using a first-order Taylor series approximation of the DOA vector with the one initially estimated. The problem of finding the new direction vector is formulated as the closed form of a generalized eigenvalue problem, which allows one to readily solve it. Thus, it can reach relatively low computational complexity and provide satisfactory estimation performance. Let the number of signals be K. After obtaining initial DOA estimates, the whole possible DOA searching range forms $2K$ smaller search ranges, the minimum value of cost function is searched in each smaller range to get each signal's DOA, rather than finding the K peaks in the whole range, respectively. Therefore, the NSIF estimator can conquer the effect of ambiguous peaks with initial DOA estimates, and reduce the search complexity greatly. This is more suitable for fine DOA searching in actual operations. Furthermore, the estimated electrical azimuth and elevation angles automatically pair without additional calculation. However, the proposed hybrid scheme provides comparable performance with conventional subspace-based estimators, respectively, but requires much less computational cost than the latter, particular in the large array scenario. Simulation results are provided to demonstrate the effectiveness of the proposed hybrid scheme under a large-scale URA.

2 PROBLEM FORMULATION

2.1 *System model*

As illustrated in Figure 1, consider a large-scale URA with $M \times N$ antenna elements located on the $x-y$ plane at the positions $((m-1)d, (n-1)d)$ for $m = 1, 2, \cdots, M$ and $n = 1, 2, \cdots, N$, where $d = 0.5\lambda$ denotes the distance between two neighboring elements and λ is the wavelength of the carrier signal (Wang et al., 2015). Assume that there are K far-field narrowband uncorrelated signal sources. Without loss of generality, the antenna element in origin is set as the reference point. With respect to the antenna at the origin of the axes, the received data matrix $\bar{X}(t) \in \pounds^{M \times N}$ is given by

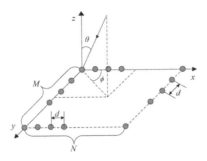

Figure 1.   The structure of a large-scale URA.

$$\overline{X}(t) = \sum_{k=1}^{K} \overline{A}(\theta_k, \ \phi_k) s_k(t) + \overline{N}(t), \ t = 1, 2, \cdots, L \tag{1}$$

where $s_k(t)$ is the signal transmitted by the kth user, L is the number of available snapshots, and $\overline{N}(t)$ represents the noise matrix. The steering matrix $\overline{A}(\theta_k, \phi_k) \in \pounds^{M \times N}$ represents the nominal array response, where the $(m,n)th$ element of $\overline{A}(\theta_k, \phi_k)$ is $[A(\theta_k, \phi_k)]_{m,n}$ $= e^{j\frac{2\pi d}{\lambda}[(m-1)\sin\phi_k \sin\theta_k + (n-1)\cos\phi_k \sin\theta_k]}$ and the kth signal impinging on the array from the azimuth angle $\phi_k \in [-90°, \ 90°]$ and the elevation angle $\theta_k \in [0, 90°]$. Electrical angles v and u are related to physical azimuth and elevation angles by $v = \frac{2\pi d}{\lambda}\cos\phi \sin\theta$ and $u = \frac{2\pi d}{\lambda}\sin\phi \sin\theta$, respectively. Then, $\overline{A}(\theta_k, \phi_k) = a(u_k)a^T(v_k)$, where $a(u_k) = [1, \ e^{ju_k}, \cdots, \ e^{j(M-1)u_k}]^T$ and $a(v_k) = [1, \ e^{jv_k}, \cdots, \ e^{j(N-1)v_k}]^T$. In order to facilitate the expression, the vectorized operation is performed to convert $\overline{A}(\theta_k, \phi_k) \in \pounds^{M \times N}$ into the equivalent model $A(u, v) = [a(u_1, v_1) \ a(u_2, v_2) \cdots, \ a(u_K, v_K)]$ with size $MN \times K$, where $a(u_k, \ v_k) = a(v_k) \otimes a(u_k)$ and $\otimes$ denotes the Kronecker product. Finally, the received signal $x(t) \in \pounds^{MN \times 1}$ can be expressed as

$$x(t) = A(u, v)s(t) + n(t), \ t = 1, 2, \cdots, L \tag{2}$$

where $s(t) = [s_1(t), s_2(t), \cdots, s_K(t)]^T$ is the transmit signals vector. The noise vector $n(t)$ is modeled as a multivariate circular complex Gaussian random process, temporally white, with zero mean and covariance $\sigma_n^2 I_{MN}$, where I_{MN} denotes an identity matrix of size $MN \times MN$.

Meanwhile, we note that the corresponding inverse relations for obtaining elevation and azimuth angles are

$$\theta_k = \sin^{-1}(\tfrac{\lambda}{2\pi d}\sqrt{u_k^2 + v_k^2}) \tag{3.a}$$

$$\phi_k = \tan^{-1}(u_k/v_k) \tag{3.b}$$

2.2   2D-MUSIC estimation

With the further assumption that $s(t)$ and $n(t)$ are uncorrelated, the received data autocorrelation matrix is given by $R = E\{x(t)x^H(t)\} = A(u, v)R_s A^H(u, v) + \sigma_n^2 I_{MN}$, where $E\{\bullet\}$ denotes the expectation operator and the signal correlation matrix $R_s = E\{s(t)s^H(t)\}$. Denote the EVD of R by

$$\mathbf{R} = \mathbf{E}_s \Lambda_s \mathbf{E}_s^H + \mathbf{E}_n \Lambda_n \mathbf{E}_n^H \tag{4}$$

where $\mathbf{E}_s \in \pounds^{MN \times K}$ is called the signal subspace and is composed of K eigenvectors associated with the arriving signals. $\mathbf{E}_n \in \pounds^{MN \times (MN-K)}$ is called the noise subspace and is composed of $(MN - K)$ eigenvectors associated with the noise. The diagonal matrices Λ_s and Λ_n have diagonal elements associated with the signal and noise eigenvalues, respectively, where the eigenvalues are arranged in decreasing order.

By exploiting the orthogonal between the signal and noise subspace, the well-known 2D MUSIC method (Chen et al., 1993) searches the continuous array direction vector over the area of $\{v,\ u\}$ or $\{\theta,\ \phi\}$ to find the K maxima of the following cost function

$$P(u,\ v) = \underset{u,v}{\text{Max}} \ [\mathbf{a}^H(u,\ v)\mathbf{E}_n\mathbf{E}_n^H\mathbf{a}(u,\ v)]^{-1} \tag{5}$$

where $\mathbf{a}(u,\ v) = \mathbf{a}(v) \otimes \mathbf{a}(u)$ is the spatial scanning direction vector with $\{u, v \in (-\pi, \pi)\}$ or $\{\theta \in (0°,\ 90°), \phi \in (-90°, 90°)\}$. Then, the K largest peaks of function (5) denote the DOA of K signal sources. However, such spectral search processes may be unaffordable for some real-time implementations, which can be explained by the following fact. To obtain each search point, the Frobenius norm $\|\mathbf{a}^H(\theta,\ \phi)\mathbf{E}_n\|^2$ or $\|\mathbf{E}_n^H\mathbf{a}(\theta,\ \phi)\|^2$ has to be computed. This drawback becomes particularly apparent in joint estimation of azimuth and elevation since we have to search over two dimensions.

3 PROPOSED HYBRID ESTIMATION SCHEME

This section presents an efficient hybrid DOA scheme for a large-scale URA. However, it is known that the conventional subspace-based methods perform EVD whose complexity is forbidden for massive MIMO systems (Cheng et al., 2015). The proposed scheme first finds initial DOA estimates through 2D-DFT and then develops the efficient fine DOA estimator by utilizing an iterative process, which avoids conventional searching processes and direct decomposing of the full-dimensional autocorrelation matrix.

3.1 *Initial DOA estimation via 2 D-DFT*

In this subsection, we show how to estimate initial 2D-DOA information from a given $\bar{\mathbf{X}}(t)$ defined in (1). DFT is widely used for conventional non-parametric spectrum analysis because of its simplicity. But, it suffers from lower resolution than the other parametric methods. Nevertheless, the 2D-DFT method can be applied to acquire coarse initial DOA estimates for the large-scale URA structure.

Define the 2D-DFT of the steering matrix $\bar{\mathbf{A}}(\theta_k, \phi_k) = \bar{\mathbf{A}}_k$ as $\widetilde{\mathbf{A}}_k$ as $\widetilde{\mathbf{A}}_k = \mathbf{F}_M \bar{\mathbf{A}}_k \mathbf{F}_N$, where $\mathbf{F}_M$ and $\mathbf{F}_N$ are the two normalized DFT matrices, whose the $(p,\ q)$th elements are $[\mathbf{F}_M]_{pq} = \frac{1}{\sqrt{M}}e^{-j\frac{2\pi}{M}pq}$ and $[\mathbf{F}_N]_{pq} = \frac{1}{\sqrt{N}}e^{-j\frac{2\pi}{N}pq}$, respectively. Meanwhile, the (p,q) element of $\widetilde{\mathbf{A}}_k$ is

$$
\begin{aligned}
[\widetilde{\mathbf{A}}_k]_{p,q} &= \frac{1}{\sqrt{MN}} \sum_{n=0}^{N-1}\sum_{m=0}^{M-1} e^{-j(\frac{2\pi nq}{N}+\frac{2\pi pm}{M} - \frac{2\pi d}{\lambda}(m\sin\phi_k\sin\theta_k + n\cos\phi_k\sin\theta_k))} \\
&= \frac{1}{\sqrt{MN}} e^{-j\frac{M-1}{2}(\frac{2\pi p}{M} - \frac{2\pi d}{\lambda}\sin\phi_k\sin\theta_k)} e^{-j\frac{N-1}{2}(\frac{2\pi q}{N} - \frac{2\pi d}{\lambda}\cos\phi_k\sin\theta_k)} \\
&\quad g\frac{\sin[\frac{M}{2}(\frac{2\pi p}{M} - \frac{2\pi d}{\lambda}\sin\phi_k\sin\theta_k)]}{\sin[\frac{1}{2}(\frac{2\pi p}{M} - \frac{2\pi d}{\lambda}\sin\phi_k\cos\theta_k)]} g\frac{\sin[\frac{N}{2}(\frac{2\pi q}{N} - \frac{2\pi d}{\lambda}\cos\phi_k\sin\theta_k)]}{\sin[\frac{1}{2}(\frac{2\pi q}{N} - \frac{2\pi d}{\lambda}\cos\phi_k\sin\theta_k)]}
\end{aligned} \tag{6}
$$

36

If the URA has an infinite number of antennas, i.e., $M, N \to \infty$, there always exist integers (p_k, q_k) that satisfy $p_k = \frac{Md}{\lambda} \sin \phi_k \sin \theta_k$, $q_k = \frac{Nd}{\lambda} \cos \phi_k \sin \theta_k$, and all channel powers will concentrate on a single 2D-DFT point (p_k, q_k), formulating the ideal sparsity. Most power of $\widetilde{\mathbf{A}}_k$ concentrates around (p_k, q_k), where $\llcorner p_k = \frac{Md}{\lambda} \sin \phi_k \sin \theta_k \lrcorner$, $q_k = \llcorner \frac{Nd}{\lambda} \cos \phi_k \sin \theta_k \lrcorner$, and $\llcorner x \lrcorner$ is a round operator that returns the nearest integer to x. Specifically, $\widetilde{\mathbf{A}}_k$ only has one nonzero point (p_k, q_k) as $M \to \infty$, $N \to \infty$. If $\frac{Md}{\lambda} \sin \phi_k \sin \theta_k$ equals some integer p_k and $\frac{Nd}{\lambda} \cos \phi_k \sin \theta_k$ equals some integer q_k, then $\setminus \widetilde{\mathbf{A}}_k$ has only one nonzero element $[\widetilde{\mathbf{A}}_k]_{p_k, q_k} = \sqrt{MN}$, which means that all powers are concentrated on the point (p_k, q_k). However, a practical array aperture cannot be infinitely large, even if $\{M, N\}$ could be as great as hundreds or thousands in a large-scale URA. However, for most other cases, $\frac{Md}{\lambda} \sin \phi_k \sin \theta_k$ and $\frac{Nd}{\lambda} \cos \phi_k \sin \theta_k$ are not integers, while the signal power will leak from the central point ($\llcorner \frac{Md}{\lambda} \sin \phi_k \sin \theta_k \lrcorner$, $\llcorner \frac{Nd}{\lambda} \cos \phi_k \sin \theta_k \lrcorner$) to others. In fact, (6) is composed of sine function such that the leakage of signal power is inversely proportional to M and N. Hence when $\{M, N\}$ are sufficiently large, $\widetilde{\mathbf{A}}_k$ is still a sparse matrix with most power concentrated around ($\frac{Md}{\lambda} \sin \phi_k \sin \theta_k, \frac{Nd}{\lambda} \cos \phi_k \sin \theta_k$). The 2D-DFT of $\overline{\mathbf{X}}(t)$ is $\mathbf{Y} = \mathbf{F}_M \overline{\mathbf{X}}(t) \mathbf{F}_N$ with the $(p, q)th$ element $[\mathbf{Y}]_{pq} = \sum_{k=1}^{K} [\widetilde{\mathbf{A}}(\theta_k, \phi_k)]_{pq} s_k(t) + [\mathbf{F}_M \overline{\mathbf{N}}(t) \mathbf{F}_N]_{pq}$. After locating the K largest peaks of magnitude spectrum $|\mathbf{Y}|$, denoted as $\{\hat{p}_k^0, \hat{q}_k^0\}_{k=1}^{K}$, we can compute the coarse initial DOA estimates $\{\hat{\theta}_k^0, \hat{\phi}_k^0\}_{k=1}^{K}$ as follows:

$$\hat{\theta}_k^0 = \sin^{-1}\left[\frac{\lambda}{2\pi d}\sqrt{\left(\frac{2\pi}{M}\hat{p}_k^0\right)^2 + \left(\frac{2\pi}{N}\hat{q}_k^0\right)^2}\right] \tag{7.a}$$

$$\hat{\phi}_k^0 = \tan^{-1}\left[\frac{2\hat{p}_k^0}{M} \Big/ \frac{2\hat{q}_k^0}{N}\right] \tag{7.b}$$

It is noted that the resolution of $\{\hat{p}_k^0, \hat{q}_k^0\}$ by directly applying 2D-DFT is still limited by half of the DFT interval, in other words, $\{1/(2M), 1/(2N)\}$. After the initial DOA $\{\hat{\theta}_k^0, \hat{\phi}_k^0\}$ is obtained, the searching regions of electrical angles from $u \in [-\pi, \pi]$ and $v \in [-\pi, \pi]$ reduce to $u \in [-\frac{\pi}{M}, \frac{\pi}{M}]$ and $v \in [-\frac{\pi}{N}, \frac{\pi}{N}]$, respectively. Hence, the DOA of the kth signal can be immediately obtained from the power-concentrated position in the 2D-DFT of $\overline{\mathbf{X}}(t)$. Nevertheless, it is noted that the proposed iterative fine estimator only needs to search within a much smaller region.

3.2 Fine DOA estimation via iterative searching

A. S-MUSIC for electrical angles estimation

Since 2D-MUSIC requires exhaustive 2D searching, it is normally inefficient due to high computational cost, and therefore this section uses the S-MUSIC method (Li et al., 2016) to handle the problem. The main shortcoming of this separate method is that the working array aperture is reduced, which will result in the loss of degrees of freedom. Its number of resolvable signals is only Min$\{M, N\} - 1$, rather than the well-known $MN - 1$. To this end, we first address $\{v_k\}_{k=1}^{K}$ estimation and then rearrange the received signal matrix to estimate $\{u_k\}_{k=1}^{K}$. However, this method does not need pair matching because the initial DOA has been given by the 2D-DFT estimation.

A.1 Electrical angle v estimation
In order to simplify the notation, we rewrite $\mathbf{a}(u_k) = [1, z_{uk}, \cdots, z_{uk}^{M-1}]^T$ with $z_{uk} = e^{ju_k}$. Then, the array steering matrix $\mathbf{A}(u, v)$ becomes

37

$$\mathbf{A}(u,v) = \begin{bmatrix} \mathbf{a}(v_1) & \mathbf{a}(v_2) & L & \mathbf{a}(v_K) \\ z_{u1}\mathbf{a}(v_1) & z_{u2}\mathbf{a}(v_2) & L & z_{uK}\mathbf{a}(v_K) \\ M & M & O & M \\ z_{u1}^{M-1}\mathbf{a}(v_1) & z_{u2}^{M-1}\mathbf{a}(v_2) & L & z_{uK}^{M-1}\mathbf{a}(v_K) \end{bmatrix} \tag{8}$$

We partition the received data matrix $\mathbf{X} = [\mathbf{x}(1)\ \mathbf{x}(2)\cdots,\ \mathbf{x}(L)] \in \pounds^{MN \times L}$, $\mathbf{x}(t)$ defined in (2) for $t = 1, 2, \cdots, L$, into M submatrices $\mathbf{X} = [\mathbf{X}_1^T\ \mathbf{X}_2^T\cdots,\ \mathbf{X}_M^T]^T$, where $\mathbf{X}_m = \mathbf{A}_m\mathbf{S} + \mathbf{N}_m$ with $\mathbf{A}_m = \mathbf{A}_v\Phi_u^{m-1}$, $\mathbf{A}_v = [\mathbf{a}(v_1),\ \mathbf{a}(v_2), \cdots,\ \mathbf{a}(v_K)]$, $\Phi_u^{m-1} = diag\{[z_{u1}^{m-1},\ z_{u2}^{m-1}, \cdots,\ z_{uK}^{m-1}]\}$, and $diag\{ \bullet \}$ stands for the diagonalization operation to transform a vector to a diagonal matrix.

In order to independently estimate electrical angles $\{v_k\}_{k=1}^K$, a new matrix can be constructed as $\mathbf{X}_v = [\mathbf{X}_1,\ \mathbf{X}_2, \cdots, \mathbf{X}_M]$, which contains the information of v. According to the structure of $\mathbf{X}$, $\mathbf{X}_v \in \pounds^{N \times ML}$ can be rewritten as $\mathbf{X}_v = \mathbf{A}_v\mathbf{S}_v + \mathbf{N}_M$. It is noted that $\mathbf{S}_v = [\mathbf{S},\ \Phi_u\mathbf{S}, \cdots, \Phi_u^{M-1}\mathbf{S}]$ can be seen as the signal matrix corresponding to the new matrix $\mathbf{X}_v$. The correlation matrix of $\mathbf{X}_v$ is $\mathbf{R}_v = E\{\mathbf{X}_v\mathbf{X}_v^H\} = \mathbf{A}_v\mathbf{R}_{vs}\mathbf{A}_v^H + \sigma_n^2\mathbf{I}_N$, where $\mathbf{A}_v = [\mathbf{a}(v_1)\ \mathbf{a}(v_2)\cdots,\ \mathbf{a}(v_K)]$ and $\mathbf{R}_{vs} = E\{\mathbf{S}_v\mathbf{S}_v^H\}$ is the source covariance matrix corresponding to $\mathbf{X}_v$. Performing EVD of $\mathbf{R}_v$, we can get the noise subspace $\mathbf{E}_{nv}$ corresponding to $\mathbf{X}_v$. Let $\mathbf{P}_{nv} = \mathbf{E}_{nv}\mathbf{E}_{nv}^H$; we construct the following MUSIC spatial spectrum function for v estimation:

$$P(v) = \underset{v}{\text{Max}}\ [\mathbf{a}^H(v)\mathbf{P}_{nv}\mathbf{a}(v)]^{-1} \tag{9}$$

Searching $v \in [-\pi, +\pi]$, we can obtain K largest peaks of $P(v)$. The corresponding $\{\hat{v}_1,\ \hat{v}_2, \cdots, \hat{v}_K\}$ are taken as the estimates of the electrical angle v.

A.2 Electrical angle u estimation
Define a new $MN \times K$ matrix $\mathbf{A}'(v, u) = [\mathbf{a}'_1(v, u),\ \mathbf{a}'_2(v, u), \cdots,\ \mathbf{a}'_K(v, u)]$, where $\mathbf{a}'_k(v, u) = \mathbf{a}(u_k) \otimes \mathbf{a}(v_k)$. Let $\mathbf{a}(v_k) = [1,\ z_{vk}, \cdots, z_{vk}^{N-1}]$ with $z_{vk} = e^{jv_k}$; then, we have

$$\mathbf{A}'(v,u) = \begin{bmatrix} \mathbf{a}(u_1) & \mathbf{a}(u_2) & L & \mathbf{a}(u_K) \\ z_{v1}\mathbf{a}(u_1) & z_{v2}\mathbf{a}(u_2) & L & z_{vK}\mathbf{a}(u_K) \\ M & M & O & M \\ z_{v1}^{N-1}\mathbf{a}(u_1) & z_{v2}^{N-1}\mathbf{a}(u_2) & L & z_{vK}^{N-1}\mathbf{a}(u_K) \end{bmatrix} \tag{10}$$

Obviously, there exists a transformation matrix $\mathbf{T}$ with size $MN \times MN$ corresponding to the finite number of row interchanged operations such that $\mathbf{A}'(u, v) = \mathbf{T}\mathbf{A}(u, v)$, where

$$[\mathbf{T}]_{i,j} = \begin{cases} 1, i = l + (h-1)M,\ j = h + (l-1)N \\ 0, \text{others} \end{cases} \tag{11}$$

under $l = 1, 2, \cdots, M$ and $h = 1, 2, \cdots, N$. Using the structure of the matrix $\mathbf{A}'(v, u)$, we introduce a virtual $MN \times L$ data matrix $\mathbf{X}'^{\text{@TX}}$. Divide the matrix $\mathbf{X}'$ into N submatrices

$$\mathbf{X}' = [(\mathbf{X}'_1)^T\ (\mathbf{X}'_2)^T\cdots,\ (\mathbf{X}'_N)^T]^T \tag{12}$$

where $\mathbf{X}'_n = \mathbf{A}'_n \mathbf{S} + \mathbf{N}'_n$ with $\mathbf{A}'_n = \mathbf{A}_u \Phi_v^{n-1}$ $\mathbf{A}_u = [\mathbf{a}(u_1), \mathbf{a}(u_2), \cdots, \mathbf{a}(u_K)]$ $\Phi_v^{n-1} = diag\{[z_{v1}^{n-1}, z_{v2}^{n-1}, \cdots, z_{vK}^{n-1}]\}$ for $n = 1, 2, \cdots, N$. In order to estimate $\{u_k\}_{k=1}^K$ independently, we construct a new $M \times NL$ matrix $\mathbf{X}_u = [\mathbf{X}'_1, \mathbf{X}'_2, \cdots, \mathbf{X}'_N]$, which contains the information of u. According to the structure of $\mathbf{X}'$ in (12), $\mathbf{X}_u \in \pounds^{M \times NL}$ can be expressed as $\mathbf{X}_u = \mathbf{A}_u \mathbf{S}_u + \mathbf{N}_N$. Note that $\mathbf{S}_u = [\mathbf{I}_K \mathbf{S}, \Phi_v \mathbf{S}, \cdots, \Phi_v^{N-1} \mathbf{S}]$ and $\mathbf{N}_u$ can be seen as the virtual signal matrix and the noise matrix, respectively. The correlation matrix of $\mathbf{X}_u$ is $\mathbf{R}_u = E\{\mathbf{X}_u \mathbf{X}_u^H\} = \mathbf{A}_u \mathbf{R}_{us} \mathbf{A}_u^H + \sigma_n^2 \mathbf{I}_M$, where $\mathbf{R}_{us} = E\{\mathbf{S}_u \mathbf{S}_u^H\}$ is the source correlation matrix corresponding to $\mathbf{X}_u$. Performing the EVD of $\mathbf{X}_u$, we obtain the noise subspace matrix $\mathbf{E}_{nu}$ of $\mathbf{R}_u$. Thus, the MUSIC spatial spectrum function for u estimation is given by

$$P(u) = \underset{u}{\text{Max}} \, [\mathbf{a}^H(u)\mathbf{P}_{nu}\mathbf{a}(u)]^{-1} \tag{13}$$

where $\mathbf{P}_{nu} = \mathbf{E}_{nu}\mathbf{E}_{nu}^H$. Searching $u \in [-\pi, +\pi]$, we obtain the K largest peaks of $P(u)$ corresponding to the estimated u.

B. Noise subspace iterative fine estimator

We know that the performance of the MUSIC method is governed by the grid size while implementing the high-resolution DOA estimation. At the same time, the drawback of the spectral search process becomes particularly apparent in joint estimation of azimuth and elevation since we have to search over two dimensions. In this subsection, we apply the NSIF method (Chang, 2015) to (9) and (13) for electrical angle estimation in a large-scale URA. The basic idea is to present an iterative search method for new direction vectors by minimizing the cost function using a first-order Taylor series approximation of the direction vector with the one initially estimated. The problem of finding the new direction vector is formulated as the closed form of a generalized eigenvalue problem, which allows one to readily solve it. It is noted that maximizing the mean output power with respect to the direction vector is equivalent to the cost functions of the null spectrum $f(v) = \underset{v}{\text{Min}} \, \mathbf{a}^H(v)\mathbf{P}_{nv}\mathbf{a}(v)$ and $f(u) = \text{Min} \, \mathbf{a}^H(u)\mathbf{P}_{nu}\mathbf{a}(u)$.

First, $\mathbf{a}^u(v_k)$ of the kth signal can be corrected in each iteration processing for the fine electrical angle v estimation. At the ith iteration, a first-order expansion may be used to approximate $\mathbf{a}(v_k)$ as $\mathbf{a}(v_k) \simeq \mathbf{B}_k^i \mathbf{c}_k^i$, where $\mathbf{c}_k^i = [1, v_k - \hat{v}_k^i]^T$ and $\mathbf{B}_k^i = [\mathbf{a}(\hat{v}_k^i), \mathbf{a}^{(1)}(\hat{v}_k^i)]$ with $\mathbf{a}^{(1)}(\hat{v}_k^i) = \partial\mathbf{a}(v)/\partial v|_{v=\hat{v}_k^i}$. As $\hat{v}_k^i \to v_k$, $\|\mathbf{B}_k^i \mathbf{c}_k^i - \mathbf{a}(v_k)\|$ tends to zero. However, $\mathbf{c}_k^i$ is unknown, though $\mathbf{B}_k^i$ is known. Therefore, we will find a new direction vector with the form $\mathbf{B}_k^i \mathbf{g}_k^i$, where $\mathbf{g}_k^i$ is a 2×1 parameter vector. When using $\mathbf{B}_k^i \mathbf{g}_k^i$ as the direction vector of the kth signal, the array output power is $P_k^i = [(\mathbf{B}_k^i \mathbf{g}_k^i)^H \mathbf{P}_{nv} \mathbf{B}_k^i \mathbf{g}_k^i]^{-1}$. The constrained optimization problem is given by

$$\underset{\mathbf{g}_k}{\text{Min}} \, (\mathbf{B}_k^i \mathbf{g}_k^i)^H \mathbf{P}_{nv} \mathbf{B}_k^i \mathbf{g}_k^i, \text{subject to} \|\mathbf{B}_k^i \mathbf{g}_k^i\|^2 = 1 \tag{14}$$

Specifically, it leads us to formulate the unconstrained minimization problem as $\underset{\mathbf{g}_k}{\text{Min}} f(\eta_k^i, \mathbf{g}_k^i)$, where $f(\eta_k^i, \mathbf{g}_k^i) = (\mathbf{B}_k^i \mathbf{g}_k^i)^H \mathbf{P}_\Psi \mathbf{B}_k^i \mathbf{g}_k^i + \eta_k^i(1 - (\mathbf{B}_k^i \mathbf{g}_k^i)^H \mathbf{B}_k^i \mathbf{g}_k^i)$ and η_k^i is the Lagrange multiplier. Taking the derivative of $f(\eta_k^i, \mathbf{g}_k^i)$ with respect to $\mathbf{g}_k^i$ and setting it to zero, we have

$$(\mathbf{B}_k^i)^H \mathbf{P}_{nv} \mathbf{B}_k^i \mathbf{g}_k^i = \eta_k^i (\mathbf{B}_k^i)^H \mathbf{B}_k^i \mathbf{g}_k^i \tag{15}$$

Substituting (15) into P_k^i, it follows that $P_k^i = (\eta_k^i)^{-1}$. The optimal solution $\mathbf{g}_k^i$ to (14) corresponds to the eigenvector associated with the smallest eigenvalue of the generalized eigenvalue problem (15). And, the new direction vector $\mathbf{w}_k^i = \mathbf{B}_k^i \mathbf{g}_k^i$. However, owing to the finite samples effect, $\mathbf{w}_k^i$ is not always proportional to $\mathbf{a}(v_k)$. Therefore, we define a normalized vector $\hat{\mathbf{w}}_k^i = (w_{k,1}^i)^{-1} \mathbf{w}_k^i$, where $w_{k,1}^i$ is the first entry of $\mathbf{w}_k^i$. The electrical angle v_k at the ith iteration is estimated via

$$\hat{v}_k^i = \frac{1}{N-1} \sum_{n=2}^{N} \arg\{\hat{w}_{k,n}^i\} \tag{16}$$

where $\hat{w}_{k,n}^i$ is the nth entry of $\hat{\mathbf{w}}_k^i$ and $\arg\{\bullet\}$ is the phase angle operation. If $|\hat{v}_k^i - \hat{v}_k^{i-1}| < \varepsilon$ is very small, we consider variance rate of cost function to achieve stabilization, where ε is a terminating error value. So it is unnecessary to increase the number of iterations, otherwise the iterative processing will continue.

Similarly, the NSIF method also can be applied on $f(u) = \underset{u}{\text{Min}} \ \mathbf{a}^H(u)\mathbf{P}_{nu}\mathbf{a}(u)$ for obtaining $\{\hat{u}_k\}_{k=1}^{K}$. Finally, we convert $\{\hat{u}_k, \hat{v}_k\}$ to $\{\hat{\theta}_k, \hat{\phi}_k\}$ using (3).

3.3 Computation complexity analysis

In this subsection, the number of complex multiplications (CM) of the 2D-MUSIC (Chen et al., 1993), 2D-OP (Chang et al., 2014), U-ESPRIT (Wang et al., 2015), RD-MUSIC (Zhang et al., 2010), S-MUSIC (Li et al., 2016), 2D-DFT, and proposed DFT-NSIF scheme are evaluated. Assume that K sources and the number of array elements for a URA is $M \times N$. The computational complexities of EVD and matrix inversion with a $MN \times MN$ correlation matrix $\hat{\mathbf{R}}$ are about $12M^3N^3$ and $2M^3N^3$ CM, respectively (Golub & Van Loan, 1996). For the 2D-MUSIC and 2D-OP methods, let F_u and F_v be the number of searching grids within the search region of u and v, respectively. For the S-MUSIC method, let F_{su} and F_{sv} be the number of searching grids within the search region of u and v, respectively. Let $F_{su,k}$ and $F_{sv,k}$ be the iteration number of the kth signal for the DFT-NSIF. Briefly, the required CM are listed in Table 1.

4 SIMULATION RESULTS

This section reports simulation results demonstrating the effectiveness of the proposed hybrid schemes for DOA estimation. Consider an URA with half-wavelength

Table 1. Computational complexities.

Estimators	Total complex multiplications (CM)
2D-MUSIC	$12(MN)^3 + (MN)^2(MN - K) + F_uF_v[(MN)^2 + MN]$
2D-OP	$2K^3 + 2K^2MN + KMN(MN - K) + 3.25(MN - K)^3$
	$\quad + 2(MN - K)^2K + (MN - K)K^2 + F_uF_v[(MN)^2 + MN]$
U-ESPRIT	$3(MN)^3 + 2\{3K^3 + 0.5[(M - 1)^2M + (M - 1)M^2]$
	$\quad + 0.5MN^2(M - 1) + 0.5MNK(MN - N) + 0.5K^3 + 0.75K^2(MN - N)\}$
RD-MUSIC	$12(MN)^3 + (MN)^2(MN - K) + F_{ru}[M^2 + M] + K[3N + 4]$
S-MUSIC	$12M^3 + M^2(M - K) + F_{su}[M^2 + M] + 12N^3$
	$\quad + N^2(N - K) + F_{sv}[N^2 + N]$
2D-DFT	$M^2N + MN^2$
DFT-NSIF	$12M^3 + M^2(M - K) + \sum_{k=1}^{K} F_{su,k}[2M^2 + 2M] + M^2N + 12N^3$
	$\quad + N^2(N - K) + \sum_{k=1}^{K} F_{sv,k}[2N^2 + 2N] + MN^2$

spacing. The symbols used for DOA estimation are modulated with binary phase shift keying modulation and average received signal power from all source signals is the same. In the simulation, the BS is equipped with $M \times N = 60 \times 70$ URA of $d = \frac{\lambda}{2}$. There are twelve equal power signals $(K = 12)$ impinging on the array from $\theta = \{54.48°, 51.67°, 45.13°, 40.47°, 33.31°, 23.17°, 26.56°, 30.98°, 36.41°, 58.37°, 62.21°, 73.83°\}$-$\phi = \{-84.35°, -7897°, 7361°, -67.34°, 56.88°, -27.2\ 1°, 26.5\ 6°, 29.05°, 32.6\ 1°, 49.76°, 42.7°, -51.34°\}$. Let the searching grids of the 2D-MUSIC and 2D-OP methods be 10^{-6} within the search region of u and v, respectively. Meanwhile, let the terminating error value of iterative fine estimator be $\varepsilon = 10^{-6}$. The noise is the zero-mean white Gaussian process. The signal-to-noise ratio (SNR) is defined as the ratio of the power of the source signal to that of the additive noise. For comparison, the results of the 2D-MUSIC (Chen et al., 1993), 2D-OP (Chang et al., 2014), U-ESPRIT (Wang et al., 2015), RD-MUSIC (Zhang et al., 2010), S-MUSIC (Li et al., 2016), and 2D-DFT methods are also provided. As an index of evaluation, all the results are computed over $\Pi = 10^3$ Monte Carlo trials and the root mean square error (RMSE) is defined as $\text{RMSE} = [\frac{1}{\Pi K} \sum_{\rho=1}^{\Pi} \sum_{k=1}^{K} (\hat{\theta}_{k,\rho} - \theta_k)^2 + (\hat{\phi}_{k,\rho} - \phi_k)^2]^{0.5}$ where $\hat{\theta}_{k,\rho}$ is the estimated elevation angle θ_k at the ρth trial, which is similar for $\hat{\phi}_{k,\rho}$. Each of the simulation results presented is after $L = 50$ snapshots processed with independent noise sample for each run.

Figure 2 depicts leakage-existing magnitude spectrum and peak power position of the 2D-DFT coarse estimator under SNR = 10dB. To introduce proper order in the spatial spectrum, one can use fftshift function (Matlab), which arranges the spatial spectrum in order: negative spectrum, zero spectrum, and positive spectrum. It is shown that the power will 'leak' from the $(\lfloor \frac{Md}{\lambda} \sin \phi_k \sin \theta_k \rceil, \lfloor \frac{Nd}{\lambda} \cos \phi_k \sin \theta_k \rceil)^{th}$ 2D-DFT point to the nearby points in Figure 2. Hence, the coarse DOA of the active signals can be immediately obtained from the power-concentrated positions in the 2D-DFT of $\bar{\mathbf{X}}(t)$. It is noted that the $\{p_k, q_k\}_{k=1}^{K}$ can be immediately estimated from the $K = 12$ largest peaks of 2D magnitude spectrum $|\mathbf{Y}|$. The results show that the peak power position of 2D magnitude spectrum $|\mathbf{Y}|$ is still useful to indicate an index of initial DOA estimates.

Figure 3 presents RMSE versus number of snapshots under SNR = 10dB. Clearly, increasing number of snapshots induces performance improvement for all estimation methods, except the 2D-DFT estimator. Figure 4 shows RMSE versus SNR. Observe that the DFT-NSIF method has better performance than the 2D-DFT, RD-MUSIC, and U-ESPRIT methods, and has very close performance to the 2D-MUSIC, 2D-OP, and S-MUSIC methods. However, the estimation accuracy of the 2D-MUSIC, 2D-OP, and S-MUSIC methods is governed by searching grid size whereas the DFT-NSIF method does not suffer from this limitation.

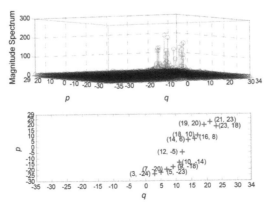

Figure 2.   Magnitude spectrum and peak power position of 2D-DFT estimator with leakage-existing.

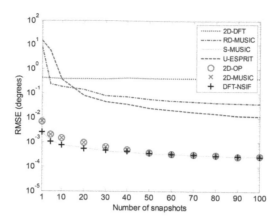

Figure 3.　RMSE versus number of snapshots.

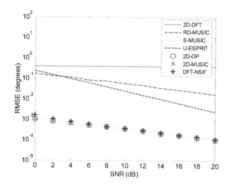

Figure 4.　RMSE versus SNR.

The proposed advantage of the iterative fine estimator is the reduction of computational complexity by reducing number of search while maintaining comparable performance. Let the searching grids of the 2D-MUSIC and 2D-OP methods be 10^{-6} within the search region of u and v, respectively, the number of searching is $F_u F_v$ and $F_u F_v = 6283186^2$. For the S-MUSIC method and the RD-MUSIC method, the required searching numbers are $F_{su} + F_{sv} = 2(6283186)$ and $F_{su} = 6283186$, respectively. Meanwhile, let the terminating error value of iterative fine estimator be $\varepsilon = 10^{-6}$. Briefly, the total searching number $F_N = \sum_{k=1}^{K} F_{su,k} + F_{sv,k}$ of the iterative fine estimators with $K = 12$ is listed in Table 2. It is noted that the presented estimator is adaptively corrected from the previous iteration stage. Again, it is clear that the proposed hybrid scheme has higher searching efficiency.

Table 2.　Total searching numbers of the NSIF.

SNR(dB)	0	2	4	6	8	10	12	14	16	18	20
F_N	114	108	104	101	99	98	98	98	97	97	97

5 CONCLUSIONS

This article has presented an efficient hybrid 2D-DOA estimation scheme for a large-scale URA. It was first found that the low resolution 2D-DFT spectrum becomes effective in providing good initial DOA estimation information. Next, it also simplifies 2D global searching in the 2D-MUSIC method to $2K$ times 1D local searching through rearranging the received signal matrix. Unlike the conventional subspace-based estimators, it avoids direct computing EVD form full-dimensional sample data autocorrelation matrix. Additionally, the proposed scheme without pairing procedure can provide a comparable angle estimating performance with the conventional subspace-based methods. While maintaining estimation accuracy, the ability to reduce the computational load of the proposed scheme for 2D DOA estimation is especially profound for the case of large-scale arrays. Numerical results validate the effectiveness of the proposed hybrid scheme.

ACKNOWLEDGMENT

This research was supported by the Ministry of Science and Technology under grant number MOST 108-2221-E-275-001.

REFERENCES

Chen, Y. M., Lee, J. H., & Yeh, C.C. (1993) "Two-dimensional angle-of-arrival estimation for uniform planar arrays with sensor position errors." IEE Proceedings Part F: Radar and Signal Processing, 140 (1),37–42.

Chang, A. C., Shen, C. C., & Chang, K. S. (2014) "DOA and DOD estimation using orthogonal projection approach for bistatic MIMO radars." *IEICE Trans. Fundamentals*, E97-A(5), 1121–1124.

Chang, A. C. (2015) Blind CFO estimation based on decision directed MUSIC technique for interleaved uplink OFDMA, *J. the Chinese Institute of Engineers*, 38(5),573–582.

Cheng, L., Wu, Y. C., Zhang, J., & Liu, L. (2015) Subspace identification for DOA estimation in massive/full-dimension MIMO systems: Bad data mitigation and automatic source enumeration. *IEEE Trans. Signal Processing*, 63(22),5897–5909.

Fan, D., Gao, F., Wang, G., Zhong, Z., & Nallanathan, A. (2017) "Angle domain signal processing-aided channel estimation for indoor 60-GHz TDD/FDD massive MIMO systems" *IEEE J. Selected Areas in Communications*, 35(9),1948–1961.

Golub, G. H., & Van Loan, C. H. (1996) *Matrix Computations*. Baltimore, MD: The Johns Hopkins University Press.

Hu, A., Lv, T., Gao, H., Zhang, Z., & Yang, S. (2014) "An ESPRIT-based approach for 2-D localization of incoherently distributed sources in massive MIMO systems." *IEEE J. Selected Topics in Signal Processing*, 8(5),996–1011.

Li, L., Chen, F., & Dai, J. (2016) "Separate DOD and DOA estimation for bistatic MIMO radar." *International Journal of Antennas and Propagation*, 2016, 1–11.

Shu, F., Qin, Y., Liu, T., Gui, L., Zhang, Y., Li, J., & Han, Z. (2018) "Low-complexity and high-resolution DOA estimation for hybrid analog and digital massive MIMO receive array." *IEEE Trans. Communications*, 66(6),2487–2501.

Wang, T., Ai, B., He, R., & Zhong, Z., 2015. "Two-dimension direction-of-arrival estimation for massive MIMO systems." *IEEE Access*, 3, 2122–2128.

Zhang, X., Xu, L., Xu, L., & Xu, D. (2010) "Direction of departure (DOD) and direction of arrival (DOA) estimation in MIMO radar with reduced-dimension MUSIC." *IEEE Communications Letters*, 14(12),1161–1163.

Communication science & Information technology

Innovation in Design, Communication and Engineering – Lam et al. (eds)

Applying rough set theory to analyze the antecedents of customer satisfaction for homestay service quality in Kinmen

Chien-Hua Wang* & Song-Bo Wang
School of Business, Lingnan Normal University, Guangdong Province, China

Jich-Yan Tsai
School of Computer Science and Engineering, Yulin Normal University, Guangxi Province, China

Sheng-Hsing Liu
Department of Information Management, Yuan Ze University, Taoyuan, Taiwan

ABSTRACT: The homestay plays a very important role in the tourism and leisure industry. It not only enhances local economic development, but also allows deep learning of the local culture. This study mainly discusses the service quality and the satisfaction of homestay in Kinmen. First, the SERVQUAL scale was used as methodology to a design questionnaire to evaluate customer perception of the service quality of homestays. There were in total 265 effective questionnaires from customers. Second, rough set theory (RST) was used to identify the satisfaction of homestay users and analyze service quality of homestays in Kinmen. And the decision rules are referred to as a combination of linguistic values and corresponding satisfaction. The aim was to demonstrate the usefulness and practicability of the proposed methods for big data analysis to make appropriate marketing strategies.

1 INTRODUCTION

Driven by the trend of sightseeing and tourism and consumers' attention to leisure life, the industry must understand and meet the needs of consumers while facing the development of the tourism industry. In tourism-related industries, the hotel industry has become an important part. However, accommodation-only services can no longer meet the needs of tourists. In order to give visitors more time to learn more about different local cultures, many warm and friendly people started to operate homestays. With the evolution of the times and people's emphasis on the quality of accommodation, the economic type of the homestay is different from the service that provided only accommodation in the past. Emerging B&Bs combine local ecological environment or unique customs and traditions. In addition to improving local economic development and promoting local specialties, it also provides visitors with a place to visit, allowing visitors more time and opportunities to experience the local culture and life.

According to statistics of the Tourism Bureau in Dec. 2018, the total number of legal homestays in Kinmen included 325 B&Bs and 1563 total rooms. It ranked seventh. Although located on the outlying islands, Kinmen is located between Taiwan and Xiamen, and its homestay industry is thriving. In view of the maturity of the Kinmen tourism industry, this study aimed at the service quality of the homestay industry in the Kinmen. This study uses the 2019 Summer Music Season to take visitors to the festival as a research object, and then gave reference suggestions to local hoteliers operators.

*Corresponding author: wangthuck@gmail.com

In recent years, there has been considerable literature suggesting that the service quality of homestays has a critical impact on tourist loyalty. The study found that transportation convenience, homestay characteristics, word-of-mouth, substantial benefits, management, environmental services, basic facilities, ancillary services, reasonable prices, custom diet, and so on, are factors that affect cunstomer satisfaction with the homestay (Wu, 2009; Mei et al., 2012; Liu et al., 2016; Su & Wang, 2015).

Different from previous researches, statistical models are used here. Thus, this study adopts the rough set theory, to which can be applied qualitative factor variable analysis at the same time. From the data generation rules, it can effectively deal with uncertain or imprecise expression and makes it easy for users to understand. Further, the results of this study can simultaneously identify multiple overall rules for consumer satisfaction from qualitative and quantitative factors (Lin et al., 2013).

Therefore, the purpose of this study is to use the conventional rough set theory to deal with incomplete or imprecise information faced by customers, to explore the influencing factors of the satisfaction of the homestay, and to analyze the influencing factors of customer satisfaction of different groups. Based on the SERVQUAL scale (Parasuraman et al., 1994), this study is based on appropriate modification to meet the service attributes of the homestay industry. Through the questionnaire survey, service attributes provided by Kinmen B&B are mainly discussed. It is hoped that the results of this study will improve and promote the service attributes for the homestay industry.

2 LITERATURE REVIEWS

2.1 *The service quality of homestay*

Both the homestay and the hotel provide a place for visitors to stay, but the feeling of being a visitor in a homestay is different from that of a restaurant or hotel (Chiou et al., 2010). According to the management method promulgated in 2001, the B&B is a self-use residential vacant room that combines local human landscape, natural ecology, environmental resources, and agriculture, forestry, animal husbandry, and other production activities to operate as a family sideline, providing accommodation for travelers to experience rural life. Morrison et al. (1996) believe that most of the homestays are privately operated; not many people can be accommodated at a time; there is a certain degree of communication with the host; and there is an opportunity to learn about the place where the special environment is located. Therefore, this study defines that operators of the B&B are the operators in Kinmen, using their own homes to provide 15 or fewer rooms for tourists to stay, and arrange for visitors to enjoy the local natural ecological environment and beautiful geological landscape and historic cultural monuments during leisure time.

In terms of service quality, the SERVQUAL scale has become the basis of most researches, and its content can be used to measure the relevant attribute of service quality (Gilbert & Wong, 2003). Thus, this study used the SERVQUAL scale as a tool to measure service quality of homestays. The corresponding five attributes are:

(1) Reliability: the ability of the service staff to provide the promised service reliably and accurately.
(2) Tangibility: the external presentation of venues, physical equipment, and service personnel.
(3) Responsiveness: service personnel assist customers and are able to provide instant service.
(4) Assurance: service staff have the expertise, courtesy, and ability to win customers trust.
(5) Empathy: service staff care for customers and other persons.

2.2 *Rough set theory*

The rough set theory, proposed by Pawlak in 1982, can serve as a new mathematical tool for dealing with data-classification problems. It adopts the concept of equivalence classes to partition training instances according to some criteria. Two kinds of partitions are formed in the mining process: lower approximations and upper approximations, from which both certain and possible rules can easily be derived. It is widely used in knowledge extraction, information system analysis, and AI, and can also be applied to the correlation between multidimensional attribute data.

The biggest advantage of this method is that it can process analytical information attributes, including analyzing approximate models; simplifying data, logic rules and deriving decisions; and analyzing uncertainty information. For information uncertainty, it can have a certain analysis model with an approximate concept (Chiu et al., 2008)

Since the analysis using the rough set theory does not set any assumptions, the application of the rough set theory is more flexible, and because the rough set theory can mine the relationship between data attributes, the rough set theory can be applied to many different fields.

3 RESEARCH METHOD

3.1 *Rough set theory*

(1) Information system and decision table. The presentation of rough set theory is in the form of information systems, and the analysis process is also based on the system. An information system can be viewed as a table of finite data. Each row in the table lists all the information about the object, with each column representing the different characteristic attributes of the object, and each row representing a different object.

IS = (U, A) is called information system, where U = {x1, x2, ..., xn} is the universe finite set of objects, and A = {a1, a2, ..., am} is the set of attributes.

A decision table is the causal relationship between the decisions made under which the data is satisfied, and is presented in a table, as shown in Table 1 below.

Table 1.   The decision table for survey results of homestay.

	Attributes		Decision
Object	Homestay facilities	Toiletries	Overall satisfaction
x1	4	4	4
x2	4	4	4
x3	4	4	5
x4	4	4	4
x5	4	5	5

(2) Equivalence relation. When analyzing a set of data, the object and objects between them become indistinguishable because they contain the same information on certain attributes, called indiscernibility relationship, and belong to the intersection of the same classification. The equivalence relation is illustrated by the data in Table 1, and the equivalence relation of some attributes can be expressed as the equations (1)–(3). If there is a problem with inconsistent classification, the problem can be handled by upper and lower approximations.

$$Ind(a_1) = \{4 : \{x_1, \ x_2, \ x_3, \ x_4, \ x_5\}\} \tag{1}$$

$$Ind(a_2) = \{4 : \{x_1, \ x_2, \ x_3, \ x_4\}, \ 5 : \{x_5\}\} \tag{2}$$

$$Ind(a_1, \ a_2) = \{\{4, \ 4\} : \{x_1, \ x_2, \ x_3, \ x_4\}, \ \{4,5\} : \{x_5\}\} \tag{3}$$

(3) The approximation relation of set. The method analyzes data according to two basic concepts, namely lower and upper approximations of a set. Let X be an arbitrary subset of the universe U, and A be an arbitrary subset of attribute set. The lower and upper approximations for A on X are:

$$\overline{A}(X) = \left\{ x_i \in U : [x_i]_{Ind(A)} \cap X \neq \emptyset \right\} \tag{4}$$

$$\underline{A}(X) = \left\{ x_i \in U : [x_i]_{Ind(A)} \subseteq X \right\} \tag{5}$$

(4) Core and reduction of attributes. The set of attributes that are indispensable in the data is called "core"; and the set of attributes that can be used as a whole under the same basic set is called "reduction." A rough set is a set that represents the attributed approximation by reduction, and uses the delete-unnecessary attribute to find a smaller set that is similar or identical to the original set's classification. Thus, in the process of attribute reduction, the extra attributes will be deleted but will not cause a reduction in classification ability.

(5) Accuracy of approximation and approximation quality. Furthermore, we can accept the concept of upper approximations and lower approximations to established data classification rules, and measure the certainty of data classification by the accuracy of the approximate set. Namely, if a set exists with boundary set in equivalence relation of A, it means that the set is still unable to definitely classify the objects belonging to the set or not belonging to the set under the information A. Therefore, the more elements in the boundary set, the lower the accuracy of the approximate set. Thus, the accuracy of the approximate set X is defined as:

$$\mu_a(X) = card(\underline{A}(X))/card(\overline{A}(X)) \tag{6}$$

(6) Decision rules. The decision rule is an orderly way to describe the new decision table formed by deletion of the attribute. However, the decision rule of the decision table is defined as: If Φ then Ψ, Φ represents the attribute, and Ψ represents the decision. After the decision rules are derived, the decision rules can also be reduced, so that the entire analysis processes can be more efficient.

3.2 Measurement method and questionnaire design

The questionnaire used in this study is based on the SERVQUAL scale to design 22 quality attributes related to homestay service. The questionnaire design of this study consists of three parts. The first part is accommodation information. The second part is the customers' feelings about the importance and satisfaction of the property of homestay. Questions in parts 1 and 2 are inferred by a five-point Likert scale: "Very satisfied," "Satisfied," "Neutral," "Dissatisfied," and "Very dissatisfied." The third part is the personal basic information of the respondent, including gender, age, and occupation. The questionnaire questions corresponding to the five dimensions of the SERVQUAL scale are shown in Table 2.

4 EMPIRICAL RESULTS AND ANALYSIS

In this study, the ROSE2 software was used to analyze customer satisfaction. The sample data included basic customer information, reason for accommodation, accommodation frequency, stay time, cost range, and satisfaction information on the 22 properties. In addition, the first 31 items are attributes and the 32nd item is a decision. Firstly, calculate the importance of the customer sample to each service attribute. Secondly, three service attributes are deleted such

Table 2.   The dimensions of SERVQUAL and questionnaire items.

Dimension	Item
Tangibility	B&B style/interior and equipment
	Room cleaning and comfort
	Free Internet service
	Toiletries
	Diet
	Self-service leisure facilities
Reliability	Correct any losses quickly
	Reasonable and accurate billing amount
	Deliver the promised service on time and correctly
Responsiveness	Provide speedy customer services
	Provide consulting services and clearly answer customer questions
	Meet customer characteristics or addition needs
	Convenience of booking service
Assurance	Make customers feel comfortable and at ease
	Operator attitude
	Operator experience and professionalism
	Safety facilities
Empathy	Understand customer needs
	Actively serve customers
	Simple accommodation
	Transportation service
	Other services

Table 3.   The table for lower and upper approximation.

	Very dissatisfied	dissatisfied	Neutral	satisfied	Very satisfied
Lower approximation	0	4	47	117	88
Upper approximation	0	4	47	117	88
Object numbers	0	4	47	117	88
Accuracy of approximation	-1	1	1	1	1
Accuracy of class	1				

as diet, simple accommodation, and other services which had less than four scores. Finally, only 28 attributes are remained for analysis.

A total of 265 valid questionnaires were recovered, during the summer music season in 2019. The sampling method of this questionnaire used the purposive sampling to collect the questionnaire, and the respondents were informed that the questionnaire was researching service attributes of the homestay, and were asked to fill out the questionnaire carefully.

According to the 265 samples, the accuracy of the decision rules and the overall approximate accuracy rate are 1 as can be seen in Table 3. This indicates that the data analyzed from the questionnaire data have good correctness and explanation ability, proving the rough set applicable to this study. The study obtained 113 kinds of reduction results after calculation, but there is no core, so this study uses all attributes for decision analysis. After calculation, a total of 39 decision rules were obtained, among which 10 items with higher percentage of attributes in each category and more representative decision rules were obtained. The results are shown in Table 4.

In the decision rules summarized in Table 4, the "neutral" situation is felt to represent overall satisfaction, and the decision rules are selected to be more representative for decisions 2 and 3. Decision rule 2: if the customer is male and the satisfaction of the service attributes such as free Internet service is "neutral," then the overall satisfaction "neutral" has 9 people, accounting for all those who feel "neutral" at 19.15%. Decision rules 3: if the customer is male, the

Table 4. Decision rules.

The set of attributes	Class	Strength	Percentage
Rule 1. (Sa03 = 2)	2	2	50.00%
Rule 2. (Gender = 1) & (Sa03 = 3)	3	9	19.15%
Rule 3. (Gender = 1) & (Sa06 = 3) & (Sa09 = 3) & (Sa13 = 3)	3	9	19.15%
Rule 4. (Basic03 = 3) & (Sa01 = 4) & (Sa11 = 4) & (Sa16 = 4)	4	23	19.66%
Rule 5. (Basic01 = 1) & (Basic03 = 3) & (Sa11 = 4) & (Sa21 = 3)	4	16	13.68%
Rule 6. (Basic01 = 1) & (Basic04 = 2) & (Sa01 = 4) & (Sa14 = 4) & (Sa16 = 4)	4	31	24.50%
Rule 7. (Age = 1) & (Sa01 = 4) & (Sa16 = 4) & (Sa21 = 4)	4	12	10.26%
Rule 8. (Basic01 = 1) & (Sa03 = 5) & (Sa12 = 5) & (Sa16 = 5)	5	25	24.41%
Rule 9. (Sa02 = 5) & (Sa11 = 5) & (Sa14 = 5) & (Sa16 = 5) & (Sa21 = 5)	5	19	21.59%
Rule 10. (Basic02 = 2) & (Sa11 = 5) & (Sa16 = 5) & (Sa18= 5)	5	11	12.50%

satisfaction of the service attributes such as free Internet service, deliver the promised service on time and correctlyand the convenience of booking service are all "neutral", then the overall satisfaction "neutral" has 9 people, accounting for all those who feel "neutral" at 19.15%.

Then, for customers who are "satisfied" overall, the decision rules with higher coverage are four groups of 4, 5, 6, and 7. Here, rule 6 is more representative, which illustrates and analyzes. Decision rule 6: if the accommodation reason for tourism, the average cost range is 3000–5999, and the B&B style/interior and equipment, make customers feel comfortable and at ease, operator experience and professionalism service attributes are satisfied, then the overall satisfaction "satisfied" has 31 people, accounting for all those who feel "satisfied" at 24.50%.

Finally, for those who are "very satisfied" overall, the decision rules with higher coverage are three groups of 8, 9, and 10. Here, rule 8 is more representative,, which illustrates and analyzes. Decision Rule 8: if the accommodation reason for tourism, and free Internet service, meet customer characteristics or addition needs, operator experience and professionalism service attributes are satisfied, then the overall satisfaction "very satisfied" has 25 people, accounting for all those who feel "very satisfied" at 24.41%.

5   CONCLUSION AND SUGGESTION

This study uses the rough set to find out the decision rules of the customers when they utilize a homestay, and analyzes the main factors that affect overall satisfaction of customers, in order to provide considerations for future B&Bs to increase profit when marketing strategies. According to the rules, the B&B style/interior and equipment, operator experience and professionalism, and transportation services are important factors influencing consumer satisfaction. If the consumer is not satisfied with these three service attributes, these attributes will affect and reduce overall satisfaction. Conversely, if the consumer is satisfied with these three service attributes, there is positive impact on the overall satisfaction of the homestay. In addition, it the homestay does not provide consulting services and clearly answer customer questions, it is likely to have a negative impact on consumers.

It is recommended that the hoteliers who have visiting groups of tourists who are staying on average for four to five days upgrade the B&B style/interior and equipment, provide consulting services, and clearly answer customer questions, demonstrate operator experience and professionalism and transportation services, and so on, affect the service attributes set. For customers who stay at the homestay two or three times a year, they can simultaneously provide consulting services and clearly answer customer questions, operator experience and professionalism, understand the needs of customers, and provide transportation services, and so on. Travel-oriented service providers can simultaneously enhance free Internet service, the B&B style/interior and equipment, provide consulting services and clearly answer customer questions, meet customer characteristics or addition needs, make customers feel comfortable and at ease, demonstrate operator experience and professionalism, and so on from the service attributes set. These related factors can effectively improve the overall satisfaction of consumers with the homestay.

REFERENCES

K. C. Chiou, S. S. Lee, T. C. Liao, 2010. A study on the application of performance evaluation matrix in home stay service quality. Journal of SME Development. 16 73–110.

H. P. Chiu, J. H. Chang, W. S. Wu, 2008. Fuzzy rough set approach to association analysis of customers and purchasing patterns. Journal of Information Management. 8 59–76.

D. Gilbert, R. K. C. Wong, 2003. Customer expectations and airline services: a Hong Kong based study. Tourism Management. 24 519–532.

C. J. Lin, T. M. Yeh, C. Y. Hsieh, 2013. The comparison of using analytical hierarchy process, rough set theory and grey relational analysis in wine evaluation. Journal of Grey System. 16 85–94.

F. Y. Liu, Y. C. Chen, S. C. Tan, M. J. Lee, Y. L. Huang, 2016. Service quality world-of-mouth of B&B industry a case study of Penghu Firework Festival. International Journal of Tourism Leisure Cultural Creative Fashion Design. 1 1–10.

K. C. Mei, Z. W. Zhu, Y. H. Hsieh, W. T. Chang, 2012. A multilevel analysis factors influencing guest satisfaction and revisit intention in B&B. Journal of Rural Tourism Research. 6 29–44.

A. Morrison, P. Pearce, G. Moscardo, N. Nadkarni, J. O'Leary, 1996. Specialist accommodation: definition, markets served, and roles in tourism development. Journal of Travel Research. 35 18–26.

L. M. Su, C. H. Wang, 2015. Using customer value map to study service quality of hotels in Taiwan. 2015 The 18th Conference on Interdisciplinary and multifunctional business management. Taipei 604–617.

A. Parasuraman, V. A. Zeithaml, L. L. Berry, 1994. Alternative scales for measuring service quality: a comparative assessment based on psychometric and diagnostic criteria. Journal of Retailing. 70 201–230.

Z. Pawlak, 1982. Rough set. International Journal of Computer and Information Sciences. 11 341–356.

Mechanical & Automation engineering

Innovation in Design, Communication and Engineering – Lam et al. (eds)
© 2020 Taylor & Francis Group, London, ISBN 978-0-367-17777-5

Squeeze film behaviors of Rabinowitsch fluid lubricated long partial journal bearings

Kuang-Yuan Kung
Department of Mechatronic Engineering, Institute of Mechanical and electrical Engineering, Lingnan Normal University, Guangdong Zhanjiang, China

Lian-Jong Mou, Shan-Chi Yuan & Cheng-Hsing Hsu*
Department of Mechanical Engineering, Chung Yuan Christian University, Taiwan, ROC

Jun-Liang Chen
Department of Mechanical Engineering, Chiayi, Taiwan, ROC

ABSTRACT: In this research preliminary study of squeeze film behaviors of long-partial journal bearing using Rabinowitsch fluids as lubricant are discussed. Different from Non-Newtonian fluid commonly used in many miniaturized industrial applications (such as Fluid Dynamic Bearing, linear motion system for Optical encoder or Micro-Electro-Mechanical-System). With the help of the Dimensional analysis and modifying Reynolds stress equation; mechanical performance like pressure, load capacity, K factor, eccentricity and response time of the Rabinowitsch fluid-filled long-partial journal bearing are discussed. Result find load capacity is strongly depend on variation of K factor, which affect change slope of strain rate. Changing value of K form positive, null to negative Fluids be categorized as dilatant fluids, Newtonian fluid, pseudo-plastic accordingly.

1 INTRODUCTION

Squeeze films phenomenon developed from two vertically approaching parts has many import-ant application; including wet plates clutch (core disk to separator disk), journal bearings (shaft to bearing), gear meshing (pinion to driven), medical artificial joints (cartilage to tibia), cam pairing (cam to follower), piston ping bearing (pin to bearing) (Bhushan 1999: 939-990). It can be easily find from past studies (Khonsari et al. 2017: 263-298) that in squeeze film applications non-Newtonian fluid is more practical than the Newtonian fluid.

Models build by Rheology researchers studies in non-Newtonian fluids can be describe from: power-law, couple stress and cubic shear stress fluids etc. In 00's (Lin 1997) studies on couple stress fluid model in finite journal bearings shows that the enhancement on load-carrying capacity with Newtonian fluids. (Naduvinamani et al. 2001) on short porous journal bearing, and (Mokhiamer et al. 1999) on finite journal bearing by conducting dimensional analysis on different geometries. Process they took on both research serve as good guideline for dimensional analysis process used in squeeze film studies. (Gupta et al. 1988) and (Ramanaiah et al. 1978) found that flow rate and load-capacity are two most inferential indexes in performance of Hydrostatics thrust bearing. (Shukla et al. 1974) discussing power law fluid model used in step bearings by introducing k empirical parameter for Rabinowitsch fluids. (Buckholz 1986) tried and compare multiple mathematical approach on plane slider bearings. As well as (Jianming et al. 1989) by

*Corresponding author: chhsu88@gmail.com

numerical method. Lastly, (Singh et al. 1982) analysis Elastohydrostatic lubricant in circular plate thrust bearing by power law and formula changing.

Nowadays, the (Rabinowitsch 1929) fluid model has been repeatedly used in study of the non-linear behaviors for non-Newtonian lubricants. This model describes the relationship between cubic shear stress and the shear rate as:

$$\tau_{xy} + k\tau_{xy}^{3} = \mu \frac{\partial u}{\partial y} \tag{1}$$

where μ represents the initial viscosity and k is the empirical parameter. Apparently; when $k = 0$ is called the Newtonian fluid, when $k < 0$ is called the dilatant fluids, and when $k > 0$ is called the pseudo-plastic fluids.

Earliest theory and the empirical equation of the Rabinowitsch fluid model have been presented by(WADA et al. 1971). Followed by many studies likes, (Lin 2012) for parallel annular disks, (Naduvinamani et al. 2001) for circular stepped plates, (Singh 2011) for pivoted curved slider bearings, (Hsu et al. 2014) for conical bearings, and in (Hung 2009) for parallel plates. Demonstrate the usability and practical properties of Rabinowitsch fluid model.

Researchers above propose many different kinds of geometry and plenty of in-detailed results. However, there is no study had done by using Rabinowitsch fluid model on the squeeze film between the long partial journal bearings. Therefor in this paper, this geometry is choose and proposed. Aiming to provide more information and utilizing squeeze film characteristic for biomedical or Mechanical engineers' aid.

The analysis of squeeze film behaviors about long partial journal bearing are carried out by solving a non-linear modified Reynolds equation with Rabinowitsch model. And applying perturbation method on Pressure differential terms. The solution are composed by two perturbed pressure sub-solutions into a closed-form equation. Then load–carrying capacity and the respond time can be solved accordingly.

2 ANALYSIS

Figure 1 shows the geometrical of a long-partial journal bearing lubricated by Rabinowitsch fluid. The film thickness is h, the radius of journal is R and the eccentricity is ε. The equations of continuity, momentum equation and the stress equation refer from Rabinowitsch fluid model are showed as follow:

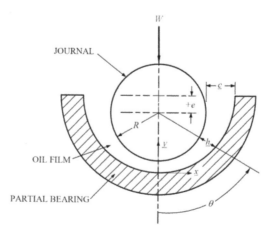

Figure 1.   Geometrical of a long partial journal bearing.

continuity:
$$\frac{\partial u}{\partial x} + \frac{\partial v}{\partial y} = 0 \tag{2}$$

momentum:
$$\frac{\partial \tau_{xy}}{\partial y} = \frac{\partial p}{\partial x} \tag{3}$$

B.C.:
$$\frac{\partial p}{\partial z} = 0 \tag{4}$$

stress equation:
$$\tau_{xy} + k\tau_{xy}^3 = \mu \frac{\partial u}{\partial y} \tag{5}$$

The boundary conditions applied in geometry shown with Figure 1 are

$$u = 0 \text{ at } y = 0 \text{ and } u = 0 \text{ at } y = h \tag{6}$$

$$u = 0 \text{ at } y = 0 \tag{7a}$$

$$v = \frac{\partial h}{\partial t} \text{ at } y = h \tag{7b}$$

By substituting equation (3) to differentiated equation (5) then perform integration on equation (5) with the boundary conditions (6) and (7) the solution of velocity of the squeeze film can be described as (8)

$$u = \frac{1}{\mu} \left[\begin{array}{l} \frac{1}{2}(y^2 - yh)\frac{\partial p}{\partial x} \\ +k(\frac{1}{4}y^4 - \frac{1}{2}hy^3 + \frac{3}{8}h^2y^2 - \frac{1}{8}h^3y)\left(\frac{\partial p}{\partial x}\right)^3 \end{array} \right] \tag{8}$$

Substituting the equation (8) into (2), with B.C (7b) after integration acquire a modified Reynolds equation as (9)

$$\frac{\partial}{\partial x}\left\{ \left[\frac{1}{12\mu}\frac{\partial p}{\partial x}h^3 + \frac{k}{80\mu}(\frac{\partial p}{\partial x})h^5 \right] \right\} = \frac{\partial h}{\partial t} \tag{9}$$

According to (Hamrock 1994). boundary conditions for squeeze film lubricant can be described as (10) and (11):

$$\frac{dp}{d\theta} = 0 \text{ at } \theta = 0 \tag{10}$$

$$p = 0 \text{ at } \theta = \pm\frac{\pi}{2} \tag{11}$$

Applying dimensionless analysis for equation (9), (10) and (11) transforming them into (12), (13) and (14) respectly.

59

$$\frac{\partial}{\partial \theta}\left\{\left[h^{*3}\frac{\partial p^*}{\partial \theta} + K\frac{3}{20}h^{*5}\left(\frac{\partial p^*}{\partial \theta}\right)^3\right]\right\} = -12\cos\theta \qquad (12)$$

$$\frac{dp^*}{d\theta} = 0 \text{ at } \theta = 0 \qquad (13)$$

$$p^* = 0 \text{ at } \theta = \pm\frac{\pi}{2} \qquad (14)$$

Where K is the non-linear dimensionless factor a transformed k from the Rabinowitsch fluid model.

When K is said to Newtonian fluid. For $K>0$ become the dilatant fluids. For $K<0$ become the pseudo-plastic fluids.

Perturbation method is used to solve the equation (12), by first assuming represent of equation (15a)

$$p^* = p_0^* + K \cdot p_1^* + O(K^2) \qquad (15a)$$

Since $K^2 \ll 1$, equation (15a) can conveniently simplify to $O(K^2) = 0$ and keeping the first order term as (15b):

$$p^* = p_0^* + K \cdot p_1^* + O(K^2) \qquad (15b)$$

Substituting (15b) into the nonlinear modified Reynolds equation (12), it can be rearranged and split into two Ordinary Differential Equations (16), (17) as functions of p_0^* and p_1^*.

$$\frac{\partial}{\partial \theta}\left\{h^*\frac{\partial p_0^*}{\partial \theta}\right\} = -12\cos\theta \qquad (16)$$

$$\frac{\partial}{\partial \theta}\left\{h^{*3}\left[\frac{\partial p_1^*}{\partial \theta} + \frac{3}{20}h^{*5}\left(\frac{\partial p_0^*}{\partial \theta}\right)^3\right]\right\} = 0 \qquad (17)$$

Solving equation (16) and (17) with the boundary conditions (13) and (14), the sub-solutions of and can be acquired. Based on this result, approximation film pressure can be expressed by (15b) as below:

$$p^* = -12\int\frac{\sin\theta}{h^{*3}}d\theta + K \cdot \frac{1296}{5}\int\frac{\sin^3\theta}{h^{*7}}d\theta \qquad (18)$$

Then the load-carrying capacity per unit length is expressed as:

$$W = \int p\cos\theta \cdot Rd\theta \qquad (19)$$

By dimensional analysis the load-carrying capacity in a dimensionless form is derived.

$$p^* = -12\int\frac{\sin\theta}{h^{*3}}d\theta + K \cdot \frac{1296}{5}\int\frac{\sin^3\theta}{h^{*7}}d\theta \qquad (20)$$

The differential equation of the dimensionless response time as a function of eccentricity is expressed as:

60

$$\frac{d\varepsilon}{dt^*} = \frac{1}{W^*} \tag{21}$$

From (20) (21) the dimensionless response time is described as:

$$t^* = \frac{Wc^2}{\mu R^3} t \tag{22}$$

Assuming initial position of journal was in central of the bearing thus initial condition for eccentricity ratio is.

$$\varepsilon = 0 \text{ at } t^* = 0 \tag{23}$$

After substituting (20), (21) into (22) and integrating, the solution of response time is obtained as (24).

$$t^* = \int_{\varepsilon=0}^{\varepsilon} \left[12 \cdot \int_{\theta=-\pi/2}^{\theta=+\pi/2} \frac{\sin^2\theta}{h^{*3}} \, d\theta \right] d\varepsilon + k \cdot \int_{\varepsilon=0}^{\varepsilon} \left[-\frac{1296}{5} \int_{\theta=-\pi/2}^{\theta=+\pi/2} \frac{\sin^4\theta}{h^{*7}} \, d\theta \right] d\varepsilon \tag{24}$$

3 RESULTS AND DISCUSSION

From solutions above (18), (20), (24) the non-Newtonian characteristics of squeeze film performances are all related to nonlinear dimensionless factor K.

Figure 2 shows a decreasing tendency for load capacity W^* with increasing value of K factor. As the Rabinowitsch fluids factor decrease, load capacity increase as long as ε increase.

Figure 3 illustrates dimensionless film pressure distribution p^* across θ (angle). This distribution represent the amount of loading force been spread in squeezed film, these curves show a similar trend as (Ayyappa et al. 2015). As expected, highest film pressure are presence at the center of the partial bearing (θ=0). Film pressure increase as value decrease, the maximum film pressure appear at θ=0 for k=-0.0014 gain 21% against Newtonian case. This increase in value indicate a good performance improvement by using Rabinowitsch-modeled fluid as bearing lubricant.

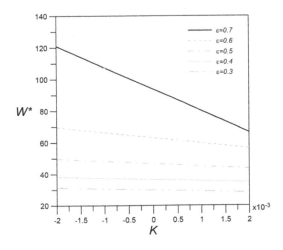

Figure 2. W^* with K for different ε.

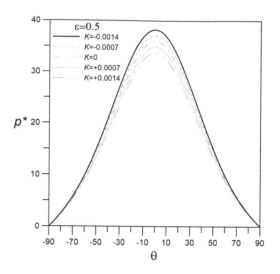

Figure 3. p^* distribution across θ for different K when $\varepsilon=0.5$.

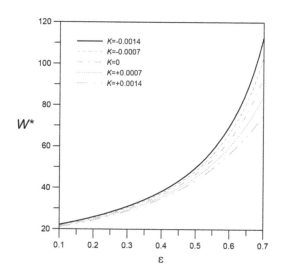

Figure 4. W^* with ε for different K factor.

Figure 4 Load capacity W^* change with of eccentricity ε.

Under dilatant region ($K=-0.0007$, $K=-0.0014$) yield 22% more load capacity then Newtonian fluid ($K=0$). Conversely, for pseudo-plastic region ($K=0.0007$, $K=0.0014$) the load capacity W^* decreases as value of K increase.

These relative values of curves is roughly on prat with (Lin et al. 2013) did, but unlike shallow two step trend Lin has, load capacity have smoother increase in value form $\varepsilon=0.5$ to 0.7. Showing a characteristic of uniformed load capacity for possible better utilization.

Figure 5 change of the dimensionless response time with eccentricity

Compared with Newtonian fluid ($K=0$), dilatant fluids ($K=-0.0007$, $K=-0.0014$) has slower response time (whose curves sets above the Newtonian fluid). For the pseudo-plastic fluid ($K=0.0007$, $K=0.0014$) are sets below Newtonian's.

In all, change in response time caused by Rabinowitsch-modeled fluid are not very obvious difference from the Newtonian's.

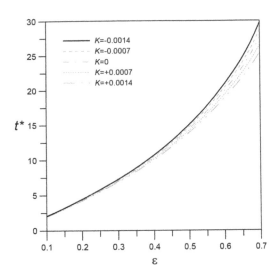

Figure 5.   With ε for different K factor.

4   CONCLUDING REMARKS AND CONCLUSION

The research of the non-Newtonian squeeze film characteristics on the long partial journal bearing with the Rabinowitsch fluid model has been successfully solved in this paper. According to the results and discussions above, conclude that.

A nonlinear modified Reynolds equation of long partial journal bearing was derived by substituting the cubic stress equation. The perturbation method was used for solving first order closed-form solutions witch describing squeeze film characteristics.

Considering all squeeze film behaviors as whole load capacity, response time. Dilatant fluids hold strong advantage over Newtonian and pseudo-plastic fluids.

There is in-general 22% performance improvement by introducing Rabinowitsch-modeled non-Newtonian fluid in long partial journal bearings as lubricant.

Nomenclature:

c radial clearance
e eccentricity x/R
h film thickness,
h^* dimensionless film thickness, $h/c = 1 - \varepsilon \cos \theta$
k is the empirical parameter for Rabinowitsch fluids
K nonlinear dimensionless factor responsible for the properties of Rabinowitsch fluids,
$\quad K = k \frac{\mu^2 R^2 (d\varepsilon/dt)^2}{c^2}$
p film pressure
p^* dimensionless film pressure, $p^* = pc^2/\mu R^2 (d\varepsilon/dt)$
R radius of the journal
t response time
t^* dimensionless response time, $t* = Wc^2 t/\mu R^3$
u,v fluid velocity components in the x, y directions
W load-carrying capacity per unit length of the bearing
W^* dimensionless load-carrying capacity per unit length of the bearing, $W* = Wc^2/\mu R^3 (d\varepsilon/dt)$
x,y rectangular coordinate
ε eccentricity ratio, e/c
θ circumferential coordinate
μ initial viscosity of the lubricant
τ shear stress

REFERENCES

Bhushan, Bharat. 1999. John Wiley & Sons.

Khonsari, Michael M, and E Richard Booser. 2017. John Wiley & Sons.

Lin, Jaw-Ren. 1997. Wear, 206: 171–178.

Naduvinamani, NB, PS Hiremath, and G Gurubasavaraj. 2001. Tribology International, 34: 739–747.

Mokhiamer, UM, WA Crosby, and HA El-Gamal. 1999. Wear, 224: 194–201.

Gupta, RS, and LG Sharma. 1988. Wear, 125: 257–269.

Ramanaiah, G, and Priti Sarkar. 1978. Wear, 48: 309–316.

Shukla, JB, and M Isa. 1974. Wear, 30: 51–71.

Buckholz, RH. 1986. Journal of Tribology, 108: 86–91.

Jianming, Wang, and Jin Gaobing. 1989. Wear, 129: 1–11.

Singh, Chandan, TS Nailwal, and Prawal Sinha. 1982. Journal of Lubrication Technology, 104: 243–247.

Rabinowitsch, B. 1929. Zeitschrift für physikalische Chemie, 145: 1–26.

WADA, Sanae, and Hirotsugu HAYAsHI. 1971. Bulletin of JSME, 14: 279–286.

Lin, Jaw-Ren. 2012. Tribology international, 52: 190–194.

Singh, Udaya P. 2011. In.

Hsu, Cheng-Hsing, Jaw-Ren Lin, Lian-Jong Mou, and Chia-Chuan Kuo. 2014. Industrial Lubrication and Tribology, 66: 373–378.

Hung, CR. 2009. Education Specialization P, 97: 87–97.

Hamrock, Bernard J. 1994. Inc. Hightstown, NJ, 8520.

Ayyappa, GH, NB Naduvinamani, A Siddangouda, and SN Biradar. 2015. Tribology in Industry, 37.

Lin, JR, PJ Li, and TC Hung. 2013. FDMP: Fluid Dynamics & Materials Processing, 9: 419–434.

Innovation in Design, Communication and Engineering – Lam et al. (eds)
© 2020 Taylor & Francis Group, London, ISBN 978-0-367-17777-5

Simulating the load torque of rotary valve in MWD with magnetic powder brake

Long Wang & Yue Shen
College of Science, China University of Petroleum, Qingdao, Shandong, P.R. China

Jia Jia
School of Petroleum Engineering, China University of Petroleum, Qingdao, Shandong, P.R. China

Yan Ling, Ling-tan Zhang & Ling-zhi Wei
College of Science, China University of Petroleum, Qingdao, Shandong, P.R. China

ABSTRACT: The load torque of the rotary valve which is used in the measurement while drilling (MWD) tool has the rapid change characteristics with the rotation angle or time. In the simulation experiment of the rotary valve speed control, it is necessary to simulate the variation law of the load torque of the rotary valve. The magnetic powder brake is an ideal load torque simulation component, and the time constant of the braking torque seriously affects the simulation of the dynamic load torque. By increasing the variation of excitation voltage, the steady state value of the local braking torque can be quickly reached. By establishing the mathematical model of the excitation voltage increment compensation, the variation of the excitation voltage which is expressed by the time discrete parameter can be constructed, and the braking torque loading speed will be improved. Research shows that the incremental compensation of the excitation voltage can increase the dynamic response speed of the braking torque to a certain extent and greatly reduce the response time of the braking torque reaching to the steady state. The experimental analysis shows that the acceleration effect of the braking torque can be increased by 26 times when the excitation voltage sampling period is 75ms. The dynamic simulation of the load torque of the rotary valve can be realized by compensating the excitation voltage of the magnetic powder brake incrementally.

Keywords: magnetic powder brake, time constant, dynamic response speed of braking torque, measurement while drilling (MWD), load torque of the rotary valve

1 INTRODUCTION

The magnetic powder brake is an ideal load torque simulation component that changes the braking torque by changing the current or voltage of the excitation coil. The time when the braking torque reaches the steady state under the pulsed excitation current is 4 times of its time constant, the time constant is usually several hundred milliseconds, causing a larger time delay between the steady state braking torque and the exciting current, which seriously restricts the simulation of high-speed varying load torque [1]. Measurement while drilling (MWD) is a modern drilling technology that measures and transmits downhole information in real time during drilling [2-3]. Rotary valve is an important component of MWD tools, which is used to generate continuous drilling fluid pressure waves in the drill collar to upload downhole measurement information to the ground. Because the speed control of the rotary valve is related to the quality of the drilling

---

*Corresponding author: sheny1961@163.com

fluid pressure signal generation, the control effect of the speed control strategy needs to be verified by simulation experiments. The key of the rotary valve speed control simulation experiment
is the simulation of the rotary valve load torque. When using the computer data acquisition
system to control the excitation voltage of the magnetic powder brake, it is required that the
braking torque should reach the steady state during the sampling period, and the sampling
period should be controlled within 100ms. However, the present method of improving the magnetic powder braking torque loading speed can only provide less than 3 times the acceleration
effect [4-6], which can not meet the requirements of the rotary valve speed control simulation
experiment. In this paper, a method is introduced for improving the dynamic response speed of
the magnetic powder braking torque greatly by the incremental compensation of the excitation
voltage to realize the simulation of the load torque of the rotary valve.

2 LOAD TORQUE CALCULATION MODEL OF ROTARY VALVE

The rotary valve is part of the MWD tools and is mounted inside the drill collar at the top of
the drill bit. It consists of a stator and a rotor that moves relatively to the stator. The rotor is
driven by the motor with a reducer [7]. Studies have shown that the load torque varies with
the rotation angle in a non-monotonic complex law [8], and its calculation model can be
expressed as a polynomial function of the rotation angle

$$M_r(\theta) = b + \rho Q^2(a_0 + a_1\theta + a_2\theta^2 + L + a_m\theta^m) \tag{1}$$

Where, θ is the rotation angle of the rotary valve; Q is the flow rate of the drilling fluid; ρ is
the density of the drilling fluid; b is a constant term; a_i are polynomial coefficients,
$i = 0, 1, 2, \cdots, m$; m is the number of polynomial term.

Set the steady moving speed of the rotary valve as n_s, and convert the rotation angle into
a function of time, $\theta = 2\pi n_s t = k_a t$, then the Eq. (1) can be derived as the calculation model of
load torque with time

$$M_r(t) = b + \rho Q^2[a_0 + a_1 k_a t + a_2(k_a t)^2 + L + a_m(k_a t)^m] \tag{2}$$

3 INCREMENTAL COMPENSATION METHOD OF EXCITATION VOLTAGE TO IMPROVE THE DYNAMIC RESPONSE SPEED OF BRAKING TORQUE

3.1 *Mathematical model*

The steady-state braking torque generated by the magnetic powder brake is in linear relationship with the excitation voltage, and the steady-state braking torque can be expressed as

$$M_s = kV \tag{3}$$

Where, k is the coefficient, V is the excitation voltage.

If the braking torque generated by the magnetic powder brake is indicated by $M_m(t)$. When
the excitation voltage changes, the braking torque has a certain time delay in response to the
excitation voltage, and the time constant of the amount of change in the braking torque generated by the magnetic powder brake is τ_m. The value of change in the braking torque can be
expressed as

$$\Delta M_m(t) = \Delta M_s(e^{at} - e^{-t/\tau_m}) \tag{4}$$

In Eq. (4), ΔM_s is the steady-state value of braking torque increment, $0 \le t \le \Delta t$ is the action time of the excitation voltage increment, Δt is the duration of the excitation voltage increment, and a is the coefficient.

The change rate of the braking torque increment with time can be defined as the dynamic response speed of the braking torque, then

$$R(t) = \frac{d\Delta M_m(t)}{dt} = \Delta M_s(ae^{at} + \frac{1}{\tau_m}e^{-t/\tau_m}) \tag{5}$$

In the case where the duration of the excitation voltage increment is smaller, the presence of the time constant causes a decrease in the braking torque response speed. In order to solve this problem, in the duration of the excitation voltage increment, the steady state value of the braking torque is increased by increasing the excitation voltage increment, so as to compensate for the decrease of the braking torque variation caused by the time constant, which can accelerate the response speed of the braking torque.

If the expected value of the braking torque change within Δt is set as ΔM_e, then

$$\Delta M_e = \Delta M_s(e^{a\Delta t} - e^{-\Delta t/\tau_m}) \tag{6}$$

At a certain time t_0, the expected value of the braking torque change amount $\Delta M_e(t_0)$ should be equal to the change amount of the rotary valve load torque $\Delta M_r(t_0)$ at that moment, that is $\Delta M_e(t_0) = \Delta M_r(t_0)$. Therefore, the steady state value of the braking torque increment at time t_0 should be reached to $\Delta M_s(t_0)$ and can be expressed as

$$\Delta M_s(t_0) = \frac{\Delta M_r(t_0)}{e^{a\Delta t} - e^{-\Delta t/\tau_m}} \tag{7}$$

Since $M_s = kV$ and its differential is $\Delta M_s = k\Delta V$, the steady state braking torque change at a certain time t_0 should be

$$\Delta M_s(t_0) = \frac{\Delta M_r(t_0)}{e^{a\Delta t} - e^{-\Delta t/\tau_m}} - k\Delta V(t_0) \tag{8}$$

The excitation voltage increment $\Delta V(t_0)$ applied to the magnetic powder brake at time t_0 should conform to the following mathematical model

$$\Delta V(t_0) = \frac{\Delta M_r(t_0)}{k(e^{a\Delta t} - e^{-\Delta t/\tau_m})} \tag{9}$$

It can be seen from Eq. (9) that as the decrease of Δt, the increment $\Delta V(t_0)$ of the excitation voltage is gradually increased, so that the variation of the braking torque generated by the magnetic powder brake is increased. An effective compensation of the excitation voltage increment to the braking torque is realized.

3.2 Improvement and analysis of dynamic response speed of the braking torque

At the time t_0, after the excitation voltage increment compensation, the action time of the excitation voltage increment reaches the action duration, the dynamic response speed at the end of the action duration is

$$R_m(t_0 + \Delta t) = \Delta M_s(t_0)(ae^{a\Delta t} + \frac{1}{\tau_m}e^{-\Delta t/\tau_m}) \tag{10}$$

Since $\Delta M_s(t_0) = \frac{\Delta M_r(t_0)}{e^{a\Delta t} - e^{-\Delta t/\tau_m}}$, then

$$R_m(t_0 + \Delta t) = \Delta M_r(t_0)\frac{\left(ae^{a\Delta t} + \frac{1}{\tau_m}e^{-\Delta t/\tau_m}\right)}{e^{a\Delta t} - e^{-\Delta t/\tau_m}} \tag{11}$$

At time t_0, the dynamic response speed when the excitation voltage increment is not compensated at the end of the action duration is

$$R(t_0 + \Delta t) = \Delta M_r(t_0)\left(ae^{a\Delta t} + \frac{1}{\tau_m}e^{-\Delta t/\tau_m}\right) \tag{12}$$

Then, the improvement of the dynamic response speed of the braking torque can be expressed as

$$G(\Delta t) = \frac{R_m(t_0 + \Delta t)}{R(t_0 + \Delta t)} = \frac{1}{e^{a\Delta t} - e^{-\Delta t/\tau_m}} > 1 \tag{13}$$

The smaller $\Delta t/\tau_m$, the larger $G(\Delta t)$, and the bigger improvement of speed.

After the excitation voltage increment is compensated, the braking torque generated by the magnetic powder brake can reach the desired value $\Delta M_r(t_0)$ during the duration Δt, and the time taken to reach $\Delta M_r(t_0)$ without the excitation voltage increment compensation is $4\tau_m$. Since $\Delta t < 4\tau_m$, the time for reaching the expected value $\Delta M_r(t_0)$ is significantly reduced, the time occupancy rate is

$$S(\Delta t) = \frac{\Delta t}{4\tau_m} = \frac{\Delta t/\tau_m}{4} \tag{14}$$

Similarly, the smaller $\Delta t/\tau_m$, the shorter time occupancy rate of the braking torque to reach the desired value.

Let $\Delta V(t)$ and $\Delta M_r(t)$ be expressed by differential, with $dV(t) = \frac{dM_r(t)}{k(e^{a\Delta t} - e^{-\Delta t/\tau_m})}$, and integrates it to obtain the excitation voltage that should be applied at time as follows

$$V(t_0) = \int_{M_r(0)}^{M_r(t_0)} \frac{dM_r(t)}{k\left(e^{a\Delta t} - e^{-\Delta t/\tau_m}\right)} + V(0) \tag{15}$$

Since $M_r(0) = kV(0)$, $V(0) = \frac{M_r(0)}{k}$

Eq. (15) can be derived as $\quad V(t_0) = \frac{M_r(t_0) - M_r(0)}{k\left(e^{a\Delta t} - e^{-\Delta t/\tau_m}\right)} + \frac{M_r(0)}{k} \tag{16}$

Where,

$$M_r(t_0) = b + \rho Q^2[a_0 + a_1 k_a t_0 + a_2(k_a t_0)^2 + L + a_m(k_a t_0)^m], \quad M_r(0) = b + a_0 \rho Q^2.$$

Let the sampling period of the computer data acquisition system be T, and the time sequence number is N, then $t = NT$ and $\Delta t = T$. If N_0 is the time sequence number of time t_0, then $t_0 = N_0 T$, and Eq. (16) will be expressed by the time discrete parameter as

$$V(N_0 T) = \frac{M_r(N_0 T) - M_r(0)}{k(e^{aT} - e^{-T/\tau_m})} + \frac{M_r(0)}{k} \tag{17}$$

where,

$$M_r(N_0 T) = b + \rho Q^2 [a_0 + a_1 k_a N_0 T + \mathrm{L} + a_m (k_a N_0 T)^m]$$

$$M_r(0) = b + a_0 \rho Q^2.$$

4 SIMULATION EXPERIMENT OF ROTATING VALVE LOAD TORQUE

The magnetic powder brake used in the experiment is FKG-50YN and the maximum braking torque is 5N $\cdot$ m. In experiments, the response characteristic of the braking torque increment is obtained as $\Delta M_m(t) = 0.58(e^{0.001t} - e^{-2t})$, therefore $\tau_m = 500$ms, and $a = 0.001$.

The simulation experiment of the rotary valve load torque is carried out under the control of the computer data acquisition system. The Labview software is used to output the pulse voltage according to Eq. (17) to the magnetic powder brake for the incremental compensation of the excitation voltage, and the braking torque is measured under the sampling period of $T = 75$ms and $T = 100$ms, the variation of braking torque with time is shown in Figure 1. It can be seen that the theoretical calculation curve agrees with the experimental curve well in amplitude except for some delay in time. Experimental analysis shows that the time delay will be greatly shorten when the sampling period is more than 100ms, indicating a better effect of incremental compensation of the excitation voltage. Besides, the time delay in the case of sampling period of $T < 100$ms begins to increase greatly, indicating that there is a physical limit to the moving speed of the magnetic powder, so the sampling period must not be set too short. If the influence of the time delay on the rotational speed control of the magnetic powder brake is ignored at a certain extent, in order to increase the rotational speed of the magnetic powder brake as much as possible, the sampling period can be determined as $T = 75$ms, and the dynamic response speed of the braking torque is improved as follows

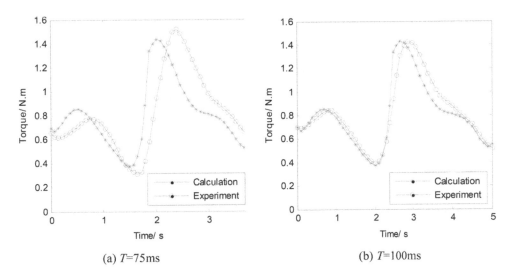

(a) T=75ms   (b) T=100ms

Figure 1. Experimental simulation of magnetic powder braking torque to load torque of the rotary valve under excitation voltage increment compensation.

$$G(T) = \frac{1}{e^{aT} - e^{-T/\tau_\mathrm{m}}} = 7.18 \qquad (18)$$

And the time occupancy rate will be

$$S(T) = T/4\tau_\mathrm{m} = 3.7\% \qquad (19)$$

From the calculation results above, the incremental compensation method of the excitation voltage can greatly increase the dynamic response speed of the braking torque, and greatly shorten the response time of the braking torque reaching to the steady state. The acceleration effect is improved by 26 times in terms of response time savings.

Set the sampling period be $T = 75\mathrm{ms}$ and the time sequence number be $N = 40$, and the braking torque of the magnetic powder brake is controlled according to Eq. (17). The variation law of the rotating torque of the rotary valve can be simulated under the condition of the speed of 5r/min.

5 CONCLUSION

(1) The excitation voltage increment compensation method can improve the dynamic response speed of the braking torque to a certain extent. In theory, the shorter the action duration, the quicker the dynamic reaction speed, but the moving speed of the magnetic powder is limited by the physical limit, so the action duration could not be too short.
(2) The incremental compensation of the excitation voltage can greatly shorten the response time of the braking torque to the steady state, and realize the dynamic simulation of the load torque of the rotary valve.

ACKNOWLEDGEMENT

This paper is supported by the National Natural Science Foundation of China (Project No.51274236).

REFERENCES

[1] Liu Jingliang, Song Ying, Liu Fei, etal. Research of magnetic powder brake loading characteristic. Aviation Precision Manufacturing Technology, 2013, 49(2): 52–56.
[2] Martin C A, Philo R M, Decker D P, etal. Innovative advances in MWD: IADC/SPE 27516, Richardson: SPE, 1994.
[3] Monroe S P. Applying digital data-encoding techniques to mud pulse telemetry: SPE 20326, Richardson: SPE, 1990.
[4] Xue Chao, Bai Guozhen. Application and research of self-adaptive fuzzy PID controller in magnetic powder brake loading system. Journal of Mechanical and Electrical Engineering, 2016, 33(2): 217–220.
[5] Yuan Yufeng, Ren Fang, Yang Zhaojian. Research on Modeling and Control Method of Magnetic Particle Brake. Machinery Design and Manufacture, 2018, (7): 13–15.
[6] Xu Kai, Bai Guozhen. A control study on the magnetic powder brake loading system. Electronic Sci and Tech, 2017, 30(10): 46–49.
[7] Malone D. Sinusoidal pressure pulse generator for measurement while drilling tools: U.S. 4847815). 1989–07–11.
[8] Shen Yue, Yin Di, Jia Jia, etal. Analysis of influence of load torque on speed of rotary valve and speed control. Science Technology and Engineering, 2018, 18(12): 71–77.

Innovation in Design, Communication and Engineering – Lam et al. (eds)
© 2020 Taylor & Francis Group, London, ISBN 978-0-367-17777-5

Study of heat transfer performance of an impingement jet on a plate heat sink

Chien-Nan Lin* & Yueh-Hung Lee
Department of Mechanical Engineering, Far East University, Xinshi, Tainan, Taiwan

ABSTRACT: This study investigates the heat transfer characteristics of an air-impinging jet on a plate heat sink. The boosted air releases from the nozzle incident on the surface of the heat sink, which is heated on the bottom side with uniform heat flux. In this study, the temperature and velocity distribution are presented for a varied Reynold number, the impingement height. An experiment facility is set up in order to examine the numerical results. The results show that a longer distance from the impinging of the nozzle covers a wider range of the fin's surface, but with a lower incident velocity than a small distance.

1   INTRODUCTION

Methods to improve the heat transfer efficiency for electric devices include: increasing the convective heat transfer coefficient, increasing the heat transfer surface area, and improving the fluid flow pattern. Generally, the most common method utilized is to employ an axial fan to take the heat from the heat-dissipating fins. Most of the fans are directly locked on the heat-dissipating fins, causing the distance between the fan and the heat-dissipating fins to close so as to infiltrate air. The surface below the fan hub forms a hollow zone without an air stream flowing through and reducing heat dissipation. An impingement jet is a device that can effectively improve heat transfer and mass transfer for the heat sink. It is widely used in various industries, such as steel plate cooling, paper drying, turbine cooling, etc., and is much more efficient than fan cooling, which meets the high-density heat dissipation requirement.

Previous works [1–3] have studied the fin arrangement on the heat sink and propose the optimum fin parameters. For the impingement jet cooling method, several studies [4–6] have shown that the geometry of the fins, the vertical distance from the nozzle to the fins, the fin material, and the Reynolds number affect the heat dissipation of the fins.

2   THEORY ANALYSIS

The physical model of the numerical simulation used in this study is a flat-plate heat sink fin, as shown in Figure 1. The dimensions of the base geometry are 49 mm (length) x 40 mm (width) x 5 mm (thickness), the dimensions of the fin width are 31 mm (height) x 40 mm (width) x 1 mm (thickness), with a fin spacing of 2 mm. The flat fins are located in the center of the computational domain; an air nozzle is placed directly above the fins. One quarter fin is adopted in this numerical simulation so as to save computer time. Thus, the symmetry boundary condition is used on both sides of the computational domain. The other sides and the upper surface are considered as pressure outlet boundaries. The bottom surface of the heat sink is a stationary wall; a 10-watt uniform heat is applied as a heat source. The impingement jet is placed at the upper surface.

---

*Corresponding author: lincn@mail.feu.edu.tw

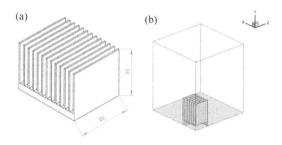

Figure 1. (a) fin geometry and (b) computational domain.

The k-ε turbulent model is utilized to couple with the Navier-Stokes equation in solving the impingement jet flow. A steady state energy equation is used to describe the heat flow of the air stream. A three-dimensional conduction equation solves the conjugated heat transfer on the surface heat sink and the air stream.

A hybrid grid system is used in this numerical simulation; the mesh is a uniform structured grid except around the nozzle jet. Careful grid tests are examined before simulation; the total grid number varies with the impingement height; for example, a total 1,200,00 grids is meshed for an impingement height of y = 100 mm.

Experimental Facility

In this study, the K-type thermocouples are used to measure the temperature, and a total of 36 positions are measured between the fins. Figure 2 shows the schematic of the experimental facility. The parameter combinations in the numerical section are the impact Reynolds number (Re = 3,000–12,000) and the impact distance (y = 60–180 mm). The uncertainty of the experimental measurement is estimated at about 20%.

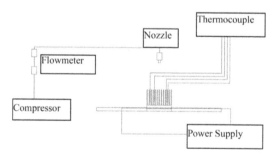

Figure 2. Schematic of experiment.

3   RESULTS AND DISCUSSION

Figure 3 presents the flow track of the air stream released from the jet incident on the fin surface of the heat sink. In this case, Re = 3,000 and the nozzle height is 60 mm above the heat sink. The air temperature gradually increases as it leaves the nozzle, and then increases quickly while touching the fin surface, finally slightly decreasing when exiting the fin area, owing to mixing with the cold air around the heat sink.

When the Re increases to 6,000, the maximum temperature of the air stream decreases from 396K to 384K, as shown in Figure 4. The flow track is almost similar to Re = 3,000, but some circulation flow appears because the velocity of the air stream is large enough to strike on the base surface of the heat sink.

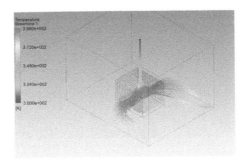

Figure 3.   The flow track of air stream Re = 3,000, L = 60 mm.

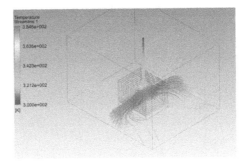

Figure 4.   The flow track of air stream Re = 6,400, L = 60 mm.

Figure 5.   The flow track of air stream Re = 3,000, L = 100 mm.

The flow track for Re = 3000 at L = 100mm shown in Figure 5 is similar to those of Figure 3 and 4. Meanwhile, the temperature distribution of the air stream and fin is showed in Figure 6. As expect, the middle zone of fin has lower temperature than side zone, owing to the air stream incident.

Figure 7 presents the thermal resistance, R, varied with impingement height, y, for a different Re. The thermal resistance decreases with increasing Re. The impingement height of the minimum thermal resistance decreases slightly with increasing Re, but the minimum is all around y = 120 mm. Note that the results are based on the numerical simulation; the experimental measurement is at about a 20% deviation from this curve.

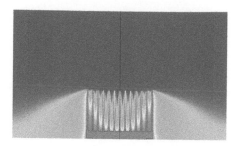

Figure 6.   The temperature distribution of the air stream and heat sink for Re = 3,000, L = 100 mm.

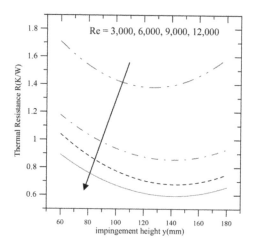

Figure 7.   The thermal resistance varied with impingement height, y, for a different Re.

4   CONCLUSION

In this study, the numerical simulation is used to investigate how the three-dimensional steady-state temperature distribution and the airflow of the jet impact the flat fins. The results of numerical simulation and experimental measurement show that the airflow impact velocity decreases with increasing impingement height, but increases with the spread zone of the fins, which results in a best impact distance around y = 120 mm, which means the thermal resistance is the lowest value.

REFERENCES

Bejan, A., and Sciubba, E. 1992. *International Journal of Heat and Mass Transfer*, 3259–3264.
Brignoni, L. A., and Garimella, S. V. 1999. *IEEE Transactions on Components and Packaging Technologies*, 399–404.
Katti, V., and Prabhu, S. V. 2008. *International Journal of Heat and Mass Transfer*, 4480–4495.
Maveety, J. G., and Hendricks, J. F. 1999. *Journal of Electronic Packaging*, 156–161.
Rammohan Rao, V., and Venkateshan, S. P. 1996. *International Journal of Heat and Mass Transfer*, 779–789.
el-Sheikh, H. A., and Garimella, S. V. 2000. *IEEE Transactions on Components and Packaging Technologies*, 300–308.

Innovation in Design, Communication and Engineering – Lam et al. (eds)
© 2020 Taylor & Francis Group, London, ISBN 978-0-367-17777-5

How can the automobile industry implement a circular economy?

Yih-Sheng Chen* & Huann-Ming Chou
Mechanical and Energy Engineering, Kun Shan University, Tainan, Taiwan

ABSTRACT: Taiwan's air quality has been deteriorating in recent years. According to a report by the Environmental Protection Agency (EPA), transportation is a major contributing factor to fine particle pollution in the country. The automobile industry should thus make every effort to reduce energy consumption as well as greenhouse gas emission. Recently, the Executive Yuan of the Republic of China set a policy goal to implement full electrification of automobiles by 2040 that runs parallel to world trends. The automobile industry should be aware of this megatrend and make big, fundamental changes, such as redesigning environmentally friendly electric vehicles, developing hot stamping technology in order to reduce vehicle weight without sacrificing safety, and innovating business models so as to accomplish the zero waste ideal of a full circular economy. The basic concept of a circular economy is the 3Rs (reduce, reuse, and recycle), but it is still necessary to add another 3Rs (redesign, refurbish, and renewable energy) for the automobile industry to implement such a circular economy. With these 6R principles, we can establish a sustainable automobile industry that is balanced in every social, economic, and environmental aspect.

Keywords: hot stamping, circular economy, automobile industry, redesign

1 INTRODUCTION

In 2016, 84.24 million automobiles were sold globally, and 439,269 of this total were sold in Taiwan. According to statistics from the motor vehicles supervision office, 21.67 million vehicles are registered in Taiwan, of which 7.92 million are automobiles. A 2015 EPA performance report indicated that the transportation sector is the single largest contributor to pollution (PM2.5), accounting for 37% of the total, and it also contributes 14.6% of the total emission of greenhouse gases in the country. It is vital and necessary to save energy and reduce emission of vehicles in order to improve the air quality in the Taiwan area (EPA of Executive Yuan, 2015).

This study aims to explore how the automobile industry can implement, from the perspective of a circular economy, a sustainable development model that takes into account the social, economic, and environmental aspects in pursuit of the well-being of humans.

In recent years, Europe, the United States, Japan, and China have gradually imposed more severe control requirements concerning corporate average fuel economy (CAFE) and CO emission standards. For example, in the United States, automobile fuel consumption per liter needs to reach 23.1 km by 2025, while CO emissions must be less than 163 g/km. In Europe, fuel consumption must reach 26.3 km/l, and CO emission shall be less than 95 g/km by 2020. Similarly, Japan has imposed a standard of 20.3 km/l for vehicles by 2020, while China has imposed a slightly stricter standard of 20 km/l for automobiles by 2020 (China Association of Automobile Manufacturers, 2016). The electrification and construction of vehicles that are lightweight without compromising safety has become an irrevocable global megatrend.

Taiwan's Executive Yuan announced its "Air Pollution Prevention Action Plan" in December 2017, which sets out its policy objectives, including the full electrification of government

---

* Corresponding author: ys66.chen@hotmail.com.tw

vehicles by 2030 and of all other automobiles by 2040. Volvo has stated that it will initiate the production of battery-operated electric vehicles or plug-in hybrid electric vehicles from 2019 onward. Toyota will work with Matsushita Battery to construct a "complete series" of cars based on electric power by 2025. Volkswagen is also currently promoting electric vehicles. It estimates that electric car sales will reach 3 million and consist of 50 variations of battery-operated electric vehicles as well as 30 types of plug-in hybrid electric vehicles by 2025. China and other countries have said that they are fully committed to build electric car recharging stations in various cities. According to a forecast by Bloomberg New Energy Balance, electric cars can compete with traditional internal combustion cars without government subsidies by 2025, and the number of electric vehicles will grow rapidly afterward. By 2040, it is estimated that 54% of global new car sales will be electric vehicles, while 33% of the 1.6 billion registered vehicles worldwide will be electric. Obviously, the electrification of automobiles will be without any question a global megatrend (Electric Vehicle Outlook, 2017).

In response to the gradual depletion of global energy resources and the impact of greenhouse effects, automobile design needs to consider the demands of environmental protection and energy conservation. Producing lightweight vehicles is also a trend in the automobile industry. However, developed countries have imposed mandatory requirements and standards for vehicle collision safety and have established relevant laws and regulations in order to ensure the structural safety of the vehicle body. To satisfy the automobile industry's requirements for weight and safety, the demand for advanced ultra-high-strength steel has risen. As a result, hot stamping technology, which combines the advantages of improved vehicle safety and saving weight, is gradually gaining popularity. Vehicle parts for hot stamping have become an indispensable choice for a lot of original equipment manufacturers (OEMs).

Another trend is the combination of artificial intelligence, big data, the Internet of Things, and cloud computing with the development of automobiles such as autonomous driving, the development of solar energy roads, etc.

2 HOW CAN THE AUTOMOBILE INDUSTRY IMPLEMENT A CIRCULAR ECONOMY?

2.1 The circular economy concept

Ellen MacArthur, a retired British navigator, founded the Ellen MacArthur Foundation in 2000 in order to promote the concept of a "circular economy." Between 2013 and 2014, the Foundation published a series of "Toward a Circular Economy" study reports that described a circular economy revolving around the accelerated transformation of economic and business philosophies and concepts. In this report, the Foundation proposed the definition of a circular economy as follows.

A circular economy is an industrial system that is restorative or regenerative by intention and design. It replaces the end-of-life concept with restoration, shifts towards the use of renewable energy, eliminates the use of toxic chemicals, which impair reuse and return to the biosphere, and aims for the elimination of waste through the superior design of materials, products, systems and business models. (Ellen MacArthur Foundation, 2014).

Based on this definition, this paper summarizes and includes the other three core concepts that can be applied to the circular economy of the automobile industry in addition to the basic concepts of the 3Rs (Li, Yu, & Liu, 2011), which can be referred to as the 6R principles, as described in Table 1.

2.2 The implementation of a circular economy by the automobile industry

In short, the automobile industry can follow the 6R principles so as to implement a circular economy. The following is a brief introduction.

Table 1.   6R principles of a circular economy for the automobile industry.

6R Principles	Description
1 Redesign	No longer using raw materials that are harmful to the environment and that cannot be recycled so as not to hinder the reuse of the products. It will also aim to achieve zero waste through the design and incorporation of high-quality materials, products, systems, and new business models.
2 Reduce	Reduce the amount and weight of steel used in car production in order to reduce energy consumption and other resources.
3 Reuse	Aim to enhance the automobile's marginal utility to the utmost in the secondhand car market.
4 Refurbish	To restore the performance of the car equivalent to a new one and guarantee its safety and performance after repairing or remanufacturing by the OEM.
5 Recycle	After the end-of-life cycle, the abandoned car shall be dismantled and recycled into the material value chain.
6 Renewable Energy	In line with the trend of increase in the production of electric vehicles, renewable energy should be developed so as to reduce the environmental pollution caused by the automobile industry.

3   REDESIGN PRINCIPLE

The requirements for the development of electric vehicles have gradually matured, and future automobiles have to be redesigned in order to achieve the objectives of being lightweight, adhering to safety standards, gaining energy conservation, and reducing the emission of greenhouse gases and PM2.5. For example, the World Steel Association (WSA) has already launched its Future Steel Vehicle research project (FSV project), which aims for a future generations of cars designed with alternative powertrains operated based on pure battery, plug-in hybrid electric power, or fuel cell electric power. The FSV project has made a major contribution to the development, implementation, and processing of advanced high-strength steel and ultra-high-strength steel in order to redesign the body-in-white with less weight (only 188 kg or a reduction in body weight of at least 39%) without sacrificing adherence to the global five-star safety standard. It will also reduce the full life-cycle carbon footprint (carbon dioxide emission) by around 50~70% (WSA, 2013).

The FSV project adopts a life-cycle assessment (LCA) approach that will reduce the carbon footprint of a car's entire life cycle because only the LCA method can properly evaluate and determine the most suitable materials that should be used in a car's body-in-white (BIW) structure in order to make sure that the savings of carbon dioxide emission by weight reduction during a vehicle's usage will not be offset by the increased carbon footprint of other non-green materials during its production and final disposal.

Regardless of the type of electric vehicles produced, the cost of the battery or fuel cell is always very high, and this is also the main reason why the construction of electric vehicles is currently not popularized. The redesign of the business model is also a priority. For example, electric car batteries can be rented out by the OEM, which would reduce the sales price and attract more potential buyers. It's also a good idea to develop a more convenient electric car rental system via the Internet so as to accommodate those people who do not have a high demand for the usage of automobiles.

4   REDUCE PRINCIPLE

Construction using a lightweight body structure without sacrificing safety is gaining rapid momentum in the development of new cars. The main method is to use advanced high-strength steel and ultra-high-strength steel to replace ordinary grades of steel. The ultra-high-strength steel can be produced by a hot stamping process line. Compared with traditional cold stamping, hot stamping possesses good formability, good dimensional stability (not easy to rebound), less mold wear and forming pass, and other advantages that can increase the tensile strength up to 1,500 Mpa after hot stamping.

Ultra-high-strength steel produced by hot stamping is mainly used for the structural components of automobiles that enhance the safety of vehicles such as front and rear bumpers, A-pillar, B-pillar, middle tunnel, front and rear girder, roof rail, side impact beams, etc. These components not only require the ability of high-absorption collision energy but also a smaller crushing margin of the collision area in order to protect the safety of passengers. Hot stamping technology is currently the only process to meet the requirements of producing ultra-high-strength steel.

Hot stamping material is mainly made from 22MnB5 of manganese boron steel, including aluminum silicon coating and non-coating material. The 22MnB5 cold-rolled sheet that contains ferrite and pearlite crystal structure is heated up to 850–950°C from room temperature after four to six minutes passing through a roller hearth heating furnace and then quickly moving into a chilled water-cooled mold for hot stamping while cooling and quenching in the mold (quenching speed is around 50~100°C per second); during a phase change, it will transform into a ultra-high-strength martensitic crystal structure, and the tensile strength can be increased from 600 MPa to 1,500 MPa, or even up to 1,700 Mpa (Karbasian & Tekkaya, 2010).

The hot stamping process was developed successfully in Sweden in the 1970s. It was first used in SAAB cars in 1984 and usually only used in the BIW structural design of luxury models in Europe in the early stage. Each car uses about 4–15% of hot stamping parts in average, with a weight reduction of 10–30 kg. The 2012 Audi A3 models already used 21.7% hot stamped parts, and the Volvo XC90 contained up to 40% of this type of material. The annual demand for hot stamping parts has grown from about 8 million pieces in 1997 to about 360 million pieces in 2015. There are about 300 hot stamping lines in the world at present, with a continuous increase of 15–20% every year. China is the largest automobile production market in the world, with the production of 28 million vehicles in 2016, and currently about 100 hot stamping production lines operate in China. It is also expected to grow rapidly in the future (Bakewell, 2016).

The main equipment of a hot stamping process line is composed of a de-coiler, leveling machine and blanking press, de-stacking robot and marking equipment, roller hearth furnace with inlet/exit rollers, hot stamping press with loading/unloading feeders, mold with rapid cooling system, exit conveyor, laser cutting machines, sand blasting and oiling device, and a full automatic control system as shown in Figure 1.

5 RECYCLE PRINCIPLE

If the body of an automobile can use an all-steel design, the car is very easy to recycle when it has to be discarded after it has been used for its full life cycle and reached marginal utility.

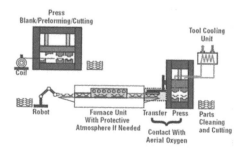

Figure 1. Hot stamping process line.

Since steel is a kind of green material, it can be easily recycled 100%. Based on the new circular economy business model, the battery and motor could be recovered and refurbished by the original manufacturer in the future and flow into the used car market. The materials that need to be discarded will also be sorted and recycled into the industrial cycle of automobile production.

6 RENEWABLE ENERGY PRINCIPLE

According to a forecast by Bloomberg New Energy Balance, approximately 530 million electric vehicles will operate worldwide by 2040. This can potentially save up to 8 million barrels of oil per day, but the required electricity consumption will reach 1,800 TWh. If the electrical power will still be generated by petroleum, natural gas, or coal, the reduction of greenhouse gas emissions and PM2.5 as a result of switching to electric vehicles will still be offset by the use of fossil products to generate electricity from the perspective of a full life-cycle assessment. Therefore, we should aggressively develop renewable energy so that we can achieve the effect of energy-saving and emission reduction in a circular economy (Electric Vehicle Outlook, 2017).

Renewable energy sources such as wind and solar energy are always subject to various natural limitations and cannot meet the demand for stable long-term operation. Although nuclear energy is not a renewable energy source, it possesses large energy reserves and can be used for long-term stable power generation in order to reduce greenhouse gas emissions effectively. However, the most serious problem from nuclear power generation at present is the disposal of nuclear waste. If not handled properly, it will be seriously harmful for human beings.

However, the Chinese Academy of Sciences has recently proposed a new method of dealing with nuclear waste – the Accelerator Driven Advanced Nuclear Energy System will eventually raise uranium resource utilization efficiency from the currently less than 1% to more than 95%. The nuclear waste will be less than 4% of spent nuclear fuel, and the half-life of nuclear waste will be reduced from hundreds of thousands of years to about 500. If the newest nuclear energy technology can be industrialized by 2030 as scheduled, it will enable nuclear waste to be recycled and turn waste into treasure, which will be of great help in achieving a true circular economy.

7 CONCLUSION

Petrochemicals and coal embedded in the earth cannot be regenerated after we extract them and use them for energy purpose. Such fuels also create problems such as air pollution and greenhouse gas emissions. Nonetheless, car manufacturing based on electrification, utilizing lightweight design and safety considerations, is an unavoidable trend. However, the external electrical power required by the automobile industry should be generated and used based on the principle of developing renewable energy instead of using petrochemical energy from the viewpoint of greenhouse gas emissions over the entire life cycle of the product. All in all, in pursuit of sustainable development, the automobile industry should strive to promote the 6R principles of the circular economy in order not to cause any additional harm to our living environment, including lower air quality and greenhouse gas emissions.

REFERENCES

Bakewell, J. 2016. Hot stamping hits heights, Automotive Manufacturing Solutions.
China Association of Automobile Manufacturers. 2016. Fuel consumption and carbon emissions management in various countries.
Electric Vehicle Outlook. 2017. Executive summary, Bloomberg New Energy Finance's annual long-term forecast of the world's electric vehicle market.

Ellen MacArthur Foundation. 2014. Toward the circular economy: Accelerating the scale-up across global supply chains.

Environmental Protection Agency (EPA) of Executive Yuan. 2015. Governance performance report.

Karbasian, H., and Tekkaya, A. E. 2010. A review on hot stamping. *Journal of Materials Processing Technology*, 210, 2103–2118.

Li, J., Yu, K., and Liu, L. 2011. Circular economy 3R principles and its development trend.

World Steel Association (WSA). 2013. Steel solutions in the green economy, Future Steel Vehicles.

Green technology & Architecture engineering

Innovation in Design, Communication and Engineering – Lam et al. (eds)
© 2020 Taylor & Francis Group, London, ISBN 978-0-367-17777-5

Modified Particle Swarm Optimization (MPSO)-based maximum power point tracking for photovoltaic system

Chiou-Jye Huang
Jiangxi University of Science and Technology, Zhanggong District, Ganzhou, China

Po-Hsien Tu & Jung-Shan Lin*
National Chi Nan University, Puli, Nantou, Taiwan

ABSTRACT: A novel concept, aiming to improve the conventional particle swarm optimization (CPSO) method for strengthening algorithm capability and improving system performance, is presented. In addition to using linear decreasing inertia weight, nonlinear adapting learning factors are employed in order to enhance tracking ability. By this intervention, the system acquires more accurate convergence and the situation of into local maximum power point is avoided. Consequently, the simulation results exhibit that the modified particle swarm optimization (MPSO) has the potential to track the global MPP with accurate rate of convergence under partial shading conditions (PSC).

1 INTRODUCTION

Rapid development of the global economy raises world demand on energy day by day. Nowadays, the principal energy resources are oil, coal, and natural gas, which, apart from being nonrenewable, also cause serious environmental pollution. Such effects are the greenhouse effect, acid rain and ozone depletion, not to mention the combustion of fossil fuels, which is the most serious one. To confront the shortage of energy and reduce fossil fuel pollution, science began to look into renewable and green energy, such as sunlight, wind, water, and geothermal heat. According to an authoritative technical report of the International Renewable Energy Agency (IRENA), from the global PV system installation numbers, total global installed capacity of solar PV reached 79.4GW in 2016, compared to the installed capacity of 55.4GW in 2015, annual growth rate has reached 43%, and is the second consecutive year to reach growth rate of more than 30%. Actual installed capacity was much higher than the initial estimate, mainly due to significantly higher-than-expected installation capacity in Mainland China, which rose to 34GW (IRENA, 2017). Among the mentioned energy resources, sunlight is the most popular, due to its advantages, as it is pure, non-polluting, and unlimited. However, fluctuation of weather conditions has proved to be a crucial problem for PV systems, causing lower power output. Therefore, enhancing the operation of the PV array in order to extract the maximum power under PSC is a primary aspiration.

In this article, the new method of improved particle swarm optimization technique claims to guarantee the maximum power point tracking of photovoltaic systems under partial shading conditions. Also, in order to deploy significant applicability, a maximum power point tracking controller with photovoltaic system has been designed. The remainder of article is organized as follows: Section 2 introduces the photovoltaic system modeling. The proposed improved particle swarm optimization algorithm is developed in Section 3. Some numerical results are also presented in Section 4. Finally, conclusions are given in Section 5.

---

* Corresponding author: jslin@ncnu.edu.tw

2 PHOTOVOLTAIC SYSTEMS

The equivalent circuit diagram of the solar cell is presented in Figure 1 (Liu, Chen, & Huang, 2015; Saravanan & Ramesh Babu, 2016). The equivalent circuit of the general model consists of light source current, diode, parallel resistor (expressing the leakage current), and series resistor, which describes the internal resistance to the current direction. In Figure 1, I_{ph} is the light generated current of the solar cell, D_j is the diode, R_p is the equivalent parallel resistance for the internal material, R_s is the equivalent series resistance of the internal material. The value of R_p is usually apical, while the value of R_s is the least. Hence, R_p and R_s are usually ignored. In addition, R_L denotes an external load, while I_{pv} and V_{pv} denote the output current and output voltage of the solar cell respectively. The I–V characteristic equations of a solar cell are presented as follows:

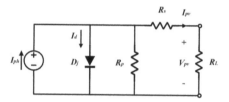

Figure 1.   Solar cell equivalent circuit.

$$I_{pv} = I_{ph} - I_{sat}\left(\exp^{\frac{qV_{pv}}{AK_bT}} - 1\right) \tag{1}$$

$$I_{ph} = GI_{scr}[1 + k_i(T - T_r)] \tag{2}$$

$$I_{sat} = I_{rr}\left(\frac{T}{T_r}\right)^3 \exp^{\frac{qV_{pv}E_q}{AK_b}\left(\frac{1}{T_r} - \frac{1}{T}\right)} \tag{3}$$

where K_b expresses the Boltzmann constant, q is the charge of an electron, A equals the ideality factor, T is the cell temperature, Isat is the reverse saturation current, I_{scr} is the short-circuit current at reference condition, k_i is the short-circuit temperature coefficient, G is the solar irradiance, E_q is the band-gap energy of material, T_r is the reference temperature, and I_{rr} is the saturation current at T_r.

3   ALGORITHM

In the proposed method, an improvement of the conventional PSO method is developed, where the magnitude values of ω, c_1 and c_2 are determined by the following linear decreasing inertia weight, linear learning factors, and nonlinear adapting learning factors methods.

In the past, Liu et al. used linear decreasing inertia weight as well as linear learning factors (Liu et al., 2012; Li et al, 2019) for PSO technique to achieve the global tracking of MPP. The two linear learning factors can be expressed as:

$$c_1(k) = c_{1\,max} - \frac{k}{k_{max}}(c_{1\,max} - c_{1\,min}) \tag{4}$$

$$c_2(k) = c_{2\,max} - \frac{k}{k_{max}}(c_{2\,max} - c_{2\,min}) \tag{5}$$

where $c_1(k)$ is the individual learning factor at kth iteration value, and $c^2(k)$ is the group learning factor at kth iteration value. In addition, $c_1$max, $c_1$min, $c_2$max and $c_2$min are the maximum and minimum values of linear learning factors.

Primarily, we allow the value of c_1 to decrease and value of c_2 to increase linearly. When the initial value of c_1 is larger, it allows particles extensive search around the area. However, when the initial value of c_2 is smaller, it does not allow particles to move to group-optimal solution position. As the number of iterations increases, gradually c_1 will decrease and c_2 will increase. The particles search for the best solution of group position. This approach reduces the chance of falling into local optimum solution.

4 SIMULATION RESULTS

In this article, the simulations were evaluated by the commercial software MATLAB/SIMU-LINK. The parameters of PV module are presented in Table 1. We employ MPSO algorithm with linear decreasing inertia weight and fixed values $c_1 = c_2 = 1.5$. This case tracks to the global MPP whose correct rate of convergence is about 50%.

The PV array was composed by six modules in series. The simulation conditions of PSC used in each module to irradiate different sunshine effect current, voltage, and power of PV array. The simulation conditions were as follows:

(1) The insolation of module 1 = 100 W/m2, (2) the insolation of module 2 = 100 W/m2, (3) the insolation of module 3 = 300 W/m2, (4) the insolation of module 4 = 500 W/m2, (5) the insolation of module 5 = 500 W/m2 and (6) the insolation of module 6 = 1000 W/m2.

Table 1. Parameters of PV module.

Symbol	Description	Values adopted
I_{sc}	Short-circuit current	5.45 A
V_{oc}	Open-circuit voltage	22.2 V
I_{mpp}	Current at MPP	4.95 A
V_{mpp}	Voltage at MPP	17.2 V

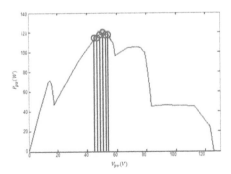

Figure 2. The simulation for MPSO including linear decreasing inertia weight and fixed c_1, c_2.

5 CONCLUSIONS

In this article, using the linear decreasing inertia weight and the nonlinear adapting learning factor to improve CPSO algorithm was developed. The simulation results clearly demonstrated MPSO algorithm's tracking performance. In comparative simulation results, when ω, c_1 and c_2 parameters were fixed, the particle swarm optimization algorithm could not guarantee the detection of the global maximum power point. However, the MPSO with linear decreasing inertia weight and nonlinear adapting learning factors regularly goes through the global maximum power point.

REFERENCES

IRENA (2017) "Renewable capacity statistics 2017."

Liu, Y.-H., Chen, J.-H., & Huang, J.-W. (2015) "A review of maximum power point tracking techniques for use in partially shaded conditions," *Renew. Sustain. Energy Rev.*,. 41, 436–453.

Saravanan, S., & Ramesh Babu, N. (2016) "Maximum power point tracking algorithms for photovoltaic system - A review," *Renewable and Sustainable Energy Reviews*, 57, 192–204.

Liu, Y. H., Huang, S. C., Huang, J. W., & Liang, W. C. (2012) "A particle swarm optimization-based maximum power point tracking algorithm for PV systems operating under partially shaded conditions," *IEEE Trans. Energy Convers.*, vol. 27, no. 4, pp. 1027–1035, 2012.

Li, H., Yang, D., Su, W., Lu, J., & Yu, X. (2019) "An Overall Distribution Particle Swarm Optimization MPPT Algorithm for Photovoltaic System Under Partial Shading," *IEEE Trans. Ind. Electron.*, vol. 66, no. 1, pp. 265–275, Jan.2019.

Innovation in Design, Communication and Engineering – Lam et al. (eds)
© 2020 Taylor & Francis Group, London, ISBN 978-0-367-17777-5

Exploring the integration of vertical greening in balcony design

Chen-Sheng Hsiung* & Huann-Ming Chou
Mechanical and Energy Engineering, Kun Shan University, Tainan, Taiwan

ABSTRACT: Vertical greening should be fully applied and promoted. It helps to provide a beautified environment, increases the amount of urban green space, provides high-quality innovation to traditionally rigid urban construction, and demonstrates the full function of an urban green ecology. Vertical greening, also known as three-dimensional greening, is a type of greening process administered in three-dimensional space and vertical to the ground. It possesses the advantage of a full usage of space and utility, in which the selected plants can utilize their functions such as climbing, twining, clinging, and drooping. The plants are groomed along the surface of a building or other structure where they climb, fix, attach, and hang on themselves, forming a vertical greening surface to enhance the rate and amount of environmental greening. Furthermore, vertical greening adds to the aesthetics of the balcony space and can create a more diversified city landscape. In addition, it demonstrates an aesthetic and artistic design, enhances people's quality of living, reduces air pollution, and beautifies and purifies the environment. This study focuses on exploring the integration of vertical greening in balcony designs. The design includes seeking breakthroughs in utilizing different characteristics of plants to create a variety of greening effects. In addition to demonstrating beautiful art pieces of living, vertical greening works alongside with environmental protection for solving the air pollution problem in that it provides a fresh, comfortable, restful, and green environment.

Keywords: vertical greening, living green wall, balcony, spatial design

1 INTRODUCTION

Putting contemporary architectural and living aesthetics together allows people to express their taste. Home, as a place to relax and to regain energy, is important as a living environment that can be combined with the nature, have ecological systems bred on a balcony, and have every simple living unit be like a garden. The purpose of this paper is to explore how changes in micro-environments can impact macro-environments.

Since the industrial revolution, owing to the growth of the population and of highly developed businesses and industries, the relationship between human beings and the natural environment has lost balance rapidly. Though industrialized technologies had obvious help to improving our living conditions, they nonetheless indirectly cause the decrease of natural resources, such as forests, that can regulate the ecosystem; these technologies also overload the natural environment, thereby making global climate change and urban heat islands important environmental issues. Because of climate change, using plants as regulative material is a possible way to protect the environment and to reduce carbon emissions, which can also beautify the living environment.

---

* Corresponding author: hsiungphoebe@gmail.com

Usually what we call "greening" comprises one of these variations: "green wall," "flower wall," "plant wall," etc. These variations have one thing in common: they all rely on particular designed containers and water. With a proper plant, these methods are also called a vertical green wall. Vertical planting, as matching design and greening, can resolve the lack of plants in cities (Lai & Tsai, 2014). The proper use of vertical greening can provide humanistic, environmental, and economic benefits. With the idea of eco-city, it can combine cities and the natural environment for continuous development (Tsai, 2013).

With the rising living conditions and the development of revenue houses, the balcony plays an important role: it satisfies the requirements for sunlight, ventilation, storage, gardening, etc., and it also improves the indoor micro-climate and provides a place for semi-outdoor activities and for household greening. It also meets the demands for leisure and social activities. Without balconies, apartments often make people feel closed in. Activities on a balcony give residents a sense of privacy and comfort, and they let people return to nature. Such activities strengthen the relationship between people and the environment, as well as the relationships among people. Balconies have positive effects on residents' physical and mental health (Peng, 2016). Balconies have, structurally, an effect of coverage, and, in terms of household design, they provide visual extension and practical usage. Unfortunately, balconies are often put too much to practical use, such as storage, and the aesthetic and eco-friendly side of balconies is neglected. Therefore, this paper hopes to introduce vertical greening to household balconies as a normal housing design feature. Also, similar kinds of balconies have already been put in use in business buildings. For example, departments in the Shin-I district of Taipei, such as Shin Kong Mitsukoshi, due to their business practices, have used terrace-like balconies. This kind of design has an extension effect, and the cloisters are similar to household balconies; vertical greening design can also be applied.

How can household balconies present vertical greening? According to many studies and analyses, the proper techniques include: hydroponic, modular, and carrier systems, cable wire, a wire mesh system, and a modular trellis panel system. The former three kinds are artificial vertical greening while the latter three are natural vertical greening. For the technique, proper kinds of plants, their features, and their shortcomings are listed in Table 1.

For the greening techniques and features listed earlier, one can refer to Figure 1 for the exposition of ivy on hydroponic and carrier system and to Figure 2 for the exposition of modular techniques.

Table 1. Different types of greening techniques and their features.

Technique	Earth	Plant	Feature	Shortcomings
Hydroponic	X	Tropical, near waterfall and riverbank	Easy-to-set-up hardware	Difficult software implementation
Modular	V	Can be planted before-hand, temperate climate plants	Many choices of plants, can change modular at any time	Need supporting splints
Carrier System	V	Vertical-growing plants	High watering ability	
Cable wire	V	Vines	Natural vertical greening system, low cost	Require longer time for plants to grow
Wire mesh system	V	Vines	Natural vertical greening system, low cost	Require longer time for plants to grow
Modular trellis panel system	V	Vines	Natural vertical greening system, low cost, hard, light but solid	Require longer time for plants to grow

Figure 1.   Examples of hydroponic and carrier systems (https://kknews.cc/home/p42v6ne.html).

Figure 2.   Examples of modular and carrier systems (Happy Hair/Tucheng Shop).

Proper plants for a vertical wall include ivy, bromeliads, dracaena, *Epipremnunm aureum*, arrowhead vine, and the striped bracket plant. Green design opens a new way and gets rid of the limitation that plants can only grow horizontally. The vertical wall technique has already been well developed in Taiwan, and we can see many such designs in different areas (Yang, 2012). Vertical gardens have been the organic wallpaper around the world; many builders use green design in order to construct a green space that fits modern urban ways of living while keeping the multifunctionality of gardening art, preservation, eco-balance, and urban cleaning, which brings humans back to nature with a tighter connection. This wave of green, organic design has been carried out on a larger scale and with better creativity than ever before, marking a new century of green coexistence (Yang, 2012). For example, the Park Lane business area (@cmpfamily) has the first large-scale green wall. That building uses 120,000 plants, using watering tubes as long as three kilometers so as to provide watering and nutrition for these plants. It creates a beautiful and ecological urban garden, and it has become the representation of Taiwan's green walls in business buildings (Figure 3). In recent years, many construction sites have put plants on their wall in order to beautify ugly sites (e.g., Figure 4). This implementation also keeps nearby residents and pedestrians away from noisy, dirty construction sites and the air pollution that construction might cause.

Figure 3.   Green building on Park Lane, Taichung.

Figure 4.   Sites from Chen-De Road, Taipei.

3   THE APPLICATION OF VERTICAL GREENING TO BALCONY SPACE DESIGN

Generally speaking, in balcony design, Taiwanese usually consider practicality rather than beauty or greening – unlike Europeans or Americans, who commonly use plants. Take Taipei as an example; community household balconies usually use potted plants as a means to beautify. However, they lack design. For present household development, if we can make horticulture design more delicate and with more variety, and put in some concept of environmental protection, then these ideas would be more practical in our daily life.

Moreover, vertical greening and the use of vertical walls should be considered green. Planting projects should overcome the thinking that design is just for beauty as before, and should be elevated to the level of green using Taiwan's local plants as materials for building green urban environments. Creating a landscape that comes close to Taiwan's natural landscape is the principle of eco-greening, which fits the principle of green building design (Research Institute of the Ministry of the Interior, 2008). This kind of thinking is necessary for Taiwan, especially Taipei, which has a particular basin landscape. This issue is discussed in the next study.

The start of vertical greening is to beautify the environment and to improve micro-climate. A small-scale, windproof green wall can provide coverage that is at the same height of humans; large-scale vertical greening covers the entire building (Lin, 2011). These functions raise the value of greening. These are the functions and effects of greening: (1) preserve energy: greening can effectively decrease the temperature on the surface, that is, it stops the indoor temperature from rising, and thereby reduces the use of air conditioners, which further preserves energy; (2) cleanse the air: plants can absorb carbon dioxide and thereby cleanse the air; they also can absorb dust; (3) reduce noise: vertical walls can absorb noises; they effectively reduce noise from outside and keep people indoors in a good mood.

4   CONCLUSION

The father of vertical greening, Patrick Blanc, puts walls and plants together and makes them an artwork (Figure 5). This gives buildings a new look, making them vigorous and giving them new value. It is because of Patrick Blanc that vertical greening on buildings is a new international trend. To put this kind of artistic use to general household balconies will also give them a new life that values greening and environmental protection.

Since the urbanization of Taipei, the city has become less and less green. Specific greening (vertical greening) will become a trend to promote continuous building, and it will take plants as one of the building elements. It will consider the effects of vertical greening in terms of urban view (Tsai, 2013). This paper has tried to apply this idea of greening techniques onto household balconies among the decade-long trend of vertical greening. Applying the concepts of eco-friendliness, green building, and environmental protection to household balconies so that they can become an eco-city or green community, and applying vertical greening and gardens to households not only can make the houses green but also makes them beautiful (Figure 6).

Figure 5.   Quai Branly Museum/Paris (Patrick Blanc) (https://kknews.cc/news/6kp3eq3.html).

Figure 6.   Example of balcony greening (https://kknews.cc/home/p42v6ne.html).

REFERENCES

Lai, K. H., & Tsai, S. Y. 2014. *Journal of Environment & Art*, 15, 1–18.

Lin, X. Y. 2011. The study of plants applying to the outdoor vegetal wall. Thesis of the Department of Landscape Architecture, Tunghai University.

Peng, C. W. 2016. The functional changes and potential requirements of balcony space in condominiums. Thesis of the Master Program in Architecture and Urban Design, National Taipei University of Technology.

Research Institute of the Ministry of the Interior. December 2008. Research report.

Tsai, M. L. 2013. A study on the public's awareness of and attitude toward vertical green architecture. Thesis of the Master Program in Architecture and Urban Design, National Taipei University of Technology.

Yang, C. C. 2012. The pattern and application of vertical walls. Thesis of the Department of Architecture and Landscape Design, Nanhua University.

Innovation in Design, Communication and Engineering – Lam et al. (eds)
© 2020 Taylor & Francis Group, London, ISBN 978-0-367-17777-5

A comparison between and lessons from the current situation at China's Erhai Lake and South America's Lake Titicaca

Ming-Fu Ho* & Huann-Ming Chou

Mechanical and Energy Engineering, Kun Shan University, Tainan, Taiwan

ABSTRACT: This study chooses two major alpine freshwater lakes, one from Yunnan, China, and the other from South America, to compare their current water environment, analyze the causes of their different current conditions, and illustrate the point that as a result of humans' lack of environmental awareness, the over-exploited lake basin resources, damaged lakeside ecological structures, and disturbed ecosystem have caused the ecological cycles of lakes to collapse. In order to attain a good balance between economic development and environmental protection, all governments with the authority over lakes and wetlands, as well as their people, should use the lakes and the basin systems with caution under the premise of protecting the virtuous lake ecological cycles and structures.

Keywords: Erhai lake, Lake titicaca, Lakes, Water environment

1 INTRODUCTION

In addition to providing livelihood opportunities and industrial water, lakes, wetlands and the basin also serve functions such as climate regulation, irrigation, shipping, aquatic products, sightseeing and biodiversity conservation. With global warming and human's increasing exploitation of natural resources, not only have we reduced the surface area of lakes, adding to pollution and worsening water and ecological issues, we have also indirectly caused frequent floods in the lakes' region and climate imbalance. All these gravely threaten the regional economic and social development. Lake pollution is now a global challenge.

In this paper, two freshwater lakes in mountainous areas with similar geographical environments have been chosen to compare the differences in their current situation as well as analyse and explore the factors that have caused these differences. This is in the hopes of inspiring the local government and the local community to reasonably develop and utilise the lakes and their basin systems without damaging the virtuous circle of the lakes' natural ecosystem.

2 THE CURRENT SITUATION AT CHINA'S ERHAI LAKE AND THE FACTORS INVOLVED

2.1 *Erhai Lake's natural conditions and river basin environmental evolution*

Located at Cangshan Donglu in China's Yunnan Province, Erhai Lake is a typical alpine fault lake situated at 1965.8 metres above sea level, 251.35 square metres in surface area, and goes as deep as 19.5 metres (according to actual measurements from Yunnan Environmental Protection Bureau in 2000) - the second largest alpine lake in Yunnan Province. The Erhai

---

* Corresponding author: meylen629@gmail.com

area has a subtropical plateau climate with obvious wet and dry seasons, derives water mainly from rainwater and a small amount of thawed snow, and has an average precipitation of 1000 to 1200 millimetres annually. There are 23 main rivers that go into the lake, but the only river that goes out of it is Xi'er River, with a basin of 2565 square metres and water transparency between 1.8 and 6.6 metres.

Before the 1970s, Erhai Lake had good water quality. But after the mid-70s, with constant development at Erhai Lake, the lake pollution situation was aggravated, and its water quality gradually deteriorated. In almost 30 years from 1981 to 2009, its water quality showed the trend of transforming from oligotrophic to eutrophic. From 1981 to 1985, the Erhai Lake water quality went from oligotrophic to oligomesotrophic. From 1986 to 1992, the water quality showed improvement, but became mesotrophic in 1988. From 1992 to 2001, Erhai Lake's water quality remained mesotrophic, but showed increasing amounts of algal blooms. And from 2004 to 2009, Erhai Lake's overall water quality was Class III, which was at the mesotrophic level (Chai, Y.F., et al., 2013). It initially maintained Class II in 2001, but following an outbreak of cyanobacteria in 2003, the water transparency fell to a historical low, with the trophic state in partial areas reaching Class IV of surface water quality. However, following the government's relevant departments' stringent implementation of water environment protection policies, the overall water quality from 2004 to 2009 was maintained at Class III.

2.2   *Analysis of Erhai Lake's environmental degradation factors*

A. Overexploitation of Watershed Resources, Pollution Overload

The excessive development of Erhai's basins led to serious soil erosion. In a 2004 research, Dong Yun-Xian et al. pointed out that Erhai Lake's forest coverage rate was low, and with the soil erosion encompassing 30% of the total surface area (Dong, Y.X., et al., 2004), the result was sediment deposition at the rivers and the entrance of the lake that affected Erhai Lake's overall ecological environment. In addition, with the increasing population in the area where men used large amounts of fertilisers and pesticides to enhance their agricultural production, which were then flushed into Erhai Lake with rain, the K, N and P ratios in the lake water increased yearly, leading to the eutrophication of the lake.

Diffuse pollution is one of the main sources of pollution at Erhai Lake. According to related research, among Erhai Lake's diffuse pollution, 65.08% TN and 42.15% TP are from village sewage Therefore, village sewage is the most important source of Erhai Lake's diffuse pollution. The village sewage can mainly be classified into three categories: animal farming wastewater, rural residents' wastewater and the tourism sector's food and accommodation wastewater. According to 2013 statistics, there were 114314 cows, 435723 pigs and 599500 people in the Erhai Lake basin (including all of the agricultural population and urban population living in rural areas), which also received tourists 9298700 times. These three types of wastewater flowed into Erhai Lake without proper management, and gradually increased its pollution load.

In the past, most of the village's garbage was dumped on riverbanks, in the village or by the lake as the people pleased. Some of them even had the habit of throwing animal carcasses into the rivers and lakeside. And every year when the rainy season came, all these were flushed into the lake, causing serious pollution to the environment of Erhai Lake. Currently, many of the villages in the basin still do not have an established sewage system, allowing wastewater from daily life as well as garbage to pour into the creeks and rivers without a care. The wastewater produced by tourists in many villages catering to the tourism sector has also put huge pressure on the environment (Zhou, Jiang, Zhang, 2016).

B. Serious Human-Caused Lakeside Ecological Disruption

To develop the "Vegetable Basket" project, the Dali Municipal Government rallied the public to build ponds, and within three years built a farming model that spanned 149.25 hectares. According to the Dali Environmental Protection Bureau's monitoring and investigation information, the TP and TN concentration in the fish ponds were respectively 20 and 10 times that of the concentrations in Erhai Lake. The digging of the ponds was a large-scale invasion of the lake's wetlands, seriously damaging the ecological structure and functions surrounding the lake. In addition, occupying the lakeside to prepare the land for fields and constructing entertainment venues as well as piers also directly damaged the natural ecological system of the lakeside, reducing Erhai Lake's ability of self-purification of its own water bodies, and ultimately worsening the water quality even more.

The "Erhai Inverted Siphon" project constructed by Xi'er River's hydropower station and annual water consumption of 162,000,000 cubic metres also used up Erhai Lake's water source beyond its means, resulting in a decline in the water level of the lake, shoreline retreat, and changes to the lakeside's biological reproduction and habitats. This not only heavily impacted on the stability of the lake's ecosystem, but also disrupted the biological diversity at Erhai Lake.

C. Ecological Imbalance at the Lake

In 1991, Erhai Lake successfully introduced the new Taihu Lake Whitebait species to its ecosystem. As whitebait feeds on zooplankton, increased production of whitebait led to a reduction of zooplankton that feeds on blue and green algae. The successful introduction of whitebait and the consequent reduction of zooplankton were one of the main reasons that caused algal blooms' explosive growth (Wu, et al., 2013). In addition, not only did whitebait require large amounts of baits, they also consumed large amounts of roes during the indigenous fish's breeding period, causing a sharp reduction in the population of Erhai Lake's indigenous fish such as the Dali Lake Carp.

From the mid-80s to the mid-90s, Erhai Lake vigorously developed the farming of fish in cages, investing large amounts of artificial feed into the lake daily, resulting in an increase in the lake's nitrogen and phosphorus concentrations. Due to the significant predation of aquatic plants, aquatic vegetation species also became simplistic, with reduced biological diversity. The algal blooms that used to compete with the aquatic plants for the same nutrition and light conditions reproduced in mass, eventually resulting in an outbreak of algal blooms. Besides, the extensive fishing using trawls with small mesh sizes and the bottom trawling method to economically catch fish stocks also influenced aquatic plants and fish resources' regeneration ability. This caused further damage to fish stocks' structure and reduced biological diversity.

2.3 *Overview of the government's remediation measures*

The first outbreak of algal blooms in 1996 forced the government to cancel trawling and fish farming in cages at the end of the year. In November 1997, the government implemented an "Anti-Phosphorus" policy that cut the total inflow of phosphorus. In 1999, the "Three Exit Three Return" policy was implemented, which consisted of exiting farming and returning the forest/lake, exiting ponds and returning the lake, as well as exiting houses and returning the wetlands. The government further made a plan for 2003 to 2020 to invest 3 billion Chinese yuan to remediate Erhai Lake, with six main measures including urban sewage treatment and sewage interception engineering, recovery and construction of Erhai Lake's ecosystem, remediation of main rivers that enter the lake, management of diffuse pollution from agriculture and the villages, maintenance of basins, as well as environmental management of Erhai Lake's basins. The above measures were instrumental towards the continuous improvement and ecological protection of the Erhai Lake environment, and Erhai Lake's water quality is now back at a Class III surface water quality level overall. In addition, following the national "11th Five-Year Plan" investment of nearly 1.5 billion yuan into the remediation of Erhai Lake, 2.927 billion were invested in the "12th Five-Year Plan," and the investment amounted up to 17.645 billion yuan in the latest "13th Five-Year Plan."

During the 2017 Spring Festival, however, Erhai Lake again showed the phenomenon of an algal bloom outbreak. Based on Dali's reception of 38.5918 million tourists in 2016, the local government suspended the operations of all food and accommodation businesses in the protected basins for investigation in order to control sewage that led to the pollution of Erhai Lake. It is clear that the environmental management of basins and basic facilities of Erhai Lake require large amounts of continuous investment. Whether future policies can be executed as they should be, is the key to Erhai Lake and its basins' environmental protection and remediation.

3 THE CURRENT SITUATION AT SOUTH AMERICA'S LAKE TITICACA AND THE FACTORS INVOLVED

3.1 *Lake Titicaca's natural conditions and current water environmental conditions*

Lake Titicaca is a tectonic lake located in the Andes on the border of Bolivia and Peru, situated at 3812 metres above sea level, 8290 square metres in surface area, and is the world's highest navigable lake. Lake Titicaca has a plateau climate alternating between rainy and dry seasons, derives its water source mainly from the alpine thaw of surrounding mountains, and has an average precipitation of 250 millimetres. There are 25 rivers that lead into the river, but the only river the leads out of the lake is Río Desaguadero, with a basin area of 58000 square metres and is as deep as 280 metres.

Lake Titicaca is surrounded by high mountains, the lakewater is clear, calm and blue, and it is possible to stand by the lake and see fish swimming around in the water. In the shallow waters, aquatic plants flourish and dark green reedmace grow around the water like elegant embroidery around the lake, adding a subtle touch to the beauty of Lake Titicaca (Guo, 2006). According to the Foster colorimetric method of water quality sampling, if the water is oligotrophic, the water is clear and blue; if the water is eutrophic leading to an overgrowth of floating algae, the water is green; and if there are too much dissolved rotted plants, the water is brown. And with the large-scale, dense growth of reedmace belonging to the reeds family on the surface of the lake, it can be deduced that Lake Titicaca is overall an oligotrophic grass-type lake.

3.2 *Analysis of Lake Titicaca's current water environmental conditions factors*

A. Low-level Exploitation of Watershed Resources, Little Pollution

Lake Titicaca is located in a plateau where the Andes has formed a natural barrier between the lake as well as human migration and development. Due to the climate and rainfall factors, the local farmers in the basin's valley agricultural district still used traditional methods to plant corns, potatoes and beans. This means that there is a lower degree of the use of mechanics, fertilisers and pesticides, and therefore the K, N and P that get flushed into the lake are less, and do not cause lake eutrophication as easily.

Just on the matter of the basin's sewage and garbage, because Lake Titicaca is surrounded by vast alpine ranches, most farmers of cows, pigs, alpacas and sheep in the 3 major villages raise their animals in the wild, unlike the livestock in Erhai Lake that are bred in captivity. Therefore, very little wastewater is produced during the farming process. There is a rough estimate of 1.6 million people that reside in urban towns and villages surrounding the lake (about 2.5 times of the Erhai Lake basin), but in 2011, they received less than 3 million visitors (about 1/ 3 of the tourist numbers that visited Erhai Lake in 2013). And with the lake's surface area being approximately 33 times greater than that of Erhai Lake, the degree of pollution to the lake resulting from the daily life of the residents in the basin as well as sewage and garbage produced by the tourism sector's food and lodging businesses is relatively lower, greatly reducing the load on the lake's water bodies.

B. Minor Human-Caused Lakeside Ecological Disruption

Lake Titicaca's surroundings are multi-level terrains, with geology akin to the Pune dessert. Other than unique mountain arid grassland everywhere, there are also many saline water concave grooves. This grassland with an abundance of saline concaves is not conducive to fish pond farming and lakeside agriculture, saving the lakes and its wetlands from extensive disruption. The lake has only a couple of piers and amusement facilities, and the natural ecosystem of the lake has not suffered serious damage.

In the Lake Titicaca waters, bulrush belonging to the reed family that are as high as three or four metres flourish. The head of the former Limnology group of Germany's "Max Planck Institute," Kaethe Seidel, discovered the following in her research: reeds are almost hollow tubes that extend to the roots; when moisture dissipates rapidly due to the sun shining on the leaves, negative pressure is produced on the inside of the reeds, sucking in air from the hollow tube and transporting it directly to the roots, then spreading the air into the mud or water in the surrounding area. Reeds are the equivalent of an aeration machine in the water, helping rivers and lakes breathe. In addition, the research also suggested that the reed community in the water and land ecotones can effectively intercept land-based sources of nutrition. The soil under the reed community's roots has obvious interception capabilities, intercepting most of the pollution sources such as nitrogen and phosphorus that go into the lake through the undercurrents (Yin, Lan & Yan, 1995). The wetland where bulrush grows in abundance is clearly one of the main factors that maintain the lake's ecosystem.

C. Ecological Balance at the Lake

Lake Titicaca currently has about hundreds of Uros residing in dzens of floating reed islands (Wang, 2017). They harvest reeds that cling onto soil and turn that into a floating reed mat, inserting four stakes at four corners of the mat and using ropes to anchor and secure it to the bottom of the lake, then stack bundles of reeds horizontally and vertically one on top of another, forming the floating island that they rely on for a living. They also use reeds to make thatched cottages, straw mats and straw boats, and cook and eat on the floating island, as well as farm livestock and build fish ponds. With the gradually increasing number of tourists, they have also started to sell a variety of unique items such as straw boats, straw hats, straw mats and other handicrafts, developing a unique "bulrush civilisation" vibe of cultural tourism. The Uros' traditions pay attention to maintaining the balance of the ecosystem. Not only do they not fish or hunt birds in excess, toilet points require a half-hour boat trip to prevent water pollution. Even though some of them have started using modern boats in line with economic developments, it is hard to detect any manmade garbage pollution on the island. The Uros basically maintain a traditional life that is at peace with the rest of the world's civilizations.

The Uros' floating grass islands are made using only bulrush, and can be said to be the manmade floating island that best meets environmental-friendly criteria. According to research, manmade floating islands effectively reduce nitrogen, phosphorus and other nutrients in the water, and have the function of purifying the water quality of the lake. Its purification principle works by using plant roots to absorb large amounts of floating objects in the water, gradually forming a biofilm on the plant roots' surface, and in the biofilm microorganisms consume and metabolise pollutants in the water into inorganic matters. Through that process, nitrogen, phosphorus and other organic matters are effectively reduced. Floating islands also provide shade from the sun, inhibiting algae's photosynthesis, reducing phytoplankton growth and effectively preventing "blooms" and improving water transparency. In addition, floating islands' plant roots also form a habitat for fish and aquatic insects, increasing biodiversity and improving the lake's ecological stability. The density of the Uros' floating islands' underwater plants may not be very high, but they do play a part in water purification.

There are a total of 51 islands at Lake Titicaca, and most of them are inhabited. The largest island (Taquile Island) has about 2000 inhabitants on the island. There are almost no modern mechanical equipment or motor vehicles. At night, they still use candles or torchlights to light their way. Only in recent years have they started using a small solar generator. Currently, the islands' inhabitants still plant corns, potatoes and other agricultural products on terraced fields using traditional methods, and farm livestock such as cows, sheep, chicken and other poultry. They lead a simple, traditional life, and it is precisely this environmental-friendly lifestyle that is naturally conducive to the lake's ecological balance.

3.3 Lessons learned

The Erhai Lake basin has experienced water environmental degradation in the past 30 years, largely due to the government and the people's low awareness of environmental protection when there was a rapid increase in the basin's residential population, resulting in excessive economic development that caused serious damage to the virtuous cycle of the lake's natural ecosystem. Following that, the Chinese government invested heavily in remediation measures. Even though the water quality has seen obvious improvements, Erhai Lake's ecosystem is still at a sensitive turning point - if remediation efforts are unable to catch up with the speed of pollution, the water environment will still worsen. On the other hand, the environment at Lake Titicaca's basin is still fairly good at the moment, but there are also many similar threats: for example, sewage from the daily life of residents inhabiting the Copacabana town in Bolivian territory flow into the lake without processing through a simple slabstone water channel. All kinds of plastic bottle waste, garbage bags and other daily life garbage can be seen everywhere by the lake. In October 2016, Peru Government officials confirmed that there were more than 10,000 scrotum frog deaths within 30 miles of the Coata River that flows past Peru's southeast city of Puno feeding into the lake. In addition, to reduce the people's poverty, Peru and Bolivia have also launched small-scale fish farming in cages at Lake Titicaca (Alejandro & Silje, 2010). China's Erhai Lake pollution model seems to quietly replicate itself at Lake Titicaca. However, the governments of Peru and Bolivia seem to have become aware of the potential threat to the lake; both parties signed an environmental protection agreement on January 7, 2016 with plans to invest approximately 500 million US dollars over the next 10 years into solving all kinds of pollution issues at the lake and maintaining the lake's ecosystem.

A similar situation to the water pollution these 30 years at Erhai Lake is, in fact, occurring at lakes all across the world. The Chinese government's painful lesson involving large amounts of funds in remediation measures serves as an example for the local governments and people inhabiting lakes' basins - if one does not take heed and fail to take the necessary cautionary measures and start implementing governance measures early, a serious price such as that of Erhai Lake's remediation will have to be paid in the future.

4 CONCLUSION

Under the double pressure of nature and human developments, lakes' surface areas are gradually decreasing, with water quality that worsens by the day and natural ecosystems that degrade continuously, severely threatening the economic and social development in their respective regions. Global warming is, of course, one of the reasons that cause the worsening of the water environment. However, having compared and analyzed the present conditions of Erhai Lake and Lake Titicaca that have experienced different degrees of economic developments as well as the factors involved, it has been found that the main reasons for the loss of the virtuous cycle at the lakes are to do with humankind. It is our lack of environmental protection awareness and excessive economic development that have caused overexploitation of the lakes' natural resources, damaging the ecological structure of the lakes and disrupting the lakes' ecological stability. It is hoped that the local governments and the people inhabiting the lakes' basins - the people who are responsible for the lakes and their wetlands - can figure out a way to reasonably develop and utilise lakes and their basins' resources without endangering the environment and the virtuous cycle of the ecosystem at the lakes. Perhaps then a balance between economic development and environmental protection can be struck.

REFERENCES

Alejandro, R., Silje, W., 2010. Cage Culture Comment: Latin America and Caribbean Sea areas, Cage Culture - Regional Comments and Global Overview. Rome: Food and Agriculture Organisation of the United Nations, 2010, 60–85.

Chai, Y.F., et al., 2013. Erhai Lake Ecological Environment's Research on Present Conditions and Existing Issues. Frontiers in Earth Science, 3, 241–252.

Dai, W.J., & Jiang, H.Q., 2008. Analysis and Evaluation of Manmade Floating Island's Efficiency at Improving Lakes' Water Quality. Journal of Taiwan Agricultural Engineering, 54(1), 26–34.

Dong, Y.X., et al., 2004. The Present Conditions of and Remediation Measures. Erhai Lake's Water Environment, 23, 101–103.

Guo, J.S., 2006. The World's Highest Freshwater Lake - Lake Titicaca. Earth, 5, 19.

Wang, S.Y., 2017. The Amazing Nature. Hohhot Cidade: Yuan Fang Publishing House.

Wu, G.G., et al., 2013. Analysis of Aquatic Plants and Phytoplankton's' Historical Changes and Influencing Factors at Erhai Lake. Acta Hydro biological Sinica, 37(5), 912–918.

Yin, C.Q., Lan, Z.W., & Yan, W.J., 1995. Preliminary Research on the Baiyangdian Water and Lane Ecotone's Effect on Intercepting Land-Based Nutritional Sources. Journal of Applied Ecology, 6(1), 76–80.

Zhou, J., Jiang, P., Zhang, H.Q., 2016. Research on the Main Challenges of Erhai Lake's Ecosystem Management and Innovative Remediation Measures Launched by the Towns and Villages. Small Town Construction, 5, 84–89.

Xuanzang theory

Innovation in Design, Communication and Engineering – Lam et al. (eds)
© 2020 Taylor & Francis Group, London, ISBN 978-0-367-17777-5

Monitoring the daily life of the elderly using the energy management system

Hui-Jen Chuang* & Chun-Chao Wang
Department of Electrical Engineering, Kao Yuan University, Kaohsiung, Taiwan

Ling-Tze Chao & Huann-Ming Chou
Department of Mechanical Engineering, Kun Shan University, Tainan, Taiwan

Tsun-I Chien
Department of Electronic Engineering, Kun Shan University, Tainan, Taiwan

Chang-Ying Chuang
Department of Electrical Engineering, National Chung Cheng University, Chiayi, Taiwan

ABSTRACT: In order to achieve the goal of assisting the elderly in managing their home appliances and of assessing the mobility of older people, this article proposes the energy management system (EMS) to monitor the activities of elderly people's daily living. The EMS consists of a general home appliance controller module and an LED light controller module that combine the energy management human–machine interface to operate various home appliance via Bluetooth wireless communication and report power consumption status of the home appliances on a real-time basis to record electricity consumption data. Equipped with an EMS APP on the mobile phone, it can help the elderly use home appliances efficiently for the benefit of energy saving. In addition, a daily activities density map observing method is proposed for the assessment of the elderly person's mobility, with which observers may understand the elderly person's activities of daily living and analyze the change of functional health status in older people for the purpose of remote monitoring and care.

Keywords: energy management system, activities of daily living

1   INTRODUCTION

Household safety is a key research theme for the development of smart home systems. The study on using technology to create an elderly-friendly living environment and protect the elderly from injury in their household living, and enable safe and independent living for the elderly for as long as possible, may help make home a better place to live for the elderly and reduce social care costs invested by the government for older people. Household safety issues include surveillance of the elderly person's mobility and functional status or daily living functions, as well as surveillance and reporting of the elderly person's falling. Regarding the monitoring of the elderly person's daily living functions in their homes, the activities of daily living (ADL) (Gale, 2009) refer to basic activities performed by individuals, on a day-to-day basis, consisting of self-care tasks (e.g., eating, bathing and showering, cloth changing, and grooming), working, household chores, and leisure and recreational activities. Isla Rippon et al. (2018); Taylor et al., 2018) argue that low ADL levels represent the potential of a certain kind

---

* Corresponding author: t20006@cc.kyu.edu.tw

of symptom or impairment of body functions; Giebel et al. (2015) find that performance of ADL is a critical indicator for determination of health status or evaluation of quality of life (QoL). According to the studies given above, a person's health is closely associated with his/her ADL performance. Therefore, in the Barthel Index, for example, activities like eating, personal hygiene, toileting, bathing, dressing and undressing, and moving, etc., are used to evaluate an aging people's movement and living function status or daily living functions. It is a subjective approach to evaluate a person's health.

Generally, the fastest and the most convenient way to simultaneously observe the living status of the elderly currently is to install webcams in a place where the elder people lives. Webcam devices, however, are not acceptable to the elderly due to privacy problems. With rapid development of information communication technology (ICT), the elderly person's hopefully feels respected when we apply ICT to monitor and care for their daily living status. For this reason, in some studies, smart meters are applied to power consumption appliances in four areas – living room, bathroom, kitchen, and bedroom – to calculate activity density of the cared for person and create activities density maps by combining and accumulating monitor data of power appliances in the respective areas. From observation of the change of activities, we may evaluate whether the person's physical heath is changed, or determine whether the person's physical functions are impaired (Yeh, 2014).

To facilitate the elderly to manage their home appliances for the benefit of energy saving, this article employs a mobile-phone energy management application to help aging people use these appliances. In this study, we develop an LED controller and a load controller for general home appliances, in combination with the said application, to turn on/off LED lights or fluorescent lamps and adjust their brightness, as well as turn on/off the load controller. From the simultaneous reporting of power consumption of various load controllers, we may create a database of home appliances' power consumption and upload the data to cloud database. In addition, by creating a daily activity density map to analyze the elderly person's activities levels, we may understand the elderly person's daily activities and assess changes of functional health status to achieve the goal of remote monitoring and care. Furthermore, families of the elderly may download power consumption data via the Internet to know how their elderly are living from time to time.

2 ENERGY MANAGEMENT SYSTEM

2.1 Lighting controller module

A lighting controller module consists of a microprocessor control unit, a dimming control circuit, an energy calculation unit, a power supply module, and a Bluetooth wireless communication module, in which the energy calculation unit reports power consumption to the control center via the Bluetooth wireless communication module. The microprocessor control unit integrates Bluetooth wireless communication module through a UART interface to receive the command from the control center, where users turn on/off LED lights or fluorescent lamps and adjust their brightness via the dimming control circuit to simultaneously report the current lighting load status. Furthermore, users may adjust the brightness required based on the amount of sunlight exposure for efficient lighting fixture management and energy saving.

2.2 Load controller module

A load controller module consists of a microprocessor control unit, a zero crossing circuit, an energy calculation unit, a power supply module, and a Bluetooth wireless communication module. When the load controller module receives a command from the remote control, the microprocessor control unit will unload it with current waveform at the zero crossing, and the energy calculation unit will report power consumption. The load controller module made for this study can be applied to the remote control or the monitor of the power consumption for ordinary electrical devices such as hair dryers, table lamps, electric fans, and so on, and can be used for the monitoring of general electrical equipment in an energy management system.

2.3 *Human machine interface in energy management system*

With an energy management system, the elderly may use mobile phones to generate a command to control lighting devices or general electrical appliances via the Bluetooth wireless communication module. The human–machine interface of the energy management system is shown in Figure 1, in which there is a control button for the LED controller module and the load controller module, and users just press the button on their mobile phones to control electrical appliances via the Bluetooth wireless communication module.

Figure 2 illustrates the actual operation of adjusting the brightness of an LED light, where the SWITCH button on the LED control interface is pressed to adjust the light's brightness; when "20%" is selected and the button "ON" is pressed, the control interface shows power consumption of 4W.

Figure 3 shows the actual operation of the load controller, in which the button LOAD_ON_2 on the load controller interface is pressed to enable the load; when the night lamp lights up, the interface shows power consumption of 3W.

Figure 1.   Energy management system.

Figure 2.   Factual operation of the LED light.

Figure 3. Factual operation of the load controller.

3 MONITORING OF THE ACTIVITIES OF DAILY LIVING

There are concerns of privacy issues when various webcams are installed in the elderly person's living environment to observe their household activities: the energy management system proposed in this article can be used to understand how their home appliances are used based on the reporting of data of various electrical devices, without invading the privacy of the elderly.

For behavior monitoring of the elderly person's living environment, the energy management system in this article is used to observe the elderly person's physical activities 24 hours a day and measure how their home appliances are used in their daily life to speculate on their activities of daily living. Monitoring the use of home appliances in the living environment, this article creates a daily activity density map for the assessment of the elderly person's mobility to understand the distribution of the sleep hours and habits across the day and activities in the living room and bathroom. It objectively analyzes the change of the functional health conditions in older people, allowing their family members to download power consumption data of the home appliances via the Internet to understand their household activities any time.

Figure 4 is a bathroom lamp density map. The system is applied to an apartment for the aged. Without interrupting the elderly person's activities, we capture the use status of lights in the daily living of an elder individual A. From the power consumption status of the bathroom light for 30 days in a certain month, we may understand the routine activities of the elderly. From the map, the elder individual A has regular daily activities; he usually gets up at 5 in the morning, except on few days during the month, and uses the bathroom the last time at 11 or 12 in the night. There were only four times he used the bathroom in the middle of the night during the whole month, and the

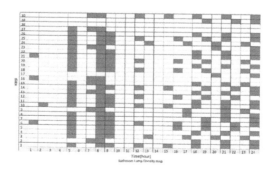

Figure 4. Bathroom lamp density map.

frequency of using the bathroom is about once every three hours. There were several days he didn't use the bathroom. According to the figure, A only used the bathroom four times in the middle of the night, suggesting that he has good quality of sleep; and the frequency of using bathroom is about once for three hours, indicating that he shows no frequent urination. Besides, there are three days that A didesn't come home for the night, suggesting that he might go out of town for a vacation.

Figure 5 shows the bedroom lamp density map, from which we may understand the elderly's living activities in the bedroom. The elder individual A, for example, almost always gets up and turns on the bedroom lamp at 5 in the morning, turns on the lamp again at 7 or 8 in the morning, and turns on the lamp at 10 or 11 at night the last time.

Figure 6 shows the living room lamp density map, from which we may understand the elderly's living activities in the living room. The elderly individual A, for example, almost always turns on the living room lamp at 5 or 6 in the afternoon and turns off the lamp at 10 or 11 at night and then returns to the bedroom to sleep.

From the bathroom, bedroom, and living room lamp density maps, we may clearly observe the time distribution of A's activities in the bathroom, bedroom, and living room and understand his activities of daily living, which can be a basis for determining whether the elderly individual's health status changes or physical functions are impaired. In addition, the families of the elderly may download the power consumption data of home appliances to understand their daily activities from time to time.

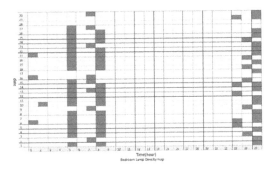

Figure 5.   Bedroom lamp density map.

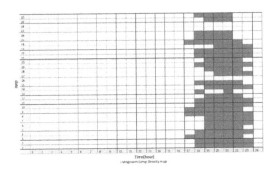

Figure 6.   Living room lamp density map.

4 CONCLUSIONS

Because an older adult's perception is deteriorating gradually, he/she is not capable of using or controlling devices or equipment efficiently little by little; however, the number of caregivers in care facilities is limited. This study develops a human-machine interface of an energy management system, in combination with all controller modules, able to remote control all home appliances. Users just need to press a configured button to control all home appliances and thereby receive reports of the daily power consumption density maps and the time distribution of living activities. From the proposed energy management system, caregivers may observe the elderly person's activities of daily living and use it as a basis for predicting whether their health status changes or their physical functions are impaired. This may help the elderly person's families, caregivers, or care facilities to take care of aging people through remote monitoring and may also achieve energy saving and carbon reduction and promote quality of care for the elderly.

REFERENCES

Gale, 2009. Activities of Daily Living Evaluation, Encyclopedia.com, https://www.encyclopedia.com/care giving/encyclopedias-almanacs-transcripts-and-maps/activities-daily-living-evaluation

Giebel, C. M., Sutcliffe, C., & Challis, D. (2015) "Activities of daily living and quality of life across different stages of dementia: a UK study." *Aging Ment. Health*, 19, 63–71.

Rippon, I., & Steptoe, A. (2018) "Is the relationship between subjective age, depressive symptoms and activities of daily living bidirectional?" *Social Science & Medicine*, 214, 41–48.

Taylor, S. A. F., Kedgley, A. E., Humphries, A., & Shaheen, A. F. (2018) "Simulated activities of daily living do not replicate functional upper limb movement or reduce movement variability." *Journal of Biomechanics*, 76, 119–128.

Yeh, M. C. (2014) "Daily life monitoring system based on a smart meter," National Taipei University of Technology, Master thesis.

Innovation in Design, Communication and Engineering – Lam et al. (eds)
© 2020 Taylor & Francis Group, London, ISBN 978-0-367-17777-5

The effectiveness of green exercise in community rehabilitation center

Tsai-Chieh Chien
Program of Mechanical and Energy Engineering, Kun Shan University, Tainan, Taiwan

Huann-Ming Chou
Department of Mechanical Engineering, Kun Shan University, Tainan, Taiwan

ABSTRACT: Large scale-studies have demonstrated that green exercise can improve the mood and self-esteem of patients with psychiatric disorders. Moreover, studies have revealed that green exercise and antidepressant drugs have similar positive effects on patients with depression. However, studies on green exercise are scarce in Taiwan, particularly on community psychiatry. Combining green exercise and the concept of recovery can provide a major opportunity for advancing depression treatment strategies. Therefore, on the basis of the aforementioned concept, this study designed a project called "recover your green mind." Participants were selected from a community rehabilitation center in Taichung City, Taiwan. Through self-determination, participants were given autonomy to choose the locations of green exercise activities. The activities were aimed at treating depression through enabling the participants to interact with and immerse themselves in natural objects such as hiking trails, plants, and animals. Qualitative interviews and the rating of perceived exertion scale were employed to evaluate the intensity of the green exercise activities and examine outcomes of the participants. The results revealed that most participants rated the green exercise activities as having a light exertion level. The qualitative interviews indicated that participants achieved favorable results in terms of mood and knowledge acquisition, suggesting the effectiveness of the project in treatment regarding community psychiatry. Long-term intensive research involving large samples should be performed in the future to study the details and stability of the effect of green exercise on Taiwanese patients with mental illness.

Keywords: effectiveness, schizophrenia, green care, green exercise, community rehabilitation center

1 BACKGROUND

Green exercise is defined as any physical activity executed in natural environments, and research confirmed such exercise to have considerable healing power (D. K. Haubenhofer et al. 2010). Strolling, biking, gardening, and even chatting in natural environments have healing power (Mind, 2007). A meta-analysis of 10 studies in the United Kingdom indicated that green exercise can improve mood and self-esteem, with such exercise being particularly effective in enhancing the self-esteem of psychiatric patients (J. Peacock et al. 2007). Moreover, a study conducted in the United Kingdom reported that green exercise has the same effect as medications for patients with moderate to severe depression (W. A. Anthony, 1993). Recovery is a transformative process through which individuals improve their attitudes, values, feelings, goals, skills, or roles. Specifically, recovery enables individuals with mental illness to develop

* Corresponding author: ball0818@seed.net.tw

satisfactory, optimistic, and fulfilling life styles. The recovery of an individual with mental illness includes finding new meaning and purpose in life (S. Abuse, 2009). The Substance Abuse and Mental Health Services Administration has adopted the National Consensus Statement on Mental Health Recovery, which defines the principles of recovery and elucidates guiding principles for facilitating the recovery of people with mental illness (R. J. Robertson et al. 2003), as presented in Table 1.

The most optimal method for facilitating the recovery of people with psychiatric disabilities is to provide them with opportunities to develop inner resources and skills in order to fulfill their recovery goals. Accordingly, the current study designed a project—designated as "recover your green mind"—that involved two 3-month green exercise activities. This project

Table 1.　The 10 principles of recovery.

Self-direction: Individuals lead, control, and manipulate their own choices. Under conditions involving a high degree of autonomy, they can determine their journey to recovery, in addition to determining how to control life resources. Accordingly, individuals can define their life goals and implement measures to achieve them.

Individualized and person-centered factors: Various factors are involved in the process of recovery, including individuals' special strengths and needs, preferences, and recovery experience (e.g., past trauma), as well as cultural characteristics under different expressions. Moreover, recovery is a long continuous journey; therefore, individuals' ultimate target must be a state of complete well-being and mental health.

Empowerment: Individuals have the right to choose and be involved in making decisions regarding their interests, including the distribution of resources. They should be educated and empowered to continue executing their recovery processes. Moreover, individuals should be encouraged to satisfy their needs and effectively voice their rights. Through empowerment, individuals can control their faith and organizational and social influences in their lives.

Holistic process: Recovery entails the improvement of an individual's life, including the individual's mind, body, spirit, and community. Recovery encompasses multiple aspects of life including housing, occupation, education, mental health and health care, contemporary and natural services, addiction treatment, spirit, creativity, social networks, community participation, and personal autonomy. Additionally, family members, health providers, organizations, systems, communities, and society play an essential role in creating and maintaining a meaningful support system for individuals.

Nonlinear process: Recovery is not a step-by-step process. It involves an individual's capability to sustain growth, experience occasional setbacks, and learn from experience. The first step of recovery self-awareness; that is, an individual's awareness of the possibility to make positive changes. Such self-awareness ensures the individual's complete devotion to the process of recovery.

Strength-based process: Recovery typically focuses on establishing and revealing a person's abilities, resilience, problem-solving strategies, and inner values. Through strength development, individuals can eliminate restrictive factors in their daily lives and thus comprehensively devote themselves to their new roles; for example, they can fully devote themselves to their roles as partners, caregivers, friends, students, and employees. Accordingly, the process of recovery also entails the development of supportive and trustworthy relationships.

Peersupport: Mutual support—including sharing related experience and knowledge and learning or teaching social skills—plays an essential role in recovery. In the process of recovery, individuals are encouraged to collaborate with other individuals; such collaboration fosters a sense of belonging or community, engenders a supportive relationship, and plays a valuable role in recovery.

Respect: Social acceptance or mutual appreciation among individuals can ensure that individuals' rights are protected and that discrimination and prejudice are eliminated, which are essential to the process of recovery. In particular, self-acceptance and the restoration of faith are vital. Respect enables individuals to completely participate in all aspects of life.

Responsibility: Individuals must assume responsibility in their care and recovery. They should be resolute in acting toward their goals. They should not only understand and assign meaning to their experiences but also identify coping strategies and healing processes to enhance their well-being.

Hope: Recovery provides a fundamental motivational message: People can overcome difficulties. Hope is internalized but can also be fostered by peers, families, friends, health care providers, and others. Specifically, hope is a catalyst for recovery.

included weekly meetings in which participants were provided autonomy to select the green exercise activity to participate in. The meetings were based on the guidelines of the Joint Commission of Taiwan, the primary objective of which is to establish a comprehensive mental health care system. According to Regulations on the Establishment and Management of Mental Rehabilitation, which was enacted on the basis of the Mental Health Act, central competent authorities should regularly evaluate rehabilitation institutions. Consequently, since 2003, the Ministry of Health and Welfare has entrusted the joint commission with the implementation of an evaluation system for mental rehabilitation. The quality of mental rehabilitation services is expected to improve through external inspection and guidance. Some of the items considered in the evaluation system are outlined as follows: human resource management, space and facilities, rehabilitation services offered, rehabilitation service quality measures, and community resource integration, as well as improvements made on the basis of suggestions from previous evaluations. Article 2.10 of the regulation states that participants should call for meetings addressing autonomy; such meetings can encourage members to be aware of their rights and surroundings, participate in public affairs, and elevate their decision-making skills. Some of the discussed items are outlined as follows: meals, facilities, equipment, community participation, arrangement of recovery activities, conventions on life, member rights, and rehabilitation institution management measures. This study also adopted the rating of perceived exertion (RPE) scale developed by Swedish psychologist Gunnar Borg (R. J. Robertson et al. 2003). The RPE scale is used to evaluate the intensity of physical activity by measuring physical sensations (e.g., heart rate, respiration, sweating, and muscle fatigue) during physical activity. The RPE scale values range from 6 to 20. However, this study combined these values into a range more compatible with the study. The revised version is presented in Table 2.

2 PURPOSE

The recover-your-green mind project comprised two exercise activities. The project involved meetings with participants to present the concept of green exercise and promote its benefits; in the meetings, the participants autonomously decide which exercise activity to engage in. The decisions regarding the exercise activities were made on the basis of group discussions. The exercise activities were held every 3 months, for half a year. This study evaluated the participants' perceptions toward the two activities through qualitative interviews and the RPE scale.

3 SAMPLE/METHOD

The first exercise activity was performed at Flying Cow Ranch(Table 3) and involved eight participants. The activity entailed grassland excursion and trail hiking processes, which the

Table 2. Rating of perceived exertion scale (revised).

Grade	Statement
1	No exertion at all
2	Extremely light
3	Very light
4	Light
5	Somewhat hard
6	Hard/Heavy
7	Very Hard
8	Extremely Hard
9	Maximal Exertion

Table 3. Description of green exercise location—Flying Cow Ranch.

Location	Description
Flying Cow Ranch	Flying Cow Ranch is located in Miaoli County.
	The ranch was established in 1975 by a group of ambitious young adults who returned to Taiwan after receiving professional training in the United States regarding dairy farming. Through the government's assistance, they developed a mountain slope—located next to Pao-an woodland of Nanho Village, Tongsiao Township, Miaoli County—covered by Acacia confusa trees into the "Central Youth dairy village" (currently known as "Flying Cow Ranch"), which serves as a venue for the professional demonstration of dairy farming processes. The ranch has sufficient manure to replenish soil fertility, enabling extensive growth of organic vegetables. Because the ranch is devoid of air or water pollution, more than 3 ha of organic land within the park is used for growing green leafy vegetables and herbs that are generally used for household consumption and used as vegetable supplies for the restaurants within the ranch. In addition to being remarkably informative, the professional demonstrations of feeding and management processes for dairy cattle at the ranch are enjoyed by visitors.
	The ranch includes a butterfly garden, veldt activity area, organic plantation area, children area, and the hacienda area. Tourists can experience the lifestyle associated with dairy farming and learn about the ecology of dairy cows. Moreover, they can personally experience the production of milk at the ranch.

participants initiated upon arrival at the destination. The participants reported feeling relaxed during the execution of these processes. Moreover, they fed some cows, sheep, and rabbits. Nevertheless, the participants were initially apprehensive about approaching the animals. After demonstrations by occupational therapists, the participants slowly started the feeding session while the guide offered related information. Subsequently, they engaged in an experiential activity that involved milking cows. As the participants lined up to practice the milking process, the guide provided explanations regarding the method of milking cows.

The second green exercise activity was performed at Lavender Cottage (Mingder branch) (Table 4). Upon passing through the front gate, the participants first arrived at a store and bakery for herbs. The tour guide educated the participants about herbs and allowed them to engage in some experiential activities. They walked along the pathway surrounding the lake and through the hallway surrounded by various herbs. Finally, they had a meal at the herb restaurant.

Table 4. Description of green exercise location—Lavender cottage.

Location	Description
Lavender Cottage	In 2001, the grand opening of Lavender Cottage was held at Xinshe District, Taichung County. The cottage was founded by Zhan Hui-jun and Lin Ting-fei. The green exercise activity was performed at the cottage's Mingder branch, which was established in December 2008. The participants were profoundly impressed by the friendly attitude of the attendants, elaborate landscape design within the area, and heartwarming and colorful activities planned.
	Tourists can acquire different experiences from these rich and diverse activities, demonstrating the appeal of this cottage.
	Natural landscapes and resources in the park such as lavender fields, the wishing tree, and fireflies are featured in different activities to attract visitors, thus increasing the number of tourists. The design of conventional activities ensures tourists' personal participation, thus rendering these activities a tradition because tourists can identify with them over time. The wishing tree, the mailboxes of happiness, culture of herbs, and various meals all constitute a pathway through which tourists can directly experience the benefits of green exercise.

4 RESULT

Results obtained from the qualitative interview of the participants on Flying Cow Ranch are presented in Table 5.

The RPE scale analysis results are presented in Figure 1.

Results obtained from the qualitative interview conducted on Lavender Cottage are presented in Table 6.

The RPE scale analysis results are presented in Figure 2.

Table 5. Results obtained from the qualitative interview on Flying Cow Ranch.

Green exercise activity	Participants' comments
Trail hiking	A: I felt relaxed and delighted; everywhere we went was fun.
	B: Even though hiking was tiring, I felt quite refreshed.
	C: Hiking the trail was interesting and the air was fresh.
	D: Hiking is beneficial to our health. I could keep on hiking, breathe fresh air, and stay healthy.
	E: Staying in the Community Rehabilitation Center every day is not pleasant. Being in close contact with nature was pleasant.
	F: The air was fresh. Strolling outdoors lifted my mood.
	G: The grassland was a comfortable place for walks.
	H: Cute animals were present along the trail.
Animal feeding	A: It was still fun, although I was afraid of feeding either the cows or goats; I kept my distance. Feeding the rabbits was quite amusing, and they were adorable.
	B: I liked the rabbits; they looked cute while eating. Today I learned that rabbits are not supposed to eat carrots.
	C: I only started approaching the cows when I noticed all the tourists feeding the cows. The goats were also adorable, although I was slightly frightened when they all rushed over as I approached them. The rabbits were my favorite because they were obedient, pettable, and felt fluffy.
	D: I liked the rabbits. I could pet and feed them.
	E: I was afraid of approaching the cows because they were excessively huge. The rabbits were more adorable, making me want to get one at home.
	F: I am a pet person.
	G: I was afraid of touching them (animals). Nevertheless, I felt less scared seeing everybody having fun feeding them.
	H: It turns out that rabbits are not supposed to eat carrots.

no exertion   extremely light

very light   light

somewhat hard   hard

very hard   extremely hard

maximal exertion

Figure 1. Result of RPE in Flycow Ranch.

Table 6. Results obtained from the qualitative interview on Lavender Cottage.

Green exercise activity	Participants' comments
Learning about herbs	A: Fish wort was smelly. Other plants smelled pleasantly. I learned a lot from the tour today. B: I was interested in the hot pepper. My interest subsided when the guide mentioned the measured heat of the pepper to be 300 heat units. C: I liked the smell of peppermint. It provided a relaxing aroma when rubbed. D: Many herbs exuded strong scents, but I found the smell of some of them to be strange. E: I learned that rosemary always comes with lamb chops. I thus learned a lot today. F: Some herbs emitted a pleasant aroma. G: It felt nice to just go out and stroll around. H: I can now differentiate lavender from common sage. The tour was not bad.
Hallway with various herbs	A: Herbs were pervasive. I could smell the aroma as I walked by. B: Although it rained today, Mingder Dam was still beautiful. C: Some parts of the pathway were damaged. It was pitiful we could not walk on all of them. D: The whole path was unique and purple, making it romantic. E: The walk was satisfactory. Strolling outdoors lifted my mood. F: The pathway around the lake was great for casual walks. I loved walking along the lake. G: Mingder Dam was indeed beautiful. H: I did not feel tired as I walked while appreciating nature.

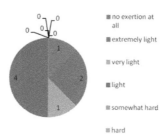

Figure 2.   Result of PRE in Ming der branch, Lavender cottage.

5   DISCUSSION

According to the results obtained from the qualitative interviews, the two green exercise activities were considerably effective in lifting the participants' moods and improving their knowledge. Most participants had positive perceptions about strolling in nature, and this finding thus validates the biophilia hypothesis (S. R. Kellert et al. 1995). Regarding the RPE scale analysis, most of the participants provided an exertion rating of "light" for the exercise activities. In terms of exercise challenge, both green exercise activities were rated to be of "just the right challenge" (E. J. Yerxa, 1990). The results reveal that the two activities not only reached their treatment objectives but also created a sense of accomplishment among the participants. Moreover, the participants were found to become more active in proposing subsequent green exercise activities during the meetings; thus, the participants' ability to provide initiatives was improved, manifesting the recovery effects.

REFERENCES

D. K. Haubenhofer, M. Elings, J. Hassink, and R. E. Hine, 2010. The development of green care in western European countries, *EXPLORE: the Journal of Science and Healing*, vol. 6, pp. 106–111.

Mind, 2007. *Ecotherapy: The Green Agenda for Mental Health*, Executive Summary: Mind.

J. Peacock, R. Hine, and J. Pretty, 2007. The mental health benefits of green exercise activities and green care, *Report for MIND*.

W. A. Anthony, 1993. Recovery from mental illness: the guiding vision of the mental health service system in the 1990s, *Psychosocial rehabilitation journal*, vol. 16, pp. 11.

S. Abuse, 2009. Mental Health Services Administration (2005) National consensus statement on mental health recovery, *Rockville, MD: US Department of Health and Human Services*.

R. J. Robertson, F. L. Goss, J. Rutkowski, B. Lenz, C. Dixon, J. Timmer, et al., 2003. Concurrent validation of the OMNI perceived exertion scale for resistance exercise, *Medicine & Science in Sports & Exercise*, vol. 35, pp. 333–341.

S. R. Kellert and E. O. Wilson, 1995. *The biophilia hypothesis*: Island Press.

E. J. Yerxa, 1990. An introduction to occupational science, a foundation for occupational therapy in the 21st century, *Occupational therapy in health care*, vol. 6, pp. 1–17.

Innovation in Design, Communication and Engineering – Lam et al. (eds)
© *2020 Taylor & Francis Group, London, ISBN 978-0-367-17777-5*

Greenhouse effect and air pollution caused by using agricultural waste as biomass power generation sources: A case study in Yunlin County

Chuan-Hsi Shih*
Program of Mechanical and Energy Engineering, Kun Shan University, Tainan, Taiwan

Huann-Ming Chou
Department of Mechanical Engineering, Kun Shan University, Tainan, Taiwan

ABSTRACT: A 2019 report by the International Energy Agency indicated that greenhouse gas emissions resulting from global energy use increased to 33.1 Gt in 2018, reaching a record high. The growth of global energy consumption in 2018 was twice the average growth during 2010 and 2018. The human race across the planet is living in an environment under constant pollution and gradual depletion of energy sources. Developing alternative clean energy sources is a crucial topic of the third industrial revolution. Moreover, clean energy development is an essential focus of energy development in Taiwan. The June report by REN21 in 2018 analyzes energy use of the global power sector in 2017. The report reveals that 26.5% of global power was generated from renewable energy. By contrast, the renewable energy power generation was merely 5% in Taiwan. To mitigate this gap, the Taiwanese government should emphasize renewable energy evaluation and invest in the development of biomass energy based on agricultural waste. The present study used straw litter in Yunlin County as a basis for developing biomass energy and explored the feasibility of straws as a power generation source and its effect on greenhouse effect and air pollution. The results can serve as a reference for relevant governmental agencies in promoting and planning biomass energy development.

Keywords: biomass energy, agricultural waste, greenhouse effect, air pollution

1 INTRODUCTION

Taiwan relies heavily on imported power generation sources for most of its residential and industrial supplies. However, it also produces abundant rice for both domestic consumption and export. The traditional method of processing rice straw involves open burning, composting, and landfilling. Government policies have begun to strongly advocate organic landfill processing; however, the on-site landfill decomposition process requires at least 20 days. During this process, errors may prevent rice straw thoroughly decomposing in the field, which may result in stifle disease in rice seedlings during the second crop season. Therefore, most farmers have adopted the burning method during this season (Tang, 2018). The on-site burning process has the greatest decomposition efficiency; however, the thick smoke it produces influences traffic safety and negatively affects air quality.

Numerous countries have made advances in research and development for the adoption of organic agricultural residues as alternative energy sources. For example, in 2018, Taiwan's produced 1,561,642 metric tons of rice, which accounted for 3,160,763 metric tons of rice straw residue. This study assumed that 14% and 17.2% of rice straw are lost in the collection process and directly landfilled back into fields, respectively. Therefore, 68.8% of rice straw is recycled (Lu & Zhang, 2010; Wang et al., 2010; Gao et al., 2002), accounting for 2,174,605 metric tons of rice straw residue collected in Taiwan annually. Therefore, if every 1 kg of agricultural residue generated 1 kWh of energy, Taiwan would be able to generate 2,175 GWh of power each year. In 2018, the amount of wind-generated energy produced in Taiwan was 1,645 GWh. Therefore, the ratio of rice straw-generated energy to wind-generated energy is 1.32:1. However, wind generators entail high production costs. In summer, when energy consumption is high, the energy generation rate of wind generators is reduced to 22% because of the absence of the northeast monsoon. Taiwan, with its abundant rice production, is capable of generating a more stable power supply though agricultural residues than it could through wind energy. To counter the energy shortages faced today, government units related to energy development should actively conduct planning and development in the field of agricultural-residue biofuel.

This study discussed the feasibility of adopting rice straw residues in Taiwan's Yunlin County as a power generation source. This study assumed that rice straw could be successfully collected and developed into solid biofuel, thereby solving the air pollution problem posed by the open burning of rice straw and providing a fixed supply of regenerative fuels to reduce Taiwan's energy shortages. Furthermore, this study explored the feasibility of adopting rice straw as an energy source. Because this study only discussed the energy conversion of agricultural residues (i.e., rice straw) into electricity, it did not discuss the financial aspects of the process. The author suggested that government units working in energy development conduct a comprehensive evaluation, support agricultural areas in recycling and producing rice straw, and adopt rice straw into their planning and development methods for generating energy.

2 LITERATURE REVIEW

The open burning of rice straw releases a substantial amount of CO_2, exacerbates global warming, and generates pollutants that include smoke (particles), sulfur oxides, nitrogen oxides, fine particulate matter (PM2.5), and organic pollutants generated through incomplete combustion. These byproducts threaten respiratory systems and severely influence the lives of people with weak body resistance, including citizens with respiratory tract diseases, older adults, and children. This study collected and compiled relevant studies, and the following were related to the open burning of rice straw. Lin (2004) analyzed monitoring data for the Yunlin and Chiayi from November 27 and 28, 2002 and revealed that on November 27 the total emissions of CO, NMHCs, NO_X, PM10, and SO_2 in five counties in central and southern Taiwan were 665.4, 153.2, 40.5, 103.5, and 10.7 metric tons, respectively. Lin inferred that severe air pollution was caused by the burning of agricultural residues. Chen (2004) indicated that open burning is generally associated with PM10, CO, NO_X, and HCs pollutant emissions. Furthermore, under certain photochemical reaction conditions, NO_X and NMHC generate the secondary pollutant O_3. Because of the transport effect, open burning held in weather conditions disadvantageous to diffusion often cause cross-region air pollution. Su (2006) revealed that the average total suspended particle concentration in the atmosphere during intensive rice straw burning periods was 254 µg/m^3, which was substantially greater than that during nonintensive burning periods (108 µg/m^3). Furthermore, the PM10/TSP ratio during intensive burning periods was 1.2 times that during nonintensive burning periods, which indicated that PM10 was the main type of particulate matter generated during burning periods. During intensive and nonintensive burning periods, polycyclic aromatic hydrocarbon concentrations of 1193 and 593 ng/m^3 were exhibited, respectively, indicating that such concentrations increased because of rice straw burning. Chang (2011) stated that the average concentration of polybrominated diphenyl ethers (PBDEs) in the atmosphere during the nonintensive burning period was 36.0–87.6 pg/m^3, whereas during the intensive period it escalated to 63.6–250 pg/m^3. Chang's results indicated that the PBDE concentration in the air surrounding rice fields during the intensive burning period was 2.85 times that during nonintensive periods. This suggested that the open burning of agricultural residues leads to significant increases in PBDE concentration.

The following studies were related to bioenergy generated through agricultural-residue reprocessing. Huang (2005) adopted various oxygen concentrations to discuss the reaction mechanisms of rice straw pyrolysis. Liu (2009) analyzed the properties (i.e., heating values) of solid biomass fuel briquettes before and after charcoalization to provide a reference for the development of solid biomass energy technology. Chen (2009) conducted a life cycle analysis on the collection and transportation of rice straw and production of rice straw-derived fuel. Xu (2011) adopted a life cycle assessment approach to evaluate various input and output scenarios of the rice straw bioenergy production chain and computed the carbon emission values of each. These values were then adopted to evaluate the latent environmental impact of each scenario. Yeh (2013) compiled the carbon and water footprints and ecological efficiency of sugar cane, sweet potato, rice straw, rapeseed, and soybean.

In summary, most scholars have discussed the influences of air pollution caused by the direct burning of agricultural residues or the environmental impacts of reprocessing agricultural residues into bioenergy. However, few studies have explored how the transportation, drying and grinding, and thermal power generation stages influence the pollutant emissions and greenhouse effects of reusing agricultural residues to generate power.

3   RESULTS

Crop residue refers to inedible agriproduct parts, including residues left in agricultural fields after harvesting. Generally, agricultural residues cannot be directly measured and are estimated using a coefficient equation (total agricultural residues = total crop production × residue-to-product ratio [RPR; also known as the straw/grain ratio]). Studies have indicated that the RPR for rice, maize, sorghum, soybean, peanut, and sugar cane are 2.024 (straw = 1.757, husk = 0.267), 2.473 (stalk = 2.0, cob = 0.273), 1.25 (stalk = 1.25), 3.5 (straw and pod), 2.777 (stalk = 2.3, shell = 0.477), and 0.39 (bagasse = 0.29, leaf = 0.1), respectively (Wei Lu et al., 2010). The Agricultural Statistical Database of the Council of Agriculture, Executive Yuan, reveals that in 2018, the total planted area and production of grain crops in Yunlin County (including rice, maize, sorghum, soybean, peanut, and sugar cane) were 69,952 hectares and 589,673 metric tons, respectively. This study adopted the RPR ratio of the aforementioned crops and inferred that in 2018, the total agricultural crop residues of Yunlin County were 921,798 metric tons. Furthermore, studies have indicated that 17.2, 21.5, 14, 2.9, 24, and 20.5% of crop residues are directly ploughed back into farmland, used for animal feed, lost in the gathering process, used as industrial raw materials, used as bioenergy, and disposed of through open burning, respectively (Lu & Zhang, 2010; Wang et al., 2010; Gao et al., 2002). Therefore, the present study assumed that 14% and 17.2% of the crop residues were lost in the gathering process and directly ploughed back into farmland, respectively; the remaining 68.8% became bioresources for generating energy. Therefore, this study estimated that in 2018, Yunlin County generated 634,197.48 metric tons of agricultural residues (Table 1).

In addition, this study assumed that the steam boiler generator scale was 6 MW and power generation efficiency was 20%. Thus, generators generated 1 kWh of electricity per 1 kg of agricultural residue expended. Furthermore, the generators could generate a total of 634.2×10^6 kWh of power. Through a literature review, the researchers compiled the burning emission coefficients of different agricultural residues (Table 2), and estimated the amount of greenhouse gasses and air pollution produced through biomass energy generation in Yunlin County. The production process of agricultural products generally includes production and gathering procedures. However, this study evaluated the air pollution emitted through adopting agricultural residues to generate energy; therefore, it did not include air pollution emitted from the production and gathering stages. After compiling the SO_2, CO, NO, NMHC, PM_{10}, and CO_2-equivalent (CO_2e) emission capacities and percentages in the transportation, drying and grinding, and power generation stages (Tables 3 and 4), the researchers established the following results:

Table 1. Analysis of agricultural residue production in Yunlin County in 2018.

Grain and crop type	Rice	Maize	Sorghum	Soybean	Peanut	Sugar cane
Planted area (hectares)	44,652	6,626	1	164	15,616	2,893
Production (metric tons)	286,426	54,071	4	265	46,229	202,678
RPR	2,024 Straw = 1.757 Husk = 0.267	2.473 Stalk = 2.0 Cob = 0.273	Stalk = 1.25	Straw and pod = 3.5	2.777 Stalk = 2.3 Shell = 0.477	0.39 Bagasse = 0.29 Leaf = 0.1
Agricultural residues	579,726.22	133,717.58	5	927.5	128,377.93	79,044.42
Collected amount	398,851.64	91,997.70	3.44	638.12	88,324.02	54.382.56
Percentage of the collected amount	62.89%	14.51%	0.00%	0.10%	13.93%	8.58%

Table 2. Burning emission coefficient of agricultural residues in the production and collection, transportation, drying and grinding, and power generation stages.

Emission coefficient (kg/ton)	SO_2	CO	NOx	NMHC	PM_{10}	CO_2e
Production and collection	0.258	0.146	0.254	0.046	0.054	206
Transportation	5.85	3.42	171	10.2	2.79	14.68
Drying and grinding	0.882	0	0.615	0	0	118.1
Power generation	1.38	0.025	5.27	0.65	1.84	1367.7

Table 3. Gas emission capacities of the transportation, drying and grinding, and power generation stages of biomass energy generation in Yunlin County in 2018.

Emission (kg)	SO_2	CO	NOx	NMHC	PM_{10}	CO_2e
Transportation	3,710,055.25	2,168,955.38	108,447,768.75	6,468,814.28	1,769,410.96	9,310,018.98
Drying and grinding	559,362.18	-	390,031.45	-	-	74,898,722.16
Power generation	875,192.52	15,854.94	3,342,220.71	412,228.36	1,166,923.36	867,391,890.77
Total	5,144,609.94	2,184,810.31	112,180,020.91	6,875,969.06	2,936,334.33	951,600,631.91

Table 4. Gas emission percentages of the transportation, drying and grinding, and power generation stages of biomass energy generation in Yunlin County in 2018.

Emission percentage (%)	SO_2	CO	NOx	NMHC	PM_{10}	CO_2e
Transportation	72.12	99.3	96.67	94.01	60,26	0.98
Drying and grinding	10.87	0	0.35	0	0	7.87
Thermal generation	17.02	0.7	2.98	5.99	39.92	91.15
Total	100	100	100	100	100	100

1. The estimated total SO_2 emissions were 5,144.61 metric tons; most (72.12%) were generated in the transportation stage (3,710.06 metric tons), followed by the power generation stage (875.19 metric tons; approximately 17.01%). The drying and grinding stage generated the least SO_2 emissions (10.87%). According to the researchers' estimation, 8.11 kg of SO_2 was emitted per 1 kWh of electricity generated.
2. The estimated total CO emissions were 2,184.81 metric tons. The CO emissions of the three stages in decreasing order were 99.27% for transportation, 0.73% for power generation, and 0% for drying and grinding. Thus, for each kWh of electricity generated, 3.45 kg of CO was emitted.
3. The estimated total NO_X emissions were 112,180.02 metric tons. The NO_X emissions of the three transportation stages in decreasing order were 96.67% for transportation, 2.98% for power generation, and 0.35% for drying and grinding. According to the researchers' estimation, 176.81 kg of NOx was generated for every 1 kWh of power generated.
4. The estimated total NMHC emissions were 6,881.04 metric tons, mainly emitted in the transportation stage (6,468.81 metric tons; 94.01%), followed by the power generation stage (412.23 metric tons; 5.99%). For each 1 kWh generated, approximately 10.85 kg of NMHC was emitted.
5. The estimated total PM_{10} emissions were 2,936.33 metric tons, most of which were emitted in the transportation stage (1,769.41 metric tons; 60.26%), followed by the power generation stage (1,166.92 metric tons; 39.74%). According to the researchers' estimation, an estimated 4.63 kg of PM_{10} was emitted per 1 kWh of power generated.
6. After greenhouse gas emissions during the production and collection stage were excluded, the total CO_2e emissions were 951,600.63 metric tons; most were released during the power generation stage (867,391.89 metric tons; 91.15% of the total), followed by the drying and grinding (6.92%) and transportation stages (0.98%). Therefore, per kWh of power generated, 1.50 kg of CO_2e was emitted. However, because the CO_2e emitted during the burning of agricultural residues originates from CO_2 absorbed by plants during photosynthesis, the CO2e in this stage is generally listed as zero. In this scenario, the primary CO_2e emission was released in the transportation and drying and grinding stages, which accounted for a greenhouse effect equivalent to 0.133 kg of $C0_2e$ emissions per 1 kWh of power generated. This value is lower than the electricity carbon emission (0.533 kg of CO_2e) announced by the Bureau of Energy, Ministry of Economic Affairs.

4 CONCLUSION AND SUGGESTIONS

In this study, estimations of greenhouse gas and air pollution emissions indicated power generation as the main emission stage for greenhouse gasses (91.15%), whereas the transportation stage is the main source of SO_2, CO, NO_X, NMHC, and PM_{10} pollution $\geq$ 60%. Therefore, biomass energy development, such as rice straw energy technology, must first be developed for regional use. This study suggests that relevant units planning to establish biomass power plants in Yunlin should do so near rice fields. This would reduce the amount of air pollution generated by vehicles during the transportation stage.

Adopting biomass power generation method effectively reduces CO_2 emissions. Additionally, the life cycle of biomass power generation creates less air and environmental pollution and greenhouse gas emissions than does fossil fuel power generation. Therefore, agricultural residue power generation has substantial development potential. Currently, biomass power generation accounts for 5% of total power generated in Taiwan, which is considerably less than the current global renewable energy average of 26.5% as well as the planned proportion of renewable energy (20%) by 2025 (National Energy Act). Thus, the development of biomass power in Taiwan has substantial room for improvement. Therefore, how government units effectively utilize domestic resources to implement biomass power generation is a critical topic.

REFERENCES

Tang, Hsueh-Jung, 2018. "Application Techniques and Precautions for Using Corruption of Rice Straw to Produce Organic Fertilizers."

Lu, W., & Zhang, T., 2010. "Life-cycle implications of using crop residues for various energy demands in China."Environmental Science & Technology, 44(10), 4026–4032.

Ya-Jing Wang, Yu-Yun Bi, Chun-Yu Gao, 2010. "Collectable Amounts and Suitability Evaluation of Straw Resource in China" Scientia Agricultura Sinica, 43(9), 1852–1859.

Gao Xiangzhao, Ma Wenqi, Ma Changbao, Zhang Fusuo, Wang Yunhua, 2002. "Analysis on the Current Status of Utilization of Crop Straw in China." Journal Huazhong (Central China) Agricultural University, 21(3), 242–247.

Chih-Yuan Lin, 2004. "Analysis of Open Burning Incidents in Yunlin and Chiayi—Influences on Air Quality and Emission Estimation." Master's thesis, Graduate Insititute of Enginering, National Taiwan University.

Chang-Hui Chen, 2004. "Influence of Rice Straw Open Burning Incidents in Taichung, Yunlin, Chiayi, and Tainan on the Development of Photochemical Smog in Kaohsiung and Pingtung." Master's thesis, Graduate Insititute of Enginering, National Taiwan University.

Yi-ling Su, 2006. "Source identification and size distribution of atmospheric polycyclic aromatic hydrocarbons during rice straw burning period." Master's thesis, Department of Environmental Engineering, Chaoyang University of Technolog.

Shun-Shiang Chang, 2011. "Atmospheric concentration and dry deposition of polybrominated diphenyl ether during the biomass open burning period."Master's thesis, Department of Environmental Engineering, National Cheng Kung University.

Chanh-Jun Huang, 2005. "Study on Pyrolysis of Rice Straw." Master's thesis, Department of Environmental Engineering, Da Yeh University.

Liu, Jen-Ching, 2009. "The Manufacture of Solid Biomass Fuel Briquettes using Agro-forestry Wastes." Master's thesis, Department of Wood Science and Design, National Pingtung University of Science and Technology.

Chen, ci-syuan, 2009. "Energy Life Cycle Analysis of Second-generation Biofuels from Rice Straw Biomass Waste." Master's thesis, Department of Environmental Engineering, National Ilan University.

Fu-Siang Syu, 2011. "Life cycle assessment of producing biocoal from rice straw via torrefaction." Master's thesis, Graduate Insititute of Enginering, National Taiwan University.

Chien-Hua Yeh, 2013. "Assessing Environmental Footprints of Biomass Energy – The Example of Liquid Biofuels." Master's thesis, Institute of Natural Resource Management, National Taipei University.

Wei Lu, Tilanzhu Zhang, 2010. "Life-Cycle Implications of Using Crop Residues for Various Energy Demands in China." Environmental Science & Technology 2010 44 (10), 4026–4032.

Agriculture Statistical Database of the Council of Agriculture, Executive Yuan (Agriculture and Food Agency).

Innovation in Design, Communication and Engineering – Lam et al. (eds)
© 2020 Taylor & Francis Group, London, London, ISBN 978-0-367-17777-5

A research on the design appearance that is based on circular economy and incorporated into green building

Pei-Ying Ou*
Ph.D Program of Mechanical and Energy Engineering, Kun Shan University, Tainan, Taiwan

Huann-Ming Chou
Department of Mechanical Engineering, Kun Shan University, Tainan, Taiwan

ABSTRACT: This article is about the investigation of possible design appearance of the circular economy concept incorporated into green building. The author wanted to integrate the circular economy concept with the creative design sector in construction industry supply chain to understand the feasible circular green building modes which can be inspired by the innovative strategies of circular economy, and to analyze the features and utilizations of cases of modular houses. This study can help architects to include practicability, creativity, and sustainability during the design of green building and let the general public recognize that circular green building is a good option for future home space; we would like to combine civil action with architects and business owners to create pleasant and friendly circular green buildings.

Keywords: circular economy, circular green building, construction industry supply chain, modular house, modular design

1 INTRODUCTION

After the economic development was accelerated by the Industrial Revolution, the technology advancement since the 21st century has also led to rising human desires. Under the economic model which encourages consumption and stimulates growth, the resources on Earth have been excessively exploited, and petrochemical energy has been gradually depleted! As a result, when human beings are enjoying the civilization achievements, they are also facing wind disasters, floods, droughts, and tsunami caused by greenhouse effect and abnormal climate and numerous climate refugees. In light of this, people with vision began advocating the development of green technology and promotion of environmental protection movement while thinking about how to achieve the proper balance under the impacts of economic development and environmental protection ideas. Via the reflection of a society with unlimited growth and in response to the implementation of zero resources waste, circular economy tries to figure out a way of life which is environmental friendly while benefitting the economic development and sustainable development.

Focused on the architectural issue, the nature of circular economy corresponds to the idea of green building, and the utilization of circular economy strategy for promoting construction industry is filled with potentials. In the book "Circular Economy", Yue-Zheng Huang pointed out that: "The construction and use of buildings in which we live and work every day account for 30% of global generation of greenhouse gases. There has been rather little progress among global construction industries as compared to other industries. In light of this, the expected benefit of transformation of construction industry can be greater than other industries,

---

*Corresponding author: ope011296@gmail.com

including reduction of material cost, and the value of recycling and reclamation of wasted construction materials (Y.Z. Huang, 2017)." The concept of circular economy is specific and easy to implement, thus the use of circular economy concept to promote green building in construction industry is a feasible strategic model. This study is aimed at developing a sustainable and feasible building mode and appearance for green building development based on the case study of modular house.

2   NATURE AND CONCEPT OF CIRCULAR ECONOMY

The nature of circular economy is ecological economy. The so-called ecological economy refers to that human beings can utilize natural resources to engage in the economic behaviors of production and consumption via the mode in compliance with natural resources and circular utilization; which means they must carry out production activities based on the concepts of ecology and ecological economics. Ecological economics emphasize that economic activities must be carried out within an ecological sustainable scope, and it is believed that only the virtuous circle within the resource capacity can allow balanced and sustainable development of ecological system. Circular economy is focused on highly efficient yet less intensive resources harvesting according to the theoretical basis of ecological economy in order to achieve the goal of zero waste of resources. With reduced and most effective use of resources, we can prevent excessive harvesting, use, or depletion of resources from affecting ecological balance and human survival.

The concept of circular economy is shown in Figure 1: there are two parts of it: the biological circle composed of green lines and the industrial circle composed of blue lines.

When the aforementioned concept of circular economy is implemented as the executive strategies, they can be summarized into the specific operations of the following aspects (Y.Z. Huang, 2017):

1. Using advanced production technology, alternative technology, and carbon reduction technology to convert the by-product of every production sector to the raw material for another process in order to enhance the asset value of product and to reduce waste of resources.
2. Establishing a system of industrial symbiosis. Forming the cooperation relationship of symbiosis and common prosperity among different industries to share economic achievements.
3. Using information technology in conjunction with management system to have total control over product life cycle. Establishing the sharing platform and prolonging product service life based on the principles of customization and sustainable service.
4. Achieving the virtuous circle of human, natural ecology, and scientific technology systems.

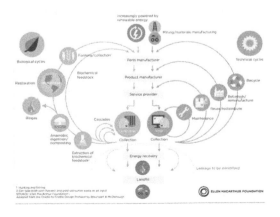

Figure 1.   Concept of circular economy.

3 GREEN BUILDING AND CIRCULAR ECONOMY

"Green Building" is also known as "Environmental Symbiosis Building", or "Ecological Building", or "Sustainable Building", meaning a building in compliance with environmental protection and ecological design (P.Y. Ou, 2015).

The green building assessment system in Taiwan, also known as the EEWH Assessment System, defines green building as the building meeting the four indicators of "Ecological, Energy Saving, Waste Reducing, and Healthy". The green building equipped with such connotation is very consistent with the circular economy concept of "High Efficiency, Less Harvesting, and Less Waste". The following section is about investigating how to combine the concept of circular economy and innovative strategies for the design in order to come up with the innovative mode of circular green building.

3.1 Connotation of Taiwan green building assessment system EEWH

"Green Building Assessment System" is developed by all countries according to the environment, climate, and requirements of each residential area for promoting green building and allowing the construction industry to control the actual connotation of green building. In order to encourage people to build energy saving, resources saving, and low pollution green building, Architecture and Building Institute of Ministry of the Interior officially announced the acceptance of application for "Green Building Mark" on September 1st, 1999. It also reviewed and updated the "Green Building Explanation and Evaluation Manual" in 2003 by listing nine major indicators: including Greenery, On-site Water Retention, Water Resource, Daily Energy Saving, CO_2 Reduction, Construction Waste Reduction, Sewage & Garbage Improvement, Biodiversity, and Indoor Environment. The passive definition of green building in the past, such as "The building consumes the least resources on Earth and produces the least amount of wastes", has been expanded to the active definition of "Ecological, Energy Saving, Waste Reducing, and Healthy Building". The assessment system based on these nine indicators is simply known as EEWH Assessment System. The indicators are as shown in Table 1 (P.Y. Ou, 2015).

Table 1. Green building assessment system.

Large indicator group	Content of indicator	
	Name	Items of assessment
Ecological	1.Biodiversity	Biological green network, small habitats, diversified plants, and soil ecology
	2.Greenery	Greenery, fixed CO_2 amount
	3.Water Retention	Water retention, storage infiltration, soft flood prevention
Energy Saving	4.Daily Energy Saving (Necessary)	Energy saving by exterior wall, air conditioning, and lighting
Waste Reducing	5.CO_2 Reduction	CO_2 emission by construction material
	6.Construction Waste Reduction	Earthwork balance, and waste reduction
	7.Indoor Environment	Sound insulation, lighting, ventilation, and construction materials
Healthy	8.Water Resources (Necessary)	Water saving equipment, and reclamation of rainwater and intermediate water
	9.Sewage & Garbage Improvement	Rainwater sewage diversion, garbage classification, and composting

3.2 *Circular green building*

Ever since the announcement of Green Building Mark in 1999, the promotion of green buildings in Taiwan has been lasting for almost 20 years. The promotion conditions include: as for the public buildings, it was stipulated in the "Green Building Promotion Program" of the Executive Yuan in 2001 that the public construction project with total construction cost of more than NTD 50 million must acquire "Candidate Green Building Certificate" first, thus the cases of public buildings granted with Green Building Mark are obviously more than private cases, and they have accounted for almost 80% of all construction projects. Among the Green Building Mark applications by private construction projects, 80% of them are residential construction projects, which account for less than 0.1% of all construction projects. This indicates that there is still a huge room for development of private projects. In addition, in terms of the green building design experiences of architects in the entire country, most architects firms did not have any experience in the design of green building (J. Yu et al. 2008). Therefore, it is a developable subject to provide a simple and convenient mode for architects. The following section is about the consideration of feasible green building design and construction modes from the perspective of strategic operation of circular economy concept.

The supply chain of building design industry can be divided into four parts of creativity forming, production, propagation, and exhibition/reception, and they can also be known as the four sectors of design, construction, sales, and consumption (P.Y. Ou et al. 2016). The relationship among these sectors of the industrial chain is as shown below (Figure 2):

The connotation and industries related to each sector are briefly described in the followings (P.Y. Ou et al. 2016):

1. Design: the business owner provides the demand, and building designer and interior designer will carry out the design to form the creativity. This sector involves industries such as: enterprise, architect, interior designer, landscape building designer, structural designer, and electromechanical piping technician.
2. Construction: after design drawing is completed, it will be submitted to the building construction unit or renovation unit to carry out the construction of building body, surrounding landscape, and interior decoration. This sector involves industries such as: civil construction material industry, construction industry, gardening landscape industry, and interior decoration industry.
3. Sales: the completed construction project can be exhibited and marketed by brokers or sales agents of real estate brokerage industry. This sector involves industries such as: real estate broker firm, consignment company, visual communication design industry, and advertisement and media industry.
4. Consumption: consumers will visit and purchase. The results of sales and reactions by consumers can be provided to business owners and architects as the references for the design of the next construction project.

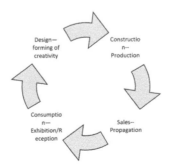

Figure 2.   Relationship of building design industrial chain. (Data source: summarized in this study).

Table 2. Green building concepts and circular economy innovative strategies. (Data source: Summarized in this study).

Industrial Chain	Relevant Industries	Green Building Concept	Circular Economic Concept	Circular Economy InnovaTive Strategy	Circurlar Green Building
Design	Landscape designer Building designer Interior designer Electromechanical technician	Ecological Energy saving Waste reducing Healthy		Resources reduction and reclamation Ecological energy saving Zero waste Building space sharing	Building renovation and revitalization
Construction	Civil construction material industry Building construction industry Gardening landscape construction industry Interior decoration industry	Ecological construction method Energy saving design Using local green construction material	Regenerate Share Optimise Loop	IoT: automatic remote sensing system calibration based on Big Data Modular establishment Circulating use of construction material 3D printing local assembly	Bionic structure Passive energy saving building Intelligent building Modular design

The sectors and industries related to building design industrial chain in conjunction with the green building concept and circular economic policies are as summarized in Table 2, which can roughly outline the possible appearance of circular green building (CTCI, 2015).

According to the analysis of the table above, the building renovation and revitalization in compliance with resources reduction and reclamation, passive energy saving design, bionic structure, smart building, and modular design can all serve as the future appearance of circular green building. The next section is about the case study of modular house.

4 MODULAR HOUSE

Modular house refers to the composition of main body and appearance of a house based on modular structures. As compared to the conventional house with concrete as the main construction material, it is equipped with modular components as shown in Figure 3 such that the pre-fabrication method can be adopted, and there are the advantages of simple assembly, adaptive to local conditions, and expansion at will. There are quite diversified forms of frequently seen modular houses in the market, including steel-framed house, container house, wooden house, and steel-framed wooden house. The following section is about the introduction of two kinds of modular houses.

Figure 3. Modular unit component. (Data source: IJIA Construction, 2017).

4.1 Connotation of Taiwan green building assessment system EEWH

Building block house is a kind of modular house based on steel frame structure. It is a house design based on the stackable and variable characteristics of building blocks. The pre-fabrication method can be adopted, and there can be special patterns of steel frames with light-weight and robust structure as shown in Figure 4. For facilitating the transportation and meeting the residential use situation, every unit of the building block house is designed based on the basic dimension of "3m×6m×3m", and every unit of wall, floor, and ceiling can be disassembled, recycled, and rebuilt. It is equipped with outstanding convenience, mobility, and environmental friendliness(IJIA Construction, 2017).

The unit above second floor will require lifting and vertical stacking, such as the lifting operation as shown in Figure 5. Considering the convenience of positioning in the air during the lifting operation, there will be the setting of LEGO-like convex and concave rings, which is the so-called lifting spot, to facilitate the future disassembly and recycling. This is another feature of the building block house. After the positioning of structure, the connection of other fundamental components and utility pipelines can be carried out. In a word, this kind of steel-framed modular house is equipped with the advantages of being robust, flame retardant, water-proof, pest resistant, and environmental friendly. Due to its lighter weight, different appearances and spatial combinations can be designed according to local conditions. And it only requires minor excavation and quick construction, so there will be less damage to the ecological environment, making it an excellent circular green building mode(IJIA Construction, 2017).

4.2 Wooden modular house

In 2014, an Italian construction company Marlegno designed a type of recyclable wooden modular house which can be pre-fabricated as shown in Figure 6. It is composed of several basic modular elements, and the different layouts can be generated by combination of these elements to meet customer needs. There are four different spatial combination patterns shown in Figure 7. Its feature is that all elements can be recycled, and the foundation is made of steel, so it is light-weight and it can be built without excavation for foundation. The living area and sleeping space are clearly defined in the spatial design as shown in Figure 8. Every room has direct visual and sensory contacts with the natural environment. Residents can directly feel the flow and variation of nature in time and space via the glass entrance and corridor of the central module.

This house was specifically designed for people in Mediterranean region. The house structure was made of XLAM structural board. The wall was 9 mm thick, the roof was 12 mm thick, and the floor was 14 mm thick. They were equipped with the features of flame retardant and seismic resistant and heat insulation by wooden fiber and hemp fiber. The façade was composed of wooden boards and fixed by dual-thread screws; the drainage ditch was made of aluminum; there were solar tiles installed on the roof to generate electricity and heat while being waterproof. With the modular design and overall pre-fabrication mode, it can be disassembled and installed via basic structural elements, such as floor, wall, window, and drainage ditch, making it very convenient to carry and to expand. The structural elements are as listed in Figure 9 (Maurizio Barberio, 2014).

Figure 4. Steel frame structure. (Data source: IJIA Construction, 2017).

Figure 5. Lifting operation. (Data source: IJIA Construction, 2017).

Figure 6. Recyclable pre-fabricated wooden modular house. (Data source: Maurizio Barberio, 2014).

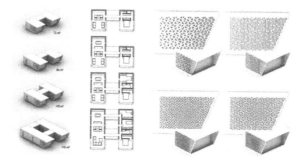

Figure 7. Spatial scheme combinations of recyclable pre-fabricated wooden modular house. (Data source: Maurizio Barberio, 2014).

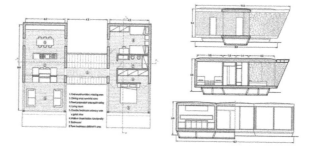

Figure 8. Indoor floor plan and elevation of recyclable and pre-fabricated wooden modular house. (Data source: Maurizio Barberio, 2014).

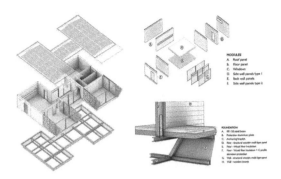

Figure 9. Structural elements of recyclable and pre-fabricated wooden modular house. (Data source: Maurizio Barberio, 2014).

If it is introduced to Taiwan with wet and hot weather, anti-mildew and anti-mite treatments should be applied to the wooden parts.

5 CONCLUSIONS

Based on aforementioned summary and analysis of concepts and innovative strategies of green building and circular economy, we have found several possible development directions for circular green building, such as: building renovation and revitalization, bionic structure, passive energy saving building, smart building, and modular design. The modular house can lead to cost reduction via the industrial serial production, and the modular design can be used for pre-fabrication, assembly, and circulating utilization; in addition to the saving of resources and construction time, it can also meet various requirements of customization. The structural elements are made of steel or wood, so every element can be recycled, re-assembled, and reused. With the energy shortage nowadays, it can be regarded as a circular green building mode meeting the ecological sustainability, which can be used as private house, holiday cottage, or themed exhibition and activity space. It provides a convenient and compact option for sustainable housing space for our residence or activities which can be moved and expanded at any time.

REFERENCES

Y.Z. Huang, 2017. Circular Economy, Commonwealth Magazine Group.
P.Y. Ou, 2015. Green Skylight: An overview of the aesthetics and eco-friendliness of a courtyard [M.S. thesis], Department of Mechanical Engineering, Kun Shan University.
J. Yu, B.Y. Chang, 2008. Analysis of Current Development Status of Green Buildings in Taiwan, *Journal of Far East University*, Vol. 2 5, No. 3.
P.Y. Ou, H.M. Chou, 2016. Output Estimates on Taiwan Architectural Design Industry, IEEE-ICAMSE 2016 - Meen, *Prior & Lam (Eds)*, pp. 330–332.
CTCI, 2015. Development Trend and Critical Issues of Circular Economy.
IJIA Construction, 2017. Building Block House.
Maurizio Barberio, 2014. Unboxed-100%Recycable Prefab Wooden House?, World Architecture Community.

Innovation in Design, Communication and Engineering – Lam et al. (eds)
© 2020 Taylor & Francis Group, London, ISBN 978-0-367-17777-5

An exploration of the personal identity dilemma and its solutions

Jung-Jung Wen*
Program of Mechanical and Energy Engineering, Kun Shan University, Tainan, Taiwan

Huann-Ming Chou
Department of Mechanical Engineering, Kun Shan University, Tainan, Taiwan

ABSTRACT: Descartes' theory "I think, therefore I am," explicitly states that the soul is a substantial entity that expresses the self, sparking the mind-body dualism debate. John Locke makes a distinction between "personal identity" and "personality," believing that consciousness and memories can be used as the standards for determining personal identity, which goes against the theory of soul's substantiality. Derek Parfit emphasizes the significance of personality and reductionist views and believes that the continuity of the mind (Relation R) is the most important. Mark Johnston points to the important distinction between personal-identity and self-identity.

The discourses of all these theories are unable to clearly define the "I" and cannot express the ultimate reality of life which transcends birth-and-death, while sustaining memory continuity. Therefore, this paper hopes to explore the Buddhist idea regarding the ultimate reality of life that there is a true, permanent, and unchanging true dweller through the use of Buddhist doctrines regarding "being," the "self," and the "repository of memories." Because of its continual nature of transcending birth and death, it is the repository of memories, which transcends the limitations of time and space. It is also the root basis of the law of causality; hence, Buddhism's Theory of Eight Consciousnesses can provide all kinds of answers lacking from the philosophical theories mentioned above.

Keywords: personal identity, being, memories, ultimate origin, the eighth consciousness

1 INTRODUCTION

Xuanzang was a Chinese Buddhist monk, his seventeen-year-long overland journey westward to India seeking the original Buddhist scriptures has been sung of over thousands of years. Looking at his life experiences, one can see that they were very different from those of ordinary people. For example, as a novice monk he was good at expounding the Mahayana Buddhism, was brave enough to travel westward to bring home Buddhist scriptures without fear of hardship and danger, and because he could not agree upon some of the sutras and treatises that had been translated into Chinese, he therefore insisted on retranslating them. He was also able to make authoritative distinctions and analyses of varying disputes between Chinese and Indian Buddhist teachings. At that time, Xuanzang was already honorably referred to as the "Deity of the Ultimate Truth" and the "Deity of Liberation." In addition, Xuanzang undertook sutra translation and composition of treatises and was able to display indomitable perseverance without trifling away his time. This paper attempts to explore the key points as to why Xuanzang's translation of sutras and authorship of treatises were capable of overcoming the three difficulties of "faithfulness, expressiveness and elegance."

---

* Corresponding author: rrwen@ms45.hinet.net

Chinese Daoist Zhuangzi didn't consider physical death to be the endpoint of one's life and asserted that the mind did not die and the soul was eternal, writing in The Nourishing the Lord of Life, "What we can point to are the faggots that have been consumed; but the fire is transmitted elsewhere, and we know not that it is over and ended." (B.M. Cai, 2018) This seems to be the basis of personal identity. Western scholar John Locke believed the self was humanity and that ego was a product of conscious thought. Personal identity was constructed from the identity of a physical entity, whereas self-identity was created from the identity of the consciousness. The ego was formed from consciousness. He also reflected on the connections between memories to prove the idea being the same person. (J. Perry, 2008) Derek Parfit believed that personal identity should be explained with reductionism. Reductionists believed the brain, body, and series of physiological and psychological events did not constitute a person. Instead, he believed people possessed these things and that the continuity of the mind (Relation R) was the most important. Reductionists asserted that (1) personal identity over a span of time was composed of certain specific facts, and (2) these particular facts can be objectively described without needing to make presuppositions about personal identity. Afterward, Parfit used the continuity of memory to supplement the insufficiencies in his theory. (D. Parfit, 1987) David Hume believed personal identity is only a feature of our perceptions, a product of our imagination. Secondly, he believed that memories were also just products of these perceptions that are based on similar relationships and causality. Therefore, memories do not produce personal identity but reveal it. (J. Perry, 2008) Mark Johnston divided human beings, hibernators, and teletransporters into three types to discuss the possibility of life after death. (M. Johnston, 2010).

This paper believes that (1) although Zhuangzi's theories discussed the perspective of eternal life, life before birth and after death were not described clearly, and discussion mostly focused on the practice of the body and mind techniques of "leaving our material form and detaching from our knowledge." For example, The Adjustments of Controversies described the technique "losing self" with "the body is like a rotten branch" and "seeming to be in a trance and to have lost all consciousness of any companion" with "one's heart is dead like ashes." In The Great and Most Honoured Master, the states of attaining the Dao, "sitting and forgetting" and the "great pervader," were also described. Therefore, its theories regarding personal identity have not provided a concrete standard for verification. (2) Western scholarly arguments mostly concentrate on the body, consciousness, and memories, but the body is always changing due to aging, illness, and death, and awareness is eliminated by the five mindless states. Even what they recognize as memories often have faults, such as transience, absent-mindedness, blocking, misattribution, suggestibility, bias, and persistence. (M.S. Sweeney, 2012) Therefore, there is yet to be any way of solving the questions of personal identity and self-identity in the previously mentioned Eastern and Western philosophies.

3   THE IMPORTANCE THAT THE DISCUSSION ON EXISTENCE HAS ON PERSONAL IDENTITY

The validation of personal identity should be based on life that has a continuous and uninterrupted substance, and the substantiality of life involves existing discussion. As for the definition of "being," empirical realization and concrete evidence should be used as the criterion for validation. Buddhist sutras seems to have already provided related arguments. First of all, in Discourse 37, Volume 2 of Samyukta-agama, the Buddha clearly states not to dispute the definition of "existence or non-existence" with people because although the world's wise men say there is "existence" based on the birth and existence of things and call the perishing of things "non-existence," they completely lack the knowledge that in this impermanent, changing "self" that arises and ceases and is subject to the five aggregates, there is still a permanent dwelling dharma that can give rise to things. Secondly, in this Agama Buddha says "existence" can be roughly divided into "three realms of existence" or subdivided into "twenty-five stages of existence"; "Non-existence" refers to what came before the emergence of these three realms and twenty-five stages of existence, which is nothing, and when the three or twenty-five stages of existence are put to an end, there will also be nothing. Moreover, things of the past that were not born, present things not existing, and future things that will not be born are also called non-existing. Therefore, Buddha's definition of existence in the Agama and modern philosophy's definition of existence are the same. In other words, in the "three realms of existence" in Buddhist doctrine that includes the five aggregates of the self and the eighteen elements of the self, there is not a permanent and indestructible and self-existing entity, and they all fall under "temporary false existence."

Based on Buddha's teachings on "knowing for oneself and realizing for oneself," the Agama states for the people of the world that in the arising, ceasing, and changing of the five aggregates there is still an "eternal, permanent, and immutable true dweller." "Eternal" refers to not arising nor ceasing, meaning it is not created by the dharma of dependent origination. In terms of time, it has existed since the beginningless eon. "Permanent" refers to eternal existence. It is adamantine and is indestructible. "Immutable" is its intrinsic attribute. Its inherent function will never be changed. "True dwelling" refers to continuous existence at all times without any instances of interruption. Therefore, it is also called the "ultimate origin." (CBETA, T02, no. 99, pp. 8, b16-28, L.C. Tsai, 2007) At the same time, he can also skip over birth and death, so it possesses the "primordial existence," the connectivity of life memories. It is a clue for providing answers to the matter of personal identity.

4 THE "SELF" SET FORTH IN THE AGAMAS HAS IMPLICATIONS FOR THE MATTER OF PERSONAL IDENTITY

All Agamas have explicit and implicit parts. The implicit parts talk about "non-self," which refers to the five aggregates, or the eighteen realms, all arising from the conditioned dharmas. They are the dharmas arising from the combination of the ultimate origin (the eighth consciousness), karmic seeds, ignorance, parents, and the four elements in a sequentially interactive way. These dharmas themselves do not possess a real self-entity, so they are not true substantial dharmas. That is to say, the "self" (the five aggregates and eighteen realms) that people believe to be true is actually the fabricated "non-self" and is the "fake-self," which will ultimately be disintegrated and annihilated. (P.S. Xiao, 2013)

The Agamas implicitly say that there exists an ultimate origin's "real self" that goes beyond birth and death and possesses connected memories, showing the existence of personal identity. For example, in Discourse 136, Volume 6 of Samyukta-agama, the Buddha explains that because all living things are concealed by ignorance, they transmigrate in the cycle of birth and death. They do not know there is the true existence of an "ultimate origin." (CBETA, T02, no. 99, pp. 42, b3-6) And this "ultimate origin" is the personal identity of every living thing. Secondly, it is also said in the Angulimala Sutra, explained with a metaphor: "For example, a village where the people are gone is called an empty village, a dried-up river is called an empty river, and a waterless bottle is called an empty bottle. They are not called empty because villages, rivers, and bottles themselves do not exist. They are called empty because inside the villages, rivers, and bottles there is void." (CBETA, T02, no. 120, pp. 527, c8-10) This metaphor was used to show that there is a real, non-ceasing "true self," which is the matter of personal identity.

5 THE REPOSITORY OF MEMORIES IS THE SPECIFIC FUNCTION OF PERSONAL IDENTITY

As opposed to scientific research, there are already Buddhist texts about the repository of memories and methods of memory extraction written long ago. For example, in Volume 1 of The Ten Epithets of the Buddha, there are records of the "wisdom of prior lifetimes." (CBETA, T17, no. 782, pp. 720, a2-5) In Volume 1 of The Long Agama Sutra, the Buddha says, "I know entirely events from countless kalpas in the past. It is owing to my skill in understanding the nature of dharmas."

At the same time, the Buddha also explicitly says that since He possesses wisdom of prior lifetimes, He could completely know any events that have occurred in the past countless kalpas. He not only knows everything in His own past lifetimes but also the entire life courses of other buddhas. (CBETA, T01, no. 1, pp. 1, b25-c12).

Moreover, in Verses on the Structure of the Eight Consciousnesses, Xuanzang Bodhisattva states, "How vast and unfathomable is the Tripitaka!" (CBETA, X55, no. 897, pp. 448, c6-11//Z 2:3, pp. 319, d6-11//R98, pp. 638, b6-11) This clearly shows that the eighth consciousness is the repository of memories because it has the functions of a storer, that which is stored, and it is the store that is the object of attachment. Therefore, the function of this repository of memories is broad and profound and cannot be restricted by boundaries.

Scientists believe the reason memories can be formed and stored is because of the creation of connections between neurons and new circuits that store memories in independent neural networks in the brain. This includes the parietal lobe, frontal lobe, medial temporal lobe, hippocampus, and neural circuits formed from the connection of neural synapses. They are all repositories of memories. During the processes of neural connections, what stood out was the variable and unfixed nature of memories. (M.S. Sweeney, 2012, Thomas Rogerson, et al. 2014) This point can provide evidence for scientists by presenting the expanded physiological phenomena through the inertial coding of the synapses and hippocampus.

Based on the above-mentioned neurophysiological phenomenon for analysis, this paper believes the aforementioned reasoning of this kind is untenable because (1) after the inertial neural circuits form when riding a bike, the hippocampus is no longer needed. Just the monitoring of the existing infralimbic cortex suffices. (Eric Burguière et al. 2013, K.S. Smith et al. 2013) This shows that the hippocampus is not where the repository of memories is. (2) From childhood to early adolescence, about 20 billion synapses are erased every day. (J. Schwartz et al. (M. Jhang, Trans., 2005)) Therefore, the features of neuroplasticity clearly show that neural networks are not the repository of memories. (3) Neurons of adult brains have about a quadrillion synaptic connections. There are too many of them, whereas genes in charge of directing connections are too few (about 35,000 types). Because of this, there is insufficient DNA for piecing together a map of connections in the brain. (J. Schwartz et al. (M. Jhang, Trans., 2005)) Hence, genes also cannot be the repository of memories. (4) When looking at the topological structure of the nervous system during processes of neural messaging in lateral inhibition of the spinal cord, which has a neural physiological function that deletes input clutter, the brain experiencing a stroke and a handicapped limb of a patient, the complete continuity of neural message mapping between the center and peripheral nervous systems cannot be seen. (T. Shou, 2003) (5) concepts belonging to Gestalt psychology are based on phenomenological and philosophical thought. Fritz Perls believes that gestalt is the entirety of experiences, and it is defined by the establishment of a replacement process characterized by the requirement for images to be perceived by each person when they appear and the gradual fading away of these images to the "background" after their requirement is satisfied. (C.J. Hu et al. 2008) Therefore, concepts of gestalt reflect the continuity of the psychological processes. It is a kind of presentation of an experience path. Certainly, it also is not the repository of memories.

This paper discovered that texts regarding the "wisdom of prior lifetimes," the "power of knowing prior lifetimes," and the withdrawal of memories indicate that the repository of memories truly exists. Because of this, it is possible for those who have these insights to extract past events many lifetimes and kalpas ago. Therefore, examining the "ultimate origin – the eighth consciousness," which acts as the repository of memories, not only requires a being having the fundamental properties of being eternal, permanent, and immutable, but also having the capabilities of the storer, that which is stored, and it is the store that is the object of attachment. It also needs to be able to transcend space and time without being affected by the limits of space-time. These are the solid foundations for the establishment of "personal identity."

6 CONCLUSION

Researching the matter of personal identity should be based on the criteria of "eternality," which is not affected by spacetime. The body-self and consciousness-self are dharmas of arising and ceasing and do not possess eternality, so they cannot be criteria for establishing the matter of personal identity. The "ultimate-origins-true-self" is not the dharma of arising and ceasing and possesses eternality, so it can become the eternal repository of life memories. Therefore it meets the criteria for establishing personal identity. Based on the principle and analysis of the Theory of the Eight Consciousnesses in Buddhist doctrine, this paper discovers that the ultimate origin, namely the eighth consciousness, is the foundational origin of personal identity research, and it should be the solution to dilemmas that Eastern and Western scholars face when researching the matter of personal identity.

REFERENCES

B.M. Cai, 2018. Living the Form, Mind, and Passions of Zhuangzi — The Path from Novice to Expert, Linking Publishing.

J. Perry, 2008. John Locke: Of identity and diversity, Personal identity, University of California Press.

D. Parfit, 1987. Reasons and Persons, Oxford University Press.

J. Perry, 2008. David Hume: Of personal identity, Personal identity, University of California Press.

M. Johnston, 2010. Surviving death, Princeton University Press.

M.S. Sweeney, 2012. Brainworks: The Mind-bending Science of How You See, What You Think, and Who You Are, F. Jheng, Trans. Taipei, Boulder Publishing. CBETA, T02, no. 99, pp. 8, b16–28.

L.C. Tsai, 2007. The Definition of Being in Agama Sutras, Journal of True Enlightenment, pp.23–47.

P.S. Xiao, 2013. Self and No-Self, The True Enlightenment Practitioners Association, Taipei.

CBETA, T02, no. 99, pp. 42, b3–6.

CBETA, T02, no. 120, pp. 527, c8–10.

CBETA, T17, no. 782, pp. 720, a2–5.

CBETA, T01, no. 1, pp. 1, b25–c12.

CBETA, X55, no. 897, pp. 448, c6-11//Z 2:3, pp. 319, d6-11//R98, pp. 638, b6-11.

Thomas Rogerson, et al. 2014. Synaptic Tagging During memory Allocation, *Nature Reviews Neuroscience*, 15(3), pp. 157–169.

Eric Burguière, et al. 2013. Optogenetic Stimulation of Leteral Orbitifronto-Striatal Pathway Suppresses Compulsive Behavior, *Science*, Vol.340, pp. 1243–1246.

K.S. Smith, A.M.Graybiel, 2013. A Dual Operator View of Habitual Behavior Reflecting Cortical and Striatal Dynamics, *Neuron*, Vol.79, No. 2, pp. 361–374.

J. Schwartz, S. Begley, The Mind and the Brain, M. Jhang, Trans., 2005, Taipei, *China Times Publishing*, pp. 113.

T. Shou, 2003. Neurobiology, Taipei, Jeou Chou Books.

C.J. Hu, J.J. Wang, 2008. A Conceptual Analysis of Gestalt Psychology, *The Journal of Psychiatric Mental Health Nursing*, Vol. 3, No. 1, Taipei.

Decomposing waste with earthworms and black soldier flies

Li-Chin Chang*
Ph.D Program of Mechanical and Energy Engineering, Kun Shan University, Tainan, Taiwan

Huann-Ming Chou
Department of Mechanical Engineering, Kun Shan University, Tainan, Taiwan

ABSTRACT: According to statistical data from Taiwan's Environmental Production Administration, ordinary garbage contains 40% kitchen waste, and these are incinerated together with
other garbage refuse (S.A. Chen, 2019). There is also concern whether food scraps have been
properly and legally recycled, and gradually its enormous amount becomes a hidden worry for the
entire Taiwan community. Waste incineration leads to air pollution, generation of harmful
dioxins, heavy metals, and so on (Y.M. Cheng, 1998). Moreover, the harmful materials sift
through soil and water sources and eventually end up on humans' dining table. The large quantity
of liquid content from kitchen waste leads to facility consuming and reduces service life of the
incinerators. The costs of all these negative effects unfortunately fall on the lap of each and every
citizen in Taiwan. Currently, the African swine fever reignited the focus on kitchen waste recycling. The complexities of food scraps dumping and recycling have been a challenge for counties as
well as cities throughout Taiwan. At the meantime the authorities concerned can consider introducing environmentally friendly insects as part of the solution. Presently, the most efficient type
from this group is the earthworms and black soldier flies. These two types of insects possess very
different appetites and together they produce high nutritional content of organic fertilizer. The
combination of these two insects can literally resolve all the problems facing farming refuse and
kitchen waste. This solution will also bring organic fertilizers back to the soil and consequently
fully materialize the agricultural circular economy. When every single-family unit can treasure the
benefits from recycling food scraps, they will be in a position to contribute to the reduction and
resolution of problems associated with climate change. In order to trace back carbon footprints, it
is very important to administrate proper preventive measures at source; the general public should
be educated to reduce food scraps. Excess food waste, even though recyclable, does not produce
positive effects on the circular economy. This article focuses on the introduction and research on
how earthworms and black soldier flies can help ordinary small size farmers, families and communities to convert agricultural waste products and kitchen waste into fertilizers in order to seek
a speedy, least workload and manpower solution to sustainable farming.

Keywords: earthworms, bblack soldier flies, organic compost, agricultural circular economy

1   UNDERSTANDING EARTHWORMS AND THE BENEFITS THEIR CASTINGS BRING TO AGRICULTURE

Earthworms exist in various species and can be roughly divided into two categories, black and red
earthworms, according to their living environment and food substances. For food substances,
black earthworms utilize inorganic matter in soil while red earthworms utilize organic matter;
therefore, it is the red earthworms that are capable of breaking down food and farming organic

---

*Corresponding author: liqin@cht.com.tw

waste. An earthworm can consume 50 to 100% (Animal husbandry and aquaculture economy, 1984) of their weight in food and produce castings comparable with its own weight daily, indicating that a kilogram of earthworms can at least consume 0.5 kilograms of waste and produce 0.5 kilograms of high quality earthworm manure—Darwin once even remarked that there is no fertile soil apart from earthworm castings. Compared with general organic fertilizers, earthworm castings are better choices as they are fertilizers with the most comprehensive effects, containing a complete range of nutrients, along with a rich content of organic matter, humus, and microflora; they do not cause harm to plants when overused, and are free from the concerns of cancer induction by nitrate salts. Hence, apart from the elimination of family biomass organic food waste and agriculture biowaste, a large advantage of earthworm farming is the production of earthworm manure.

The advantages concerning crops, and largest differences between general organic and chemical fertilizers of earthworm manure:

(1) Earthworm castings do not contain nitrate salts (F. Chin, et al. 2013). It is currently acknowledged that excess application of nitrogenous fertilizers and deficient exposure to sunlight can lead to hyper-accumulation of nitrate salts, increasing the eaters' risks of developing cancer. On the other hand, the crops grown with earthworm manure have a low content of nitrate salts.
(2) The formation of earthworm castings is stable and the production cycles are short: Earthworm manure is formed immediately after the earthworms eat; the fermentation and composting time is relatively short compared to that of general fertilizers.
(3) The elimination of organic waste produced by humans can be achieved: The beneficial bacteria within the guts of earthworms can rapidly break down organic matter and produce high quality earthworm manure.
(4) Complete and slow-release fertilizers (G.W. Dickerson, 2001): Although earthworm manure possesses higher fertilizing strengths, damage caused by excess fertilizing is less prone to occur due to its slow-release properties.
(5) Capable of residual heavy metal removal: Earthworms can not only produce organic manure, but also produce castings without detectable heavy-metal after filtering and digestion. This is not achieved through degradation of the heavy metal, but rather through their ability to accumulate heavy metal (R.C. Liang et al. 2005). Hence, researchers have employed earthworms in industrial waste degradation, and Indian scientists have used earthworms in the recovering of polluted land (S.S. Huang, 2007).
(6) A natural deodorizer which reduces plant diseases and pests: Earthworm castings contain a large amount of aerobic bacteria. A thin layer of earthworm castings applied on compost giving off a bad odor can bring about immediate odor elimination; earthworm castings are also often used to cover the ground in the pens of chicken farms to get rid of bad odors. An application of some earthworm castings can effectively prevent diseases and pests, reducing the incidence rates in plants (S.M. Tan, 2015).
(7) Heavy work and expensive equipment are not required: A normal farm can be capable of in situ possessing of organic wastes; as compost turning is not needed, machine and man power can be saved.

From the reasons above, it is clear that earthworms are the solution to agriculture cycling economics—dealing with organic waste and composting in one shot!

2 UNDERSTANDING BLACK SOLDIER FLIES AND THE BENEFITS THEY BRING TO AGRICULTURE

The shape of black soldier flies is similar to big flies. In order to efficiently learn about black soldier flies, we will make a comparison between black soldier flies and flies, which will let us clearly see the reasons black soldiers are the most perfect insects for the environment. Both of them are insects that undergo complex metamorphosis; this indicates that they go through four stages—egg, larval, pupal and adult—in their life, and they exhibit a different morphology in each stage. Their food sources are similar, but as black soldier flies stay in the larval stage for a longer period, can efficiently process organic waste, has astonishing consuming potential, does not disturb the living environment of humans, and does not spread pathogens, they are highly suitable for the handling of food and agricultural waste. Currently, as the pioneer of Taiwan, the Cleaning Squadron of Tienzhong Township, Changhua County, have used black soldier flies to process food waste (C.P. Chiu, 2019).

In the following Figure 1 and Figure 2, we will compare the life cycles of black soldier flies and flies.

When the life cycles of the two species are compared, it can be observed that there are six instars in total in the larval stage of black soldier flies, which is longer than the three instars in that of flies. The adult stage of black soldier flies is short, and the adult insects live outdoors completely; besides not disturbing humans during their adult stage, they also do not threaten the natural environment, and can process biomass waste more effectively and reliably.

The agricultural advantages of black soldier flies in processing organic waste include:

(1) A wide food range: The food that cannot be eaten by earthworms can all be taken over to be processed by black soldier flies; for example, cooked food waste, including meat products, the carcasses of livestock dead from diseases, livestock manure, and food with high

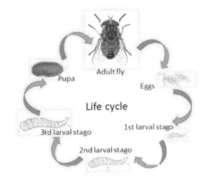

Figure 1.   The life cycle of flies.

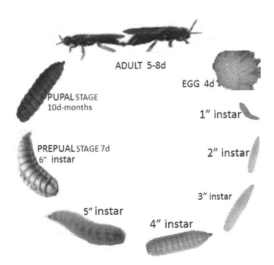

Figure 2.   The life cycle of black soldier flies.

fat content, high salt content, or mold can all serve as food for the black soldier flies (L. Ping, 2010).

(2) High production efficiency: The food waste consuming weight of a black soldier fly has been calculated by Swedish researchers to reach 0.1 gram per day (B.M.A. Dortmans et al. 2017), which is comparable to its own weight; in other words, 1 ton of food waste would require 10,000,000 black soldier flies. Experiments have shown that black soldier flies can consume 10 tons of food waste within 24 to 48 hours, efficiently and extensively dealing with organic waste.

(3) Rich in anti-bacterial peptides: Black soldier fly larvae, though dwelling in trash, manure, carcasses, rotting organic matter and living with diseases and bacteria, rarely fall ill. The black soldier flies possess a unique immune function; they possess a "Bacteria-resistant protein" with strong bacteria-killing properties. The anti-bacteria peptides are synthesized in large quantities for the eradication of invading pathogens. The anti-bacteria peptide is a natural product, and does not result in drug resistance easily. During the eating process, black soldier fly larvae are even capable of digesting and breaking down the harmful pathogens within their feces, reducing the harm to the environment (Editorial department, 2019).

(4) Rich in chitin: The exoskeleton and epidermis of black soldier fly larvae, pupae, and adults contain a rich content of chitin, which is 30-fold of that of flies (J.P Zhu et al. 2013). The exoskeletons and feces rich in chitin can turn into nutrition for plants; in addition, chitin has insect resistance properties strong enough to dissolve the tough epidermis of harmful insects and hard shells of nematode eggs. Therefore, the application of black soldier fly culturing debris on agriculture and farming can reduce the usage of chemical fertilizers and pesticides.

(5) The insect feces have high fertilizing strengths: It takes around 14 days for the black soldier flies to go from hatching to the prepupal stage. A hundred tons of food waste can be processed into 15 tons of insect bodies and 15 tons of debris after 14 days by an appropriate ratio of larva. According to the experiences of the farmers in Taiwan, using insect feces mixed with insect exoskeletons as a fertilizer can, apart from improving soil quality and implementing soil fertilizing strength, effectively prevent the reproduction of some harmful insects; it is an organic fertilizer with natural insect-repelling effects. The application of black soldier fly culturing debris on agriculture and farming can reduce the usage of chemical fertilizers and pesticides, or even completely substitute chemical fertilizers and pesticides for some crops (L. Ping, 2010).

(6) Getting rid of maggots: The larvae of black soldier flies compete with maggots for food; by doing so, wild houseflies numbers can be effectively controlled, thus decreasing fly reproduction (Editorial department, 2019).

(7) Pollination: Over these few years, bees have been dying out and disappearing in large numbers, leading to an agricultural crop decline due to the lack of pollination. Some farmers have used flies as pollinators, but they ended up becoming environmental hazards; black soldier flies could be considered for beneficial pollinator substitutes which can, at the same time, deal with agricultural waste.

3 THE MYTH OF ENVIRONMENT-PROTECTING INSECTS BEING USED AS INSECT PROTEINS

Apart from important contributions to the processing of organic waste, earthworms and black soldier flies also bring huge benefits to agriculture. Although the organic fertilizers they produce can definitely boost organic agriculture up to another level, most people still think of earthworms and black soldier flies as proteins for animal fodder supplements, feeding them to fishes, chicken, cows, pigs, and so forth, then proceeding to feed earthworms and black soldier flies with livestock manure. This can seem to be the perfect circular economy, without any waste being produced in the process, but is it really the most considerate form of circular economy to the Earth when we

Figure 3. Proteins from insects first enter the mouths of livestock, then that of humans.

think from the angle of humans and their stomachs. (Figure 3) There are many problems concerning this issue, including the needs to prevent the occurrence of mad cow disease, or evaluate risks of prion proteins in black soldier flies. Moreover, we should also pay more attention to the parts of the "Animal Welfare Act" regarding insects since mass farming of insects place them under the danger of massacre. "White Paper on Edible Insects" reminds us that, just like any high-density farming, insects may face new health threats (C.Y. Cheng, 2017). Humans determine the survival of all species for the sole purpose of satisfying human appetites, and the hard work involved in the reproduction of all species are provided just to fill human stomachs; shouldn't the diversity and appropriate number of species be considered more in order to protect the entire Earth in the best way? The rapid increase of the 70 million tons of edible meat in 1960s to 330 million tons in 2017 demonstrates a 5-fold growth (FAO, 2018/7). Huge amounts of edible meat products resulted in problems such as global warming, an increase of carbon emissions, and severe pollution of soil and water sources. Earthworms and black soldier flies definitely are capable of solving these pollution problems, but if human desires do not cease, pollution of the sources will never decrease, rendering the efforts of earthworms and black soldier flies futile as they cannot be compared to human influence. Under these circumstances, the actions can only result in a decline of meat opportunity costs, leading to excess up-taking and wasting of meat and unsolved pollution problems. In the end, vegetarian diets are the only methods that can ensure the feeding of all people living on Earth. We could also calculate the opportunity costs of the carbon emissions and livestock farming, from which we would learn that only through natural free-range farming and by returning to the land, true contributions could be made to the Earth.

4 THE DIFFICULTIES OF EARTHWORM AND BLACK SOLDIER FLY COMPOSTING, AND THE IMPROVEMENTS AND APPLICATIONS

Our researchers have experimented on the farming of earthworms and black soldier flies in order to understand the characteristics of the two insects. It was also discovered that their farming does not suit normal families due to points that should be considered, including the fear of insects, farming techniques, temperature, humidity, and all sorts of environmental sanitation problems. For improvements regarding these difficulties, establishments and management could be carried out outdoors or in fields, returning to the embrace of Mother Nature, and directly assigning the help of earthworms and black soldier flies in the production of biological composting. However, this method is more suitable for small-scale farmers and apartment complexes or buildings with a clearing. A simple method is as follows: Place a food waste bucket outdoors, and put the lid on, not too tightly, in order to attract wild black soldier flies to approach and lay eggs. This is because black soldier flies usually lay eggs in gaps; although there is a lid, the odor from the food waste bucket can help them complete the egg-laying process while reaching into the lid. It has been exhibited through several experiments by researchers that once black soldier flies have laid eggs on the bucket, the odor will continue to attract them to lay eggs even after the food waste bucket has been washed with water. When the food waste bucket is filled with food waste, the contents can be directly dumped into a pre-established composting area, then lightly covered with a thin layer of soil to make biological compost. Black soldier flies have been observed to be able to complete its life cycles underground, but will have a slightly smaller body size. This biological compost contains the chitin from the six shellings of the black soldier fly larvae in the soil and feces rich in plant nutrition. This composting method is faster and more energy-efficient than general organic composting, does not produce bad odors, will not be contaminated by mosquitoes and flies, does not produce methane, and can even directly be returned to soil to enrich the nutritional content. Also, more earthworms were discovered in the composting area; the both of them can together induce rapid forming of the compost. Compared with traditional composting, this method does not require turning, and can produce organic waste compost in different batches any time, and in any amount.

The biological composting process can be designed by oneself by taking the environment and obtainable waste into consideration, and can be developed and processed in apartment complexes, buildings, or even parks. Besides effectively dealing with the organic waste of apartment complexes and producing organic fertilizers, education can also be achieved, and a consensus can be formed; the mass should be helped to understand the efficiency of insects and their influence on the environment, and proceed to contribute to Earth.

REFERENCES

S.A. Chen, 2019. Low Recycling Rates, Fated to be Incinerated—What has Happened to Food Waste Recycling? WuoWuo Issues, pp.1.

Y.M. Cheng, 1998. Are Incinerators the Solution to Trash? Collected in Taiwan Watch, *Taiwan Watch Institute*, pp.4.

Animal husbandry and aquaculture economy, 1984. *The News and Information Center of Renmin University of China*, pp.51.

F. Chin, L.R. Su, T.M. Su, Y. Chang, Y.C. Ho, T.G. Ho, 2013. The Influence on Soil Physiochemical Properties and Cabbage Heart Nitrate Content by Earthworms Castings, *Researches and Applications of Agriculture*, No. 04.

R.C. Liang, R.S. Chen, 2005. A Feasibility Study of the Assessment of Soil Heavy Metal Pollution Through Earthworm Growth and Reproduction, National Chung Hsing University, Department of Soil and Environmental Science.

S.S. Huang, 2007. Severe Pollution in West India Assisted by 300 Thousand Earthworms, Edited: Excerpt from EBC report.

S.M. Tan, 2015. Fodder and Base Formula Selection in Soil-Free Earthworm Farming, Hebei Agricultural University.

C.P. Chiu, 2019. The Cleaning Squadron of Tienzhong Township farm black soldier flies for the use of composting food waste, reported from Changhua.

L. Ping, 2010. Optimization and Application of Process Conditions for Transforming Livestock and Poultry Manure by Using Bright Spotted Angle Water Margin, Huazhong Agricultural University.

B.M.A. Dortmans, S. Diener, B.M. Verstappen, C. Zurbrugg, 2017. Black Soldier Fly Biowaste Processing - A Step-by-Step Guide Eawag: Swiss Federal Institute of Aquatic Science and Technology, Dubendorf, Switzerland.

Editorial department, 2019. Demystifying the eating of black pupa larvae by fluid, *technology science monthly*, No. 591.

J.P Zhu, C.X. Liu, S. Yang, B.C. Wu, Y.F. Hong, 2013. The Research and Developments of Insects (black soldier flies) Resources in Fodder, Science, *Heilongjiang Animal Science and Veterinary Medicine*, issue 13 pp. 61–63.

C.Y. Cheng, 2017. "Saving the World by Eating Bugs? From food safety, fodder management, dioxin residues, to animal welfare, the European Union requires regulations", Up and downstream Italy contributing reporter, in Fishery and Farming.

FAO, 2018/7. food outlook: biannual report on global food markets, pp.7.

Innovation in Design, Communication and Engineering – Lam et al. (eds)
© 2020 Taylor & Francis Group, London, ISBN 978-0-367-17777-5

A discussion on Xuanzang's Buddhist scriptural translation approach by using Xuanzang's orthoepic translation of India's name as an example

Shu-Chen Sun* & Huann-Ming Chou
Mechanical and Energy Engineering, Kun Shan University, Tainan, Taiwan

ABSTRACT: The name of India (C. Yindu) was translated by Xuanzang based on its orthoepic pronunciation, and was henceforth widely adopted. However, as Xuanzang abandoned the existing Chinese translation of Tianzhu and invented the translation of Yindu based on other principles, this constituted a phenomenon worth studying. At the same time, the Buddhist scriptures translated by Xuanzang were being differentiated from the ancient translations and old translations, and were classified as new translations, thus the supreme translation principle adopted by Xuanzang became crucial. By researching on how Xuanzang produced the translation of Yindu's name and by referring to the example of Diamond-Cutter Perfection of Wisdom Sutra's (S. Vajracchedikā Prajñāpāramitā Sūtra; C. Jingang Jing) retranslation, it is discovered that "no forms of loss" was the supreme principle of Xuanzang's scriptural translations, on top of the principle of "five categories of untranslated terms."

Keywords: Xuanzgang, Yindu (India), Diamond-Cutter, No forms of loss

1 INTRODUCTION

The name of Yindu originated from The Great Tang Dynasty Record of the Western Regions, where Xuanzang named the main areas in the Western regions that were influenced by Buddhism. He explained the reason for choosing such names, as he used the most representative and welcomed trait of each area to determine their names. The name of Yindu was adopted ever since Xuanzang first used it in his translation. Before Xuanzang decided on the translation of "Yindu," there were already names like Tianzhu, Shendu (Shengdu) and Xiandou. Most of the Buddhist scriptures used the name Tianzhu. Why did Xuanzang abandon the existing and commonly-used names to produce the name of Yindu for the Five Regions based on orthoepic pronunciation? The underlying reasons are worth pondering.

Similarly, before Xuanzang started translating Buddhist scriptures, the Dharma Master Dao'an during the Eastern Jin Dynasty had described the problems of "five [points of] departing from the original and three [points which should remain] unchanged" when one translates barbarian things into Chinese. Also, a Dharma Master Yan Zong during the Sui Dynasty formulated "eight prerequisites and ten principles" of translation. How were Xuanzang's translations able to be distinguished from the ancient translations and old translations, and be commended as new translations? From Xuanzang's perspective, what exactly was his supreme principle of translation?

As Xuanzang was well versed in Sanskrit, Chinese, and Buddhist scriptures, he was certainly capable of becoming the main translator who could translate Buddhist texts accurately.

---

*Corresponding author: sun.shuchen@msa.hinet.net

Besides the principle of "five categories of untranslated terms" which belongs neither to literal nor liberal translation, this paper seeks to study the supreme principle of Xuanzang's scriptural translations through the cases of his translation of Yindu's name and the Diamond-Cutter Perfection of Wisdom Sutra.

2 ORTHOEPIC TRANSLATION OF "YINDU"

Xuanzang's rationale for choosing the translations of such names can be deduced from the following text in the Great Tang Dynasty Record of the Western Regions.

In a careful study we find that Tianzhu is variantly designated, causing much confusion and perplexity. Formerly it was called Shengdu, or Xiandou, but now we should call it Indu (India) according to the right pronunciation (Chen & Fan, 2003).

Since "Tianzhu" was first mentioned, how did this name come about? According to the Miscellaneous Names in Sanskrit Language (C. Fanyu zaming) compiled by Li Yan in the Tang Dynasty and the Collection of Bilingual Pairs of Chinese and Sanskrit Languages (C. Tang Fan Liangyu Shuangdui Ji) compiled by Tathāgatapāla and Guṇaviśeṣa, Tianzhu's Sanskrit pronunciation was "Xi Nu (Shang)", and the respective Sanskrit text was "Indu". However, earlier translations in the Chinese history converted this Sanskrit pronunciation into the Chinese pronunciations of "Shengdu" or "Xiandou" but not "Tianzhu," so how did such a transliteration come about, and what kind of script did it adopt? Before Persia's invasion during the sixth century B.C., also known as the Vedic period, this area was known as Sapta Sindhu. After the Persian conquest, it was known as Hapta Hindu instead, then it was again renamed as indus by the Romans. All these names refer to the same river basin (Ganga Ram Garg, 1992), and the name of this basin denotes the name of the entire area, thus the Chinese pronunciation of "Shengdu" probably came from Sindhu.

As for the name Xiandou, some believed its original name was indra-mandira, which means the residence of the King of Gods that is protected by the heaven. Due to this reason, it was known as Tianzhu. When Xuanzang arrived at this area, its name was already transformed into the Sanskrit script known as Indus. This means that all sūtra translators before the Tang Dynasty employed the Chinese translation of Tianzhu according to the Sanskrit script Indu.

According to Li Yan in the Tang Dynasty, the Sanskrit pronunciation of Indu is "Xi Nu," which was probably deduced from the Sanskrit script of Sindhu during earlier times. According to Phonetic and Semantic Dictionary for All Buddhist Sutras (C. Yiqie jing yinyi) compiled by Hui Lin from the Tang Dynasty, it was repeatedly brought up that the Sanskrit language placed all focus on the pronunciation but not the meaning. Thus, the orthoepic pronunciation mentioned by Xuanzang should be directed towards the Indu area that occurred many times in the scriptures. As he intended to correct the respective Chinese words used to represent the Sanskrit pronunciation, he decided to name it as "Yindu."

After Xuanzang decided on the name Yindu, he went on to explain another layer of meaning behind such a name: The people of India use different names for their respective countries, while people of distant places with diverse customs generally designate the land that they admire as India (Chen & Fan, 2003). What was the most representative and welcomed trait of the area? How did Xuanzang choose this as the general designation of the area? Xuanzang said: The word indu means "moon," which has many names, and this is one of them (Chen & Fan, 2003).

The meaning of Indu in Sanskrit was denoted by bright moon in the scriptures (Lin & Lin, 2005). India region then did have many designations of moon (Lin & Lin, 2005), but Xuanzang chose only to use Indu to represent the Chinese character moon, and he even linked it to the translation of the name Yindu. Modern scholars and Yijing during the Tang Dynasty hold different ideas regarding this matter. Firstly, Xuanzang placed much emphasis on orthoepic pronunciation, and the Sanskrit pronunciation of bright moon stated in the scriptures was Indu because Indu was the only word that refers plainly to the bright moon without conveying other metaphorical meanings. In addition, Xuanzang also explained the reason why he chose to use the bright moon to make an allusion to India: It means that living beings live and die in the wheel of transmigration in the long night of ignorance ceaselessly, without a rooster to announce the advent of the dawn. When the sun has sunk candles continue to give light in the night. Although the stars are shining in the sky, how can they be as brilliant as the clear moon? For this reason India was compared to the moon. Because saints and sages emerged one after another in that land to guide living beings and regulate all affairs, just as the moon shines upon all things, it was called India (Chen & Fan, 2003).

Throughout Xuanzang's entire lifetime, he firmly believed that he "would rather die going to the west than live by staying in the east," and it was also a well-known historical fact that he embarked on an arduous journey to Tianzhu to bring back the Buddhist sutras. What motivated Xuanzang to persist in the face of hardships was the hope to acquire the brightness of the moon for all sentient beings. Back then during the Tang Dynasty, there were numerous brilliant stars in the practice of Buddhist Dharma, but there was a missing bright moon that could shine light on the darkness. As Tianzhu had a succession of holy and wise men in Buddhist Dharma, it resembled a moon that shed its bright influence and led Buddhist disciples to break through their confusions. Hence, the country Tianzhu was named as Yindu which meant bright moon. Such a translation was completely aligned with Xuanzang's approach and ideology in Buddhist scriptural translation.

3 TRANSLATION OF THE DIAMOND-CUTTER PERFECTION OF WISDOM SUTRA

The following section seeks to make a comparison of the Great Perfection of Wisdom Sutra (S. Mahāprajñāpāramitā Sūtra; C. Jingang Jing): Ninth Diamond-Cutter Section translated by Xuanzang and the Diamond Perfection of Wisdom Sutra (S. Vajracchedikā Prajñāpāramitā Sūtra; C. Jingang Jing) translated by Kumārajīva based on three areas, so as to verify Xuanzang's Buddhist scriptural translation approach. Xuanzang pointed out several crucial parts in the Diamond Perfection of Wisdom Sutra translated by Kumārajīva that deviated from the Sanskrit text, including the name of the sūtra; one out of the three questions is missing; one out of the two verses is missing; and three out of the nine metaphors are missing. His retranslation was included into the 577th volume of the 600 volumes of the Great Perfection of Wisdom Sutra and was titled Ninth Diamond-Cutter Section. The following part seeks to investigate the differences in these first three areas.

3.1 *Regarding the "name of the sūtra"*

Xuanzang mentioned that the Sanskrit text "vajracchedikā prajñāpāramitā" referred to the Diamond-Cutter Perfection of Wisdom, but Kumārajīva translated it directly as the Diamond Perfection of Wisdom. Xuanzang explained the difference by stating that "the confusions of discrimination are as solid as diamond, and can only be removed by the non-discriminating wisdom expounded in this sūtra" means the confusions arising from discrimination are as unbreakable as diamond, and only the non-discriminating prajñā wisdom described in the Buddhist sūtra can remove them. This was the authentic meaning of the text, which was different from Kumārajīva's literal translation that implied a resemblance between prajna and diamond. There was a paragraph in the scriptural text that corresponded to Xuanzang's explanations on the name of the sūtra. Original Sanskrit text: prajñāpāramitā (prajñā wisdom) nāma (called) ayaṃ (this) subhūte (subhūte) dharmaparyāyaḥ (dharma-door). Xuanzang's translation: The Buddha replied Sudṛśa, "Subhūti! This discourse on dharma is now named the Diamond-Cutter Perfection of Wisdom. You should uphold it by that name." Kumārajīva's translation: The Buddha told Subhūti, "The name of this sūtra is the Diamond Perfection of Wisdom. You should uphold it by that name."

According to the meaning of the original Sanskrit text, the subject was "dharma-door." This so-called "dharma-door" conveyed the meaning of using a dharma as a gate for a certain purpose. From the perspectives of sentence structure and textual meaning, Xuanzang's translation of "This discourse on dharma is now named the Diamond-Cutter Perfection of Wisdom" was able to achieve a coherent logic. For this part of the research, we have to make a reference to the difference regarding "one out of the three questions is missing," so as to understand it thoroughly.

3.2 Regarding "one out of the three questions is missing"

Original Sanskrit text: Tat (it) katham (how) bhagavan (World Honored One) bodhisattvayāna samprasthitena (set out for the bodhisattva-vehicle) kulaputreṇa (son of good family) vā (or) kuladuhitrā (daughter of good family) vā (or) sthātavyam (abide) katham (how) pratipattavyam (progress) katham (how) cittam (heart/mind) pragrahītavyam (subdue).

Xuanzang's translation: World Honored One! For adherents set for the bodhisattva-vehicle, how should one abide? How should one practise? How should one subdue one's mind? Kumārajīva's translation: World Honored One, if a good man, or good woman, resolves his heart on Anuttarasaüyaksaübodhi, how should he dwell, how should he subdue his heart?

Clearly, three questions were brought up in the Sanskrit text: how to abide, how to practise, and how to exert (subdue) one's mind. They conveyed the meaning of abiding and practising according to certain things in order to subdue one's mind. Thus, Xuanzang's translation of the name of the sūtra as the Diamond-Cutter Perfection of Wisdom was based on the original Sanskrit text.

3.3 Regarding "one out of the two verses is missing"

Original Sanskrit text: Ye (who) māṃ (me) rūpeṇa (form) cādrākṣur (observe/see) ye (who) māṃ (me) ghoṣeṇa (voice) cānvaguḥ (seek) mithyā-prahāṇa-prasṛtā (hold erronous view) na (not/non) māṃ (me) drakṣyanti (see) te (those) janāḥ (people) Dharmato (dharma) buddho (buddha) draṣṭavyo (see/observe) dharmakāyā (dharma-body) hi nāyakāḥ (Venerable One) dharmatā (dharma-nature) ca na (not/non) vijñeyā (discerned) na (not/non) sā (it) śakyā (possible) vijānitum (cognition) [9].

Xuanzang's translation: Whoever saw me through my physical form, whoever sought me through the sound of my voice, engaged in the wrong endeavours, those people will not see me. That which ought to be observed is the Buddha's dharma-nature, and it is the Venerable One's dharma-body; but the dharma-nature being unknowable by sensory consciousness, it cannot be known by those ordinaries. Kumārajīva's translation: If one sees me in forms, If one seeks me in sounds, He practices a deviant way, and cannot see Tathāgata.

There were two verses in the original Sanskrit text that were interlinked, thus the omission of the second verse would lead to incomplete textual meaning. An important principle that Xuanzang upheld during his translation of Buddhist scriptures was to retain the precise meanings of the original text.

4 CONCLUSION

After a thorough investigation into the matter, it is revealed that by re-incorporating the three differences into the original Sanskrit text, the content of the sūtra became coherent and logical, thus readers would have less confusions and misunderstandings on the sūtra. This serves as the supreme principle of "no forms of loss." Similarly, Xuanzang translated the name of Buddhist Heaven of Tianzhu or Shengdu as Yindu according to its orthoepic pronunciation, and he translated the name into Chinese based on the Sanskrit pronunciation that refers to the bright moon in darkness. Such an approach encompassed both literal and liberal translation, and it revealed the origin of Buddhist Dharma into the Eastern land through the ingenious technique of "no forms of loss".

When Xuanzang was translating Buddhist scriptures, he was faithful to the meaning of the Sanskrit text, respectful towards the empirical ideology in the scriptures, and determined in retaining the educational connotations. As such, even though the Buddhist scriptures were presented in the Chinese language, they were able to clear up the readers' confusions. The contents of the scriptures provide a verification standard for those with true practice and actual realization, and serve as the first layer of "no forms of loss." On the other hand, by translating the Buddhist scriptural teachings and principles accurately and with no bias, translators were able to utilise theories and empirical truths to convey the authentic meaning of the original sūtras and discourses. This serves as the second layer of "no forms of loss." Although Xuanzang's translation approach placed little focus on the fluency of language, he was able to fully comprehend the principle of "conveying profound sūtra meanings through easily understandable words." This could be viewed as a pure act of remaining faithful to the source text, which is also the third layer of "no forms of loss." From the orthoepic translation of the name Yindu to the translation of Diamond-Cutter Perfection of Wisdom Sutra, "no forms of loss" served as the supreme principle in Xuanzang's scriptural translation that was faithful to the source text.

As Xuanzang's translation principle of "no forms of loss" was credible and reliable, Buddhist scriptural translation was brought into an era of new translation from the previous era of ancient translation and old translation. All the translation henceforth adopted the newly-translated nouns to ensure accurate dissemination of Buddhist teachings (Mizuno (Trans. X. R. Liu), 1996), making indelible contributions to the quality of Chinese Buddhist scriptural translation and the communication of Buddhism in China.

REFERENCES

Chen, F., & Fan, P., 2003. A New Translation of the Record of Travels to Western Lands. Sanmin Book-store, sec. printing of the first edition, Taipei.

Ganga Ram Garg, 1992. Encyclopaedia of the Hindu World. Ashk Kumar Mittal Concept Publishing Company, New Delhi.

Lin, G.M., & Lin, Y.X., 2005. Chinese-Sanskrit Dictionary. Jiafeng Publication, Vol. I, Taipei.

Mizuno, Kogen, (Trans. X.R. Liu), 1996. Buddhist Sūtras: Origin, Development, Transmission. Taipei, Dong-Da Publishing House.

Yi, J., 2016. "A Record of Buddhist Practices Sent Home from the Southern Sea," Wang Bangwei, "Chinese Literature and History," January issue 2016. How did ancient Chinese address India?.

Innovation in Design, Communication and Engineering – Lam et al. (eds)
© 2020 Taylor & Francis Group, London, ISBN 978-0-367-17777-5

The power of vows and of virtues: A discussion of the causes leading to rebirths in the form of a dragon, as set forth in the *Great Tang Records on the Western Regions*

Ta-Yuan Lu* & Huann-Ming Chou
Mechanical and Energy Engineering, Kun Shan University, Tainan, Taiwan

ABSTRACT: The dragon is a significant sacred creature commonly found in the Buddhist scriptures. Questions ranging from its modes of life through to the reasons which, according to the Buddhist law of causality, led to its rebirth as a dragon are difficult to address. However, the *Great Tang Records on the Western Regions*, an important book in cross-cultural studies authored by Dharma Master Xuanzang, mentions dragons more than 40 times. Even more valuable is its detailed accounts of how four dragon kings came to be reborn as dragons, thus providing precious records regarding the living modes of this sacred creature. Based on the reports, these four kings were born in the body of a dragon due to the power of vows and virtues from their past lifetimes. Therefore, this paper attempts to explore the causes leading to rebirths in the form of a dragon in Buddhism based on a summary of the key elements found in these four accounts.

Keywords: Xuanzang, Great Tang Records on the Western Regions, dragon, vows, virtues

1   INTRODUCTION

The *Great Tang Records on the Western Regions* (henceforth referred to as *Record*) is a transcription of Dharma Master Xuanzang's (602–664) oral narrative by his disciple Bianji at the request of Emperor Taizong. It is an irreplaceable, valuable historical document in terms of Chinese history, Indian history, and the history of Buddhism in China and India.

The number of accounts of a dragon (*nāga*) found in *Record* is 40 or more, suggesting this book pays certain attention to the creature. What's more, the dragon is a prevalent mysterious creature in ancient Chinese and Indian books. Compared with related historical accounts and folklore, some Buddhist classics depict the dragon in much more detail from its temperament and karma, as well as its stance-taking in terms of the Buddha Dharma (being either a protector or a violator). Buddha Śākyamuni once warned his disciple monks in the Āgama at his first turning of the Dharma wheel:

At that time, the World Honored One addressed the bhikṣus: "There are four things which are beyond comprehension. Which four? Sentient beings are inconceivable, worlds are inconceivable, dragon kingdoms are inconceivable, and the states of the Buddha-realm are inconceivable; the reason being, one cannot attain extinction and enter nirvāṇa from these four states." (Book Two of the Taishō Tripiṭaka, 2007).

The subtleties of these three realms are inconceivable since all living creatures vary considerably, the universe extends enormously, and the state of Buddha's kingdom is when *bodhisattvas* attain unsurpassed perfect enlightenment. As one of the living creatures, the dragon is

---

* Corresponding author: yuandragon101@gmail.com

particularly mentioned, and its realm coexists with the other three; accordingly, the dragon occupies a peculiar position in Buddhist teachings.

The image of the dragon shown in *Record* is worth exploring since Dharma Master Xuanzang possessed a great amount of knowledge and experience related to this creature. As a prominent monk in China, Master Xuanzang was deeply exposed to Chinese culture and Buddhist *sutras*; accordingly he was familiar with many accounts of dragons, not to mention his direct contact with the dragon legends in Western regions and India in his scripture-seeking journey.

What is worth noticing is that Master Xuanzang respectively described four pieces of dragon kings' *jātakas* and *nidānas* in *Record*. He mentioned particularly "why to reincarnate as a dragon" in great detail, and these "inconceivable" references can explain well why Buddha said, "dragon kingdoms are inconceivable."

2 ILL-INTENDED VOW-MAKING AND THE POWER OF MERITS: BAD DRAGONS' *JĀTAKAS* AND *NIDĀNAS*

Although each dragon's temperament varies in these approximately 40 accounts, they are referred to in general as a creature in the realm of animals with certain merits, power, and resentment, as shown in the depiction of an evil dragon in the terrain called the northern side of the Pamir Range in Book One:

Which are located at the northern side of the Pamir Range, where most of the streams flow eastward. Snow is accumulated in the valleys, which are freezing even in the spring and summer seasons, and although they sometimes melt a little they soon become frozen again. The path is dangerous and the cold wind blows with a piercing vehemence. There are frequent disasters caused by ferocious dragons that give trouble to travelers. Travelers going by this route should not wear garments of reddish-brown color, nor should they carry calabashes or shout loudly. The slightest infringement of these taboos will cause immediate disaster. A fierce wind will arise all of a sudden, sand flying in the air and pebbles raining down from the sky. Those who encounter such a catastrophe are sure to die, [or at least] it is difficult for them to escape alive. (Xuanzang and Bianji [annotated by X. L. Ji], 2007).

In this account, it is apparent that the dragon is powerful and filled with a sense of resentment – not only as a mysterious creature recognized by Buddhist practitioners but also as an intimidating creature horrifying secular business travelers. Those with a Buddhist cosmology are definitely curious about dragons' *jātakas* and *nidānas*, wondering what *nidānas* and past karma contribute to the reincarnation of dragons. Three exact references to evil dragons' *jātakas* and *nidānas* appear in *Record*: the dragon king of the Great Snow Mountains in Book One (Xuanzang and Bianji [annotated by X. L. Ji], 2007), the dragon king Gopāla in Book Two, and the dragon king Apalāla in Book Three. They all have something in common – certain merits accumulated from their past lives, and ill-intended vows to seek revenge by being reincarnated as dragons.

Before being reincarnated as *nāga* Great Snow Mountains in his next life, the novice monk found his own *arhat* teacher was served heavenly nectar far better than the food of the human world he was offered, so he felt strong resentment. Although the dragon king donor repented and provided offering, this novice monk wanted to use all the merits collected from his past lives, including the merit he earned from this present life as a monk, in order to vow he would be reincarnated as a powerful dragon king so that he could avenge the unfair share of offerings he received. His ill vow was sincere and intimidating:

The novice washed his master's alms bowl, as usual. When he discovered some remaining grains of rice in the bowl he was amazed by its fragrance and he immediately cherished a malignant feeling against his master and the nāga king, saying, "May the power of whatever good deeds I have performed appear to kill this nāga and let me be the king." The moment the novice expressed this desire the nāga king felt a headache. After listening to the sermons delivered by the arhat, the nāga king repented his misdeed and blamed himself, but the novice, deeply resentful, would not make a confession nor accept the nāga king's apology. Having returned to the monastery, the novice, by his earnest desire and the power of his good deeds, died that night and was reborn as a great nāga king with majesty and valor. He came to the lake, killed the resident nāga king, occupied the nāga palace, took possession of his subordinates, and became the master of all. (Xuanzang and Bianji [annotated by X. L. Ji], 2007).

This excerpt is very precious: it describes the two main factors of dragon reincarnation: ill-intended vows and the result of merits, both leading to immediate death. That is, this instant revenge carries sudden death as its cost. Even before this novice monk actually reincarnated as a dragon, the ill vow had already made the dragon king "feel headache."

The *nāga* Gopāla in Book Two of *Record* was originally a shepherd who was reprimanded by a king. Thus, he made offerings to the Buddhist pagodas (*Stūpas*) and then transferred this merit of offering for an ill vow:

When the Tathāgata was living in the world the nāga was a cowherd whose duty was to supply the king with milk and cream. Once he failed to fulfill his task properly and was reprimanded by the king. With a feeling of hatred and malice, he purchased some flowers to offer to the stupa of prediction, in the hope that he might be reborn as an evil nāga to devastate the country and do harm to the king. Then he went up to the rocky precipice and jumped down to kill himself. Thus he became a nāga king and lived in this cave. He desired to go out of the cave to carry out his evil wishes. (Xuanzang and Bianji [annotated by X. L. Ji], 2007).

His wish to revenge the disgrace of being reprimanded was extremely strong, so he committed suicide right after transferring his merit and making a vow. The same two factors of dragon reincarnation also are shown in this excerpt: making offering to the *Stūpas* for merits, and wishing to be a vicious dragon. However, there is a difference between the novice monk and the shepherd. The novice monk's ill vow was a prompt reaction and the merits that he relied on came from merits collected from his past lives, so he wished, "May the power of whatever good deeds I have performed appear." The shepherd seemed to know he did not have enough merits, so he particularly cultivated some merits by making offerings to the *Stūpas*, and then took his own life right after making an ill vow in front of Buddha. This shows the shepherd was quite aware of the prerequisites of dragon reincarnation. He knew to have such a reincarnation took not only an ill vow but also certain merits. Therefore, he strategically made offerings to the *Stūpas* and committed suicide after making his ill vow. Making offerings to the *Stūpas* is essential, and this illustrates the fact that both "merits" and "vows" are essential for dragon reincarnation and share equal significance.

The dragon Apalāla in Book Three of *Record* was originally a sorcerer in his previous life. He cast spells to stop a bad dragon from causing storms and was then greatly admired by the local people. Each family was willing to give him a bushel of grains as tax. Yet some farmers postponed their offering of the due grains and that irritated him, leading him to make an ill vow: to reincarnate as a bad dragon that would wreak havoc on humans in his next life:

At the time of Kāśyapa Buddha the dragon [of this spring] was born a human being named Jingqi, who was an expert in the art of exorcism and had restrained a malicious dragon from causing rainstorms. It was because of his help that the people of the country had surplus grain to store at home. Out of gratitude for the exorcist's virtuous deeds, each household of the inhabitants contributed one dou [10 liters] of grain as a gift to him. As time passed some people neglected their duty and Jingqi became angry and wished to become a malignant dragon and cause storms to spoil the seedlings of the crops. After his death he was reborn a dragon at this place and caused white water to flow from the spring and it damaged the fertility of the soil. (Xuanzang and Bianji [annotated by X. L. Ji], 2007).

In this story, the sorcerer's ill vow of wishing to become a malignant dragon is blunt, but it is not clear if he had certain merits. In the text, the favors the sorcerer did for these people by stopping the malicious dragon and rainstorms seemed to be paid back by the grain tax offered. However, if we read the text closely, the grains are local people's voluntary offerings to show the sorcerer their gratitude. In other words, the grain tax was initially a gift. As time passed, some people neglected this yearly present and greatly offended the sorcerer, who developed resentment.

This story shows us that the sorcerer defeated the malicious dragon out of good intentions and his own will. If he had done this in a menacing manner, the local people would not have been grateful for him. What causes his anger is their disrespect. It is reasonable to argue that his merits came from his lifelong defeat of the dragon for the local people. As a result, "merits" and "vows" are still two core factors in this dragon Apalāla's *jātakas* and *nidānas*.

From these three stories in *Record*, it is apparent that bad dragon reincarnation requires two necessary components: merits and vows. Those with certain merits first, such as the jealous novice monk and the sorcerer, made ill vows in order to be reincarnated in the form of a dragon, while those with ill intentions first, like the angry shepherd, had to cultivate some merits by deliberately making offerings to the *Stūpas* and then killed themselves right after making a vow.

3 GOOD VOWS AND MERITS: ANAVATAPTA LAKE DRAGON KING'S *JĀTAKAS* AND *NIDĀNAS*

Dragons in East Asian cultures possess the duality of nature (good and bad). While the image of a fierce and intimidating dragon is widely recognized, the positive image of a good dragon also exists. That is, there is a kind of good dragon – as one of the Eight Patron Deities of Buddhism. In *Record*, there appears only one good dragon with a clear record of *jātakas* and *nidānas* – the *nāga* king of Anavatapta Lake in Book One (Xuanzang and Bianji [annotated by X. L. Ji], 2007). Although the narrative is succinct, it contains profound meanings that are worth exploring in comparison with the bad dragons discussed earlier. The deeds of the Anavatapta Lake *nāga* king can be seen in Book One:

In the center of the Jambu continent is Anavatapta Lake, which is south of Fragrant Mountain and north of the Great Snow Mountains, with a circuit of eight hundred li. Its banks are adorned with gold, silver, lapis lazuli, and crystal. It is full of golden sand and its water is as pure and clean as a mirror. A bodhisattva of the eighth stage, having transformed himself into a nāga king by the power of his resolute will, makes his abode at the bottom of the lake and supplies water for the Jambu continent. (Xuanzang and Bianji [annotated by X. L. Ji], 2007).

In this account, the Anavatapta Lake *nāga* king is a good dragon, different from the bad dragons mentioned earlier. He even sustains the entire water supply of the Southern Continent of Jambudvīpa. However, his *jātakas* and *nidānas* still highlight the subjective factor of "vows," and no direct mention is made of the factor of "merits," which is manifested apparently in the stories of bad dragons.

This does not necessarily mean that a good dragon reincarnation can follow without certain merits. Quite the reverse, an Eighth Ground *bodhisattva* was already a great *bodhisattva* with much advanced practice, and he owned certain merits for sure because merits are one of the two core cultivations for those practicing the *bodhisattva* path: "cultivation of merits and wisdom." His reincarnation as a dragon is his voluntary vow taken after taking pity on the beings living in the Southern Continent of Jambudvīpa, not the result of an ill vow. After all, the high stage of an Eighth Ground *bodhisattva* is superseded only by five higher cultivation levels within the 52 stages – namely the Ninth Ground, the Tenth Ground, Virtual Enlightenment, Sublime Enlightenment, and Ultimate Buddhahood. It is impossible for an Eighth Ground *bodhisattva* to be in such a high stage without enough merits. Therefore, it is reasonable to argue that Master Xuanzang skipped this when it was common sense at that time that *bodhisattvas* had certain merits.

As a result, "good vows" and "merits" can be seen as two core factors in the *jātakas* and *nidānas* of good dragons.

4 CONCLUSION: DRAGON REINCARNATION TAKES SUFFICIENT MERITS AND VOW

From these discussion of the dragon kings' *jātakas* and *nidānas* (Xuanzang and Bianji [annotated by X. L. Ji], 2007) in accordance with the *Record*, it is noticeable that "vows" and "merits" contribute to dragon reincarnation. Whether vows are good or ill, dragon reincarnation demands both merits and vows, which can be illustrated from the story of the dragon king of the Great Snow Mountains.

Master Xuanzang tends to talk more about the aspect of "vows" in his account. However, *nāga* Gopāla's story exemplifies the importance of merits. His story shows that a person with an ill-intended wish has to make up the merits he needs first and then make a solemn vow before his death in order to ensure the vow will be realized. Also, in the story of the *nāga* king of the Great Snow Mountains, the novice monk made his horrible vow while wishing, "May the power of whatever good deeds I have performed appear." Vows alone cannot lead to dragon reincarnation, and sufficient merits are necessary.

From this viewpoint, Master Xuanzang maintains a consistent logic both directly and indirectly while completing his book. The idea that dragon reincarnation takes sufficient merits and vows can be seen as Master Xuanzang's opinion. Also, if we read the book from a broader scope, we can notice that dragons were considered real creatures in the customs of Western regions and India, and certain karma teachings derived from this environmental legend. We also can notice that contemporary Buddhist culture had a profound impact on Western regions and India.

REFERENCES

Book Two of the Taishō Tripiṭaka. 2007. Ekottaragama-sutra, 21, 657.

Xuanzang and Bianji (annotated by X. L. Ji). 2007. *The Great Tang Dynasty Record of the Western Regions*. 1st edition. 3rd print. Taipei: Xin-wen-feng Publishing Company.

Innovation in Design, Communication and Engineering – Lam et al. (eds)
© 2020 Taylor & Francis Group, London, ISBN 978-0-367-17777-5

A discussion on the Chan concept underlying the Japanese way of tea based on the essence of Buddhist realization: The tea ceremony of Sen Sōtan

Mei-Ling Chang* & Huann-Ming Chou
Mechanical and Energy Engineering, Kun Shan University, Tainan, Taiwan

ABSTRACT: Chinese practitioners travelled to India seeking the Buddha Dharma as early as the Three Kingdoms, starting with Zhu Shixing, through to the Ming Dynasty when many official monks made the journey. During this thousand plus years, approximately more than a thousand people risked their lives in search of the real truth. Their spirit is highly revered and they are an inspiration to the general public. As well, they contributed immensely toward the development of Chinese Buddhism. Journey to the West is seen as a depiction of one such Buddhist pilgrimage to India with Tang Seng representing Xuanzang. The Dharma Master Xuanzang became the model of Chinese pilgrims who went to India in search of Buddhist doctrines. This paper compares and analyzes the differences between the trips made by Master Xuanzang and various pilgrims in terms of public archives, Buddhist achievements, quality and quantity of translated Buddhist sutras, language capability, original literary works, visited countries, patronage of the emperor, etc. in an attempt to find the reason why Master Xuanzang became the symbol of those who travelled west in search of the Dharma. In addition to the widely publicized story about Journey to the West, Master Xuanzang's personal talents, his characteristic of seeking the truth via facts, his achievements in Buddhism, and language capability in both Sanskrit and Chinese are all important factors that contributed to his success. Master Xuanzang was widely respected and admired in both India and China at the time and his attainments and integrity were unsurpassed, thus he became the symbol of those who travelled west in search for Buddhist Dharma.

Keywords: Chinese seekers of Buddhism in India, Journey to the West, Dharma

1 INTRODUCTION

In the public's impression, Journey to the West, written by Ming novelist Wu Cheng'en, is a story about a westbound seeking of Buddhism monk and the Tang Monk in the story is derived from Xuanzang. In fact, Dharma Master Xuanzang has been regarded as a synonym for a westbound Buddhism seeker due to this novel's influence. However, in the history of Chinese Buddhism, he is not the only one who undertook a westbound journey for original scriptures. Generally speaking, it is Zhu Shixing who initiated these kinds of trips in the Three Kingdom period. This westbound Buddhism seeking journey reached its first peak in the Eastern Jin Dynasty. There were quite a lot of westbound monks who headed to India during the Sui and Tang Dynasty and this number reached its last peak in the Song Dynasty. Emperor Taizu of Song assigned one hundred and fifty-seven monks to India, which is the largest official mission to India sent by the court in Chinese history. The following sixty to seventy years in the Northern Song Dynasty, the number of scripture-seekers who went to India and returned was one hundred and thirty-eight. The Buddhism seeking activities remained popular, but they had little impact on Chinese Buddhism in this

* Corresponding author: mei.ling.apple@gmail.com

period. The number dropped substantially after the Song Dynasty. Some official monks were assigned to India during the Ming Dynasty, but there was little contribution. This has something to do with the fading of Buddhism in its native soil, India, since the thirteenth century. To sum up, there were thousands of westbound monks in pursuit of scriptures in Chinese history (Wei, 2009).

However, why Dharma Master Xuanzang stands out among all Chinese seekers of Buddhism in India to be regarded as a synonym for a westbound seeker is the focus of this study. Most of the past studies on the Chinese seekers of Buddhism who headed west centered on the individual seeker's contributions and his pursuing process, or on the individual's Buddhist writings, westbound routes, and related deeds (Wei, 2009; Lu, 1977). Little is known about the reason and the significance behind the general impression that equates the westbound Chinese seeker of Buddhism with Dharma Master Xuanzang. Accordingly, this paper intends to explore the reasons of such an equation by comparing Master Xuanzang with other westbound seekers from various aspects, to promote a better understanding of Master Xuanzang as well as other westbound seekers among readers, and to pay homage to them. After all, the image of Tang Monk portrayed in Journey to the West (C. Xiyou ji) as a timid good-hearted monk who needs his pupil's protection is greatly different from Master Xuanzang in the real history.

2   A COMPARISON OF SOME LEADING WESTBOUND SEEKERS OF BUDDHISM

2.1   Background of Buddhism-seeking

Buddhism originated in India and spread to China through the Silk Road. The earliest Han Buddhist canon was not a direct translation from Sanskrit. Instead, it was a translation from some ancient Central Asian languages and the translated scriptures were mostly works of oral translation. In the early phase, the Buddhist scriptures spread through the Western Regions. Sometimes, they were incomplete, had some translation distortion, or even had some contradictions. This sparked some dedicated monks' motivation to go abroad in search of original Buddhist scriptures.

2.2   Objectives and contributions

There are four outstanding well- recognised representative figures of the westbound seekers of Buddhism: Zhu Shixing of the Three Kingdom period, Faxian of the Eastern Jin Dynasty, and Xuanzang as well as Yi Jing of the Tang Dynasty (Wei, 2009).

Zhu Shixing often delivered speeches in Luoyang on the Daoxing Jing (also known as the old version of Minor Prajñā Sutra). However, the translation of this scripture was rather terse and omitted some meanings. He felt sorry that such a significant scripture was not thoroughly translated, so he vowed that he would find its original Sanskrit text to complete it. Although there was only one scripture that was brought back, this initiated a specific doctrinal study of Buddhism in China, set up a great example, and inspired succeeding seekers of Buddhism. His non-self-attitude shown in the process of his Dharma-searching is really moving.

Faxian sighed for the fact that there was a shortage of precepts for the monks to regulate their daily schedule and serve as conduct principles despite the existence of Mahāyāna and Hīnayāna scriptures that were translated into Chinese, so he decided to go to India in search of precepts. Most of the scriptures he took back were precepts. His Records of Buddhist Kingdoms (C. Foguo ji) describes the contemporary geography, produces, customs and the prevalence of Buddhism in the northwestern regions of China. This is a valuable source for the studies of histories, cultures and religions of some countries in the ancient Central Asia and South Asia.

Master Xuanzang took several prominent monks as his mentors and had gained a solid understanding of the Buddhist studies. He noticed every Buddhist theory had its own denominations and that there were some inconsistent, sometimes even contradictory, remarks, so he expected to cultivate a more comprehensive understanding of Buddhist teachings by accessing Treatise on the Seventeen Stages (S. Yogâcārabhūmiśāstra; C. Yuqie shidi lun), which encompassed the profound teaching of the Three Vehicles. He was determined to search for this scripture to settle the disputes. The Great Tang Dynasty Record of the Western Regions authored by Master Xuanzang was even an invaluable historical source, which depicts what Xuanzang saw, heard of and experienced in one hundred and thirty-eight countries and city-states in the Western Regions on his way to India. In this book, he put down a variety of things about each country in detail, such as geography, traffic, climate, produces, ethnic peoples, languages, history, religions, politics, economy, cultures and customs. It has a great reference value for the learning and studies of the ancient Central Asia and South Asia.

Yi Jing had been longing to imitate Master Xuanzang since he was little. He believed in addition to the original scriptures brought back by Xuanzang, there must have been some other treasurable Buddhist texts that waited to be discovered. Yi Jing's two works, Chronicle of Eminent Monks Who Traveled to the West seeking the Dharma (C. Datang xiyu qiufa gaoseng zhuan) and A Record of Buddhist Practices Sent Home from the Southern Sea (C. Nanhai jigui neifa zhuan) offers some treasurable information for the studies of Sino-India relationships, traffic, histories of the Chinese and Indian Buddhism, and the Southern Sea in the early Tang Dynasty.

Table 1 is a comparison among the four monks in terms of various perspectives.

Table 1. A comparison among the different aspects of westbound seekers of Buddhism.

Aspects \ Representatives	Zhu Shixing	Faxian	Xuanzang	Yi Jing
Publicity-enhancing	The prototype of Zhu Ba Jie (Pig) in *Journey to the West*	No	The prototype of the Tang Monk in *Journey to the West*	No
Performance in Buddhist studies	Before departure, he delivered speeches in Luoyang on *Daoxing Jing*.	Before departure, he received education in the temple and learned by himself.	Before departure, he took thirteen prominent monks as his teachers, read through all contemporary Chinese Buddhist theories and had gained a profound understanding of Buddhist studies. In India, he took Dharma Master Śīlabhadra, the abbot of the Nālandā Monastery, as his mentor and became one of the ten knowledgeable Tripiṭaka masters. He was well known in India, and was highly regarded as deity of the Great Vehicle and deity of liberation	Before departure, he took Monks Shan Yu and Chan master Hui Xi as his teachers and studied disciplinary precepts closely. He traveled Tang China thoroughly to enhance his Buddhist understanding. In India, he studied in the Nālandā Monastery and learned Buddhist cannons from Monk Bao Shi Zi.

(Continued)

151

Table 1. (*Continued*)

Aspects \ Representatives	Zhu Shixing	Faxian	Xuanzang	Yi Jing
The quantity and quality of translation	Quantity: He did none but he transcribed *Dapin Bore Jing* by himself.	Quantity: He translated 6 collections and 63 volumes with the aid of Monks Bao Yun and Buddhabhadr.	Quantity: He translated 75 collections and 1,335 volumes. Quality: He wanted his translation to be true to the original text as well as understandable to the public. He proposed five categories of untranslated terms. He innovated the translation style and did the translation by following a plan with the translation task allocated.	Quantity: He translated 56 collections and 230 volumes. Quality: He strictly did straight translation. His characteristic is to add notes under the original text or the translation.
Language capability	He didn't translate but transcribed scriptures.	He learned Sanskrit in Central Tianzhu for three years.	He learned some native languages and Sanskrit before departure. He joined the assembly on Dharma doctrinal debates, where many monks from five parts of India attended. He composed *Three Treatises* in Sanskrit and translated *Awakening of Faith in the* Mahāyāna and *Dao De Jing* into Sanskrit. He mastered both Chinese and Sanskrit, and can be called a language genius.	He stayed in Sumatra for half a year to learn Sanskrit. For the same purpose, he stayed in South India for a year. His work, *Sanskrit 1000-Character Text,* is a small Sanskrit-Chinese dictionary for Sanskrit learners.
Works	No	Records of Buddhist Kingdoms (Biography of Faxian)	Cheng Weishi Lun, The Great Tang Dynasty Record of the Western Regions, Three Treatises (in Sanskrit), the True Consciousness-only and Verses Delineating the Eight Consciousnesses.	A Record of Buddhist Practices Sent Home from the Southern Sea, A Record of Returning to the Southern Sea, Chronicle of Eminent Monks Who Traveled to the West seeking the Dharma and Sanskrit 1000-Character Text.
Countries visited on the journey	He made it to Khotan (Yutian in	He traveled through almost thirty countries (such as Western	He visited 110 countries, and heard of things about 28	He traveled through more than 30 countries (such as

(*Continued*)

Table 1. (*Continued*)

Aspects \ Representatives	Zhu Shixing	Faxian	Xuanzang	Yi Jing
	Xinjiang), not India.	Regions, India, Indonesia and Sri Lanka).	countries (such as the Western Regions, Afghanistan, Iran, Pakistan, India, Sri Lanka, Nepal, and Bangladesh.)	Indonesia, Malaysia, the Southern Sea, and India).
The emperor's regard	Before departure, he delivered speeches in Luoyang on *Daoxing Jing*.	Before departure, he received education in the temple and learned by himself.	Before departure, he took thirteen prominent monks as his teachers [8], read through all contemporary Chinese Buddhist theories and had gained a profound understanding of Buddhist studies. In India, he took Dharma Master Śīlabhadra, the abbot of the Nālandā Monastery, as his mentor and became one of the ten knowledgeable Tripiṭaka masters. He was well known in India, and was highly regarded as deity of the Great Vehicle and deity of liberation	Empress Wu, Zetian welcomed him in person when he returned from India. He often joined the empress's tour inspection. Tang Emperor Zhongzong set up an official translation office for him in the Jianfu Temple.

3 EXPLORATION OF THE REASONS WHY DHARMA MASTER XUANZANG BECOMES A SYNONYM FOR A WESTBOUND SEEKER OF BUDDHISM

3.1 *Journey to the West was widespread*

In modern times, the sales of products, newly-released movies and new published books all take marketing and advertisement to generate publicity. The westbound scripture-seeker is widely recognised by the public as Tang San Zang mainly due to a widespread recognition of Wu Cheng'en's novel, Journey to the West, while the other westbound seekers of Buddhism do not have any similar text as advertisement. Even if Zhu Shixing is another character (Zhu Ba Jie) in Journey to the West, he is not the protagonist in search of scriptures. The reasons behind Wu Cheng'en's choice of Xuanzang's story as the theme of his novel might be because Xuanzang's westbound seeking of Buddhism story had accumulated certain publicity through previous works and Xuanzang's firm determination by risking his own life to travel to the west. In fact, this story had become a popular topic for story-tellers since the Tang Dynasty. In addition, there had been some related plays, such as Datang Sanzang Qujing Shihua of the Southern Song Dynasty and Yang Jingxian's A Mix Play of Journey to the West of the Yuan Dynasty. Although this image is a distorted image of Master Xuanzang, this name has become a synonym for a westbound seeker of Buddhism.

3.2 His excellence in Buddhist studies and Sanskrit with Chinese as a foundation

His excellence in Buddhist studies: Before the age of 25, Master Xuanzang had taken 13 prominent monks as his mentors and studied The Abhidharma (C. Pitan), Satyasiddhiśāstra (C. Chengshi lun), Abhidharmakośabhāṣya (C. Jushe lun), Mahāyānasaṃgraha (C. She Dasheng lun), etc [8]. He had a solid understanding of Buddhist studies and grasped the true principles of Buddhist doctrines, thus earning certain reputation and status in Tang China. In India, he took Dharma Master Śīlabhadra, the abbot of the Nālandā Monastery, as his mentor. Master Xuanzang was even invited by Dharma Master Śīlabhadra to dominate the lectures held in the Nālandā Monastery, a task that only excellent monks who mastered scriptures and precepts had the honour of doing according to the customs in ancient India. He wrote the Treatise on the Complimentarity of Tenets (C. huizong lun) 3000 Verses in Sanskrit to deliver a message that the seemingly contradictory between the two Mahāyāna doctrines, Yogâcāra (C. Yuqie) and Madhyamika (C. Zhongguan), was not true as they end up with the same principle. He even wrote Zhi Ejian Lun that contained 1,600 verses to prevent some wrong theories from disseminating and established the principle of the True Consciousness-only. Thanks to Shi Ejian Lun, King Harsa held an open great assembly for doctrinal debate in Kanauj, where Xuanzang highlighted the significant principle of Mahāyāna. In five parts of India, the Mahāyāna practitioners praised him as the "deity of the Ultimate Truth" and "deity of the Great Vehicle," and the Hīnayāna practitioners extolled him as the "deity of liberation." He was greatly admired by hundreds of thousands of people over there at that time. Based on the information above, Master Xuanzang is no ordinary seeker of Buddhism from Tang China, but a well-recognised monk as one of the ten knowledgeable Tripiṭaka masters, who had a prestigious status in Indian academic circles. This achievement far exceeds that of his predecessors and successors, therefore justifying his supreme excellence in Buddhist studies.

His mastery of Sanskrit and Chinese: Before Master Xuanzang took his perilous journey to India to seek the original scriptures; he showed great initiative in learning languages. He learned Tocharian languages, which were used in Central Asia, and some aboriginal languages from some foreigners in Chang'an. He also started to learn Sanskrit in order to communicate with some prominent monks from the Western Regions (Wriggins, 1997). Later, Xuanzang traveled through several countries in the Western Regions and Tianzhu. Due to his inborn language talent, Xuanzang had a better mastery of Sanskrit than the other westbound seekers of Buddhist, so he could engage in the discussion of debates in the assembly, where many monks from five parts of India attended. In addition, he could translate texts from Sanskrit to Chinese and from Chinese to Sanskrit thanks to his incredible mastery of Chinese and Sanskrit. He translated some texts, such as Dao De Jing and Awakening of Faith in the Mahāyāna (C. Dasheng quixin Lun), into Sanskrit and sent them to India. He also made some contributions to Indian Buddhism by his Sanskrit translation of Huizong Lun, Zhi Ejian Lun and Treatise on the Three Bodies of the Buddha. These achievements derive from his outstanding language talent, which is second to none.

3.3 Demonstration of his individual talents and down-to-earth character

To sum up, thanks to his outstanding excellence in Buddhist studies, Sanskrit, and Chinese, incredible personal talents and down-to-earth character, Master Xuanzang performs better than the other westbound seekers of Buddhism in terms of the aspects of the emperor's regard, countries visited on the journey, works, and the quantity as well as quality of translation.

The Emperor's regard: Emperor Taizong of Tang absolved Xuanzang of his violation of the edict that forbade any westbound journey owing to his supreme reputation in the five parts of India. Later, Emperors Taizong and Gaozong of Tang wanted to take advantage of his knowledge, so they asked him to renounce his monk status and become a government official. Xuanzang rejected their offers, but still won their admiration and support. Emperor Taizong set up an official translation centre (National Translation Office) and wrote the preface of his translation (Yogâcārabhūmiśāstra) for him. Emperor

Gaozong paid homage to Xuanzang, and made him the abbot of the Da Ci'en Temple—a construction built to commemorate the emperor's mother, Queen Wende. The emperor even imitated Indian towers and built the Giant Wild Goose Pagoda in the west part of the temple where Xuanzang could place scriptures, statues and relics. The emperor also wrote an inscription for the Da Ci'en Temple. These all demonstrate that his talents are earnestly accepted by the emperors.

Countries visited on the journey: Due to his down-to-earth character, Xuanzang preferred field research. He visited 110 countries, heard of things from 28 countries, and made vigorous research on geography, Buddhist historic buildings, and related histories, legends, biographies and dispersion. Expert Wang, Shi Ping indicates, "The distance measured by Xuanzang's feet turns out to be exactly correct, which is admired wholeheartedly by Stein.

Works: The Great Tang Dynasty Record of the Western Regions is a prominent work for understanding the Western Regions in the seventh century as well as a significant source of the history of Chinese Buddhism. It is paramount to the studies of Indian history as well. His other works, such as Cheng Weishi Lun, Three Treatises (in Sanskrit), the True Consciousness-only and Verses Delineating the Eight Consciousnesses, are the manifestations of his outstanding mastery of the Buddhist studies, Sanskrit, and Chinese, incredible personal talents and down-to-earth character.

The quantity as well as quality of translation: Xuanzang's achievements in translation are the demonstrations par excellence of his outstanding mastery of the Buddhist studies, Sanskrit, and Chinese, incredible personal talents and down-to-earth character. He took translation seriously and did it rigorously. In terms of scripture translation, he argues that the translation should "be true to the original text as well as understandable to the public," and proposed five categories of untranslated terms five principles of non- translation. There are five conditions when the meaning of a word is not translated and phonetic translation is employed instead: when the word is with secrets and multiple significations, better understood by its past translation, absent in the target language and more respectful. What's more, when Xuanzang did the translation, he usually translated a volume first, and read out the Sanskrit text and the Chinese translation in public. If anyone present disagreed, there would be further debates and refutations among experts until they reached an agreement. Based on this agreement, the translation was modified and then finalised. He invented the translation style, made use of the official translation office to allocate the translation task and followed a plan to complete his translation, and thus the quantity and quality of his translation far surpassed the works of the other three westbound seekers of Buddhism.

4   CONCLUSION

Most of the seekers of Buddhism in India committed themselves to searching for original scriptures and they were courageous, persistent and even willing to sacrifice their lives for this cause. Four representatives, Zhu Shixing, Faxian, Xuanzang and Yi Jing, belong to this group, and their contributions to Chinese Buddhism are truly memorable. Among them, Xuanzang has become the synonym for the westbound seekers of Buddhism because the wide distribution of Journey to the West earned him certain publicity. Another reason is that his outstanding mastery of Buddhist studies, Sanskrit and Chinese as well as his incredible talents and down-to-earth character all contribute to his excellent performance in translation, works, and winning the emperor's regard. Above all, Xuanzang's outstanding excellence in Sanskrit, Chinese and the Buddhist studies enabled him to solve some essential controversial issues that existed in Indian Buddhism. This accomplishment won himself some praise and admiration from every country in Tianzhu. He was extolled as the "deity of the Great Vehicle" and the "deity of liberation," and widely recognised in five parts of India. This is the leading reason why Xuanzang is the protagonist in Journey to the West, not to mention his lived experiences of traveling to 110 countries and hearing of things about 28 countries that enriched its story contents. This is the true image of the Tang Monk as a persistent westbound seeker of Buddhism, the most symbolic figure of westbound seeker of Buddhism of all time.

REFERENCES

Hsu, M.Y., 2009. A Westbound Scripture-seeker, Faxian, and Biography of Faxian. Dharma Light Monthly, 237.

Lu, C., 1977. Xuanzang and Indian Buddhism. Modern Buddhist Scholarship Series.

Wang, B.W., 1995. Yi Jing and A Record of Buddhist Practices Sent Home from the Southern Sea. Chung Hwa Book Company.

Wei, T.R., 2009. The Chinese Monks' Westbound Scripture-seeking. Encyclopedia Knowledge, 14.

Wriggins, S. (translated by Du Mo), 1997. Xuanzang: A Buddhist Pilgrim on the Silk Road. Taipei: Human Thesaurus Publishing Group.

Innovation in Design, Communication and Engineering – Lam et al. (eds)
© 2020 Taylor & Francis Group, London, ISBN 978-0-367-17777-5

A study of Xuanzang's great ambition to "Inherit and carry forward the Tathagata's Lineage" based on his westbound route to acquire Buddhist scriptures

Hung-Chuan Yu* & Huann-Ming Chou
Mechanical and Energy Engineering, Kun Shan University, Tainan, Taiwan

ABSTRACT: During the first year of the Zhenguan era (AD 627), Master Xuanzang began his long journey to the West in search of Buddhist scriptures. He traveled more than 50,000 miles in total, and spent about 18 years exploring 110 countries, authoring the legendary *Great Tang Records on the Western Regions*. It is commonly known that Xuanzang's purpose was to resolve the different and opposing views on Buddhist doctrines that various schools espoused at that time. However, this paper presents another perspective as it studies Xuanzang's route to the West to find out that he took five years to reach the kingdom of Magádha and to enter Nālandā Monastery, the highest Buddhist educational institution in India at the time. He then became a student of Master Śīlabhadra for five years, but he did not return to China immediately after he acquired the authentic principle of *Yogâcārabhūmiśāstra* (C. *Yuqieshidi lun; Treatise on the Stages of Yogic Practice*) and many original Buddhist scriptures. Xuanzang traveled to almost everywhere in the five Indias, as he was able to uphold his lofty ambitions to acquire the Buddha Dharma and the truth, and he maintained sincere sentiments to cherish the memories of the saint. In order to acquire the ultimate truth of Buddhism and to know the cultivation sequence on the way to Buddhahood, Xuanzang did not subject his travels to a rigid schedule, as he stayed and left different places whenever necessary. Furthermore, he even participated in debates with monks from diverse schools and with non-Buddhists in order to refute erroneous thoughts and spread the truth, to defend Mahāyāna Buddhism, and to propagate the true principle of the fundamental and profound *tathāgata-garbha*'s absolute truth. Therefore, Xuanzang was definitely not just a famed practitioner who journeyed to the West seeking Buddhist scriptures,; neither was he simply a scriptural translator or traveler. Instead, he was a successful Dharma seeker and propagator.

Keywords: Xuanzang, Pilgrimage to the West for Buddhist *sutras*, Nālandā Monastery, Yogâcārabhūmiśāstra, Mahāyāna Buddhism

1 INTRODUCTION

1.1 *Background and motivation*

May 22, 2017, witnessed the launch of "The 12th Xuanzang Route Gobi Business Challenge" near the ruins of King Asoka's temple, which has a history of 1,000 years. This event was officially held in Guazhou County, Jiuquan City, Gansu Province, with 2,500 participants from 57 Chinese business schools all over the world. This race took place in the Moheyanqi Gobi, also known as "the quicksand of 800 kilometers," between Gansu Province and Xinjiang Province. Every participant walked on the ancient road Venerable Xuanzang took more than 1,000 years ago. The competition spanned four days and three

* Corresponding author: hcyuvghtpe@gmail.com

nights, covering 117 kilometers. With the rise of "One Belt, One Road," "The Xuanzang Route" has received great acclaim from the circles of education and industry for the past two years, generating the phenomenon of the "Gobi cult" – one that has drawn groups of "Gobi fellows" to carry out Xuanzang's expedition step by step with the attitudes of endurance, strength, and willpower as well as the belief in realizing dreams, action, perseverance, and transcendence. The perseverance of Xuanzang is likely to resonate with "Gobi fellows" who have undertaken this incredible journey, and who then develop new values and understandings for life. Based on the routes of his westbound scripture-seeking collected from past studies, this paper explores Xuanzang's grand aspiration and passion in his taking a vow to pursue the Dharma and truth so that readers may have a better understanding of Xuanzang's historical position in the dissemination of Buddhist teachings to the East.

1.2   The background of Xuanzang's taking a vow to travel west

At the age of 13 (AD 615), Xuanzang was exceptionally granted official permission to take tonsure. He had had a great passion for the Mahāyāna Dharma since he was little. He persevered in studying Buddhist Dharma, and he had an outstanding talent for learning them. Before he had reached 25 years of age, Xuanzang traveled around China for his Buddhist studies, and he took 13 eminent monks as his mentors. They were Jing, Yan, Kong, Hui Jing, Dao Ji, Bao Xian, Dao Zhen, Hui Xiu, Dao Shen, Dao Yu, Fa Chang, Seng Bian, and Xuan Hui. Due to them, Xuanzang was transformed into a well-learned monk. Better yet, he was eloquent and many of his contemporary scholars admired his wisdom and demeanor; as a result, Xuanzang became a noted figure, and two prominent monks, Fa Chang and Seng Bian, who were well known for understanding and studying the Great and Small Vehicles as well as for thoroughly actualizing three Vehicles, even praised him. "You can make a profound contribution to Buddhism just as an extraordinary horse can run long distances without any difficulty" (CBETA, T50, no. 2053, pp. 222, b23–24).

The deeper Xuanzang delved into Buddhist studies, the more questions he found. When he was young, Xuanzang felt sorry that the system of the Buddhist canons in China was too disorganized: the translated scriptures often lost their original meaning due to either overly literal or overly liberal translations. Meanwhile, Xuanzang noticed that every Buddhist theory had its own denominations and that inconsistent, sometimes even contradictory, remarks provoked serious antagonism at times. Many Buddhists didn't know what to follow. All of these had issues a great impact on the continuation of Buddhism in China. Since no transcription of *Yogâcārabhūmiśāstra* (C. *Yuqie shidi lun*) was available in China as evidence, Xuanzang could not settle these disputes with any citation of original canons. This also exposed the fact that Chinese Buddhism had a shortage of scripture transcriptions.

In the ninth year of the Takenori era of Emperor Gaozu of Tang (AD 626), Monk Prabhākaramitra arrived in Chang'an from India. As a disciple of Venerable Śīlabhadra, a prestigious master from Nālandā Monastery, Monk Prabhākaramitra could recite 100,000 verses of both Mahāyāna and Hīnayāna scriptures by heart. Hearing this news, Xuanzang was delighted, as if he were about to obtain a precious gem, and went to visit this eminent monk for the Dharma. Monk Prabhākaramitra inspired Xuanzang with the news that in Nālandā Monastery, Venerable Śīlabhadra, the master of the generation, taught *Yuqie shidi lun*, a scripture that encompassed the profound teaching of the Three Vehicles. He also told Xuanzang that Nālandā Monastery had remarkable teaching programs. Xuanzang then wanted to learn the Dharma in India. He accepted the responsibility of reinvigorating Chinese Buddhism and rescuing all living beings, and took a vow to find and then investigate Sanskrit original scriptures directly on his westbound scripture-seeking journey.

2 THE ROUTE XUANZANG TOOK ON THIS WESTBOUND SCRIPTURE-SEEKING JOURNEY

In the first year of the Zhenguan era of Emperor Taizong of Tang (AD 627), Xuanzang set off in Chang'an, went west through Qinzhou, Lanzhou, and Liangzhou and arrived in Dunhuang. Then, he passed Yumenguan and Yiwu, and reached Gaochang, an affiliated nation of the Tang dynasty, where he firmly declined King Gaochang's offer of a longer stay. He continued to cross the boundary of the Tang dynasty and entered the Western Regions, from which his westbound journey initiated, strictly speaking. Xuanzang went through the hazards of Lingshan and the Hot Sea, passed such cities as Suyeshui and Tanluosi, took a south turn, and traveled through the southern region of modern-day Central Asia and the northeast region of Afghanistan. He then headed east, traveled through the northern region of modern-day Pakistan, where he stayed until the end of the third year of the Zhenguan era. Afterward, it took him most of the fourth and fifth years of the Zhenguan era to head southeast along the northern region of the Indian peninsula, and he traveled through the southern area of modern-day Nepal. At the end of the fifth year of the Zhenguan era, Xuanzang arrived in Magadha and entered Nālandā Monastery, the highest seat of Buddhist learning, where he took Venerable Śīlabhadra as his mentor and stayed for about five years. After that, he spent about three years visiting Dharma masters. Xuanzang entered Bangladesh along the east side of the Ganges, headed south along the east side of the Indian peninsula, and reached Dravida, which faces modern-day Sri Lanka with a sea in between. Later, he took a northwest turn, headed west along the west side of the Indian peninsula, visited the Ajanta Caves, and once entered the hinterland of the Ganges – that is, the southeastern region of what is known today as the Chambal River. He continued west, entered modern-day Pakistan, walked northbound along the Sindhu River, and arrived in Parvata, near Chamo in the south of modern-day Kashmir. He spent about two years studying there, and then returned eastbound to Magadha for the continuation of his learning from Venerable Śīlabhadra. With time, Xuanzang mastered Buddhist studies, delivered lectures, and wrote in the Nālandā Monastery. In about the 16th year of the Zhenguan era, Xuanzang presided over the famous assembly held in Kanauj thanks to the invitations of Kings Mara and Harsa. The next year, King Harsa held a 75-day-long open assembly for Xuanzang. Then, Xuanzang bid farewell and initiated his return journey. He walked past the northern part of modern-day Pakistan and the northeast part of modern-day Afghanistan, turned eastbound, and traveled through the Wakhan Valley, which is the southern part of what is now known as the Pamir Plateau. In the spring and summer of the 18th year of the Zhenguan era, he arrived in Khotan. He came back to Chang'an the next January. The round trip spanned more than 50,000 kilometers and took about 18 years. Xuanzang traveled through 110 countries in person, creating "a miracle of westbound expedition" that stunned the world (Zhang, 1977).

3 CHARACTERISTICS OF XUANZANG'S ROUTES ON HIS WESTBOUND JOURNEY AND THE REAL PURPOSE OF HIS SEEKING OF BUDDHISM

This paper studies the characteristics of Xuanzang's routes on his westbound Buddhism-seeking journey as well as the real purpose for such a journey.

3.1 *Xuanzang traveled through 110 countries solitarily. What kind of grand ambition and motivation sustained him to complete such an incredible feat?*

The distance of the 2017 "Gobi Challenge" is far shorter than that of Xuanzang's westbound journey – that is, more than 50,000 kilometers. The place where this race was held was the difficult path where Xuanzang almost died without water. In this path, he experienced being ambushed, being betrayed, and getting lost; however, he did not crumble and still continued his journey toward Tianzhu firmly. The "Gobi Challenge" enables every participant to experience in person the hardships Xuanzang experienced in his life and mind. Through "walking on Xuanzang's path," participants undertake this ascetic trek, strengthen their minds, discover their potential, understand their abilities, and witness Xuanzang's feat. How could Xuanzang bravely fight against all odds and hazards for a sacred mission he dedicated himself to in the hope of reinvigorating Chinese Buddhism? Just as the founder of "the Xuanzang Road," Qu Xiangdong, said, "What walks on this path is not your feet, but your heart. What is tested is not the stamina of your feet, but that of your heart."

During the 2017 challenge, many participants dropped out since they couldn't withstand the harsh weather conditions and they fell short of the required physical and mental strength, although they were equipped with advanced modern technological gear and medical support. From this, we can imagine how incredible it was for Xuanzang to overcome the hardships of enduring starvation, trekking across deserts, climbing snowy mountains, experiencing storms, and being threatened by robbers without any thought of giving up. Only one idea was in his mind: "Eliminate false scriptures and obtain true scriptures. I will never return eastbound until I reach Tianzhu" (CBETA, T50, no. 2053, pp. 224, b16–18). This manifests the fact that Dharma Master Xuanzang was an eminent monk who always actualized his aspirations for the Dharma with a down-to-earth attitude and took Buddhist studies seriously and with a rigorous approach. He never took his own safety into account and was even ready to sacrifice his life in his pursuit of truth. In his mind, the seeking of Buddhism was his mission as a Buddhist disciple. Therefore, it is apparent that the fundamental reason behind his earnest choice of undertaking a journey westward in order to seek Buddhism without considering his own life originated in his unconditional regard for the truth of Buddhism, his initiative to spread Mahāyāna Buddhism as a sacred mission, and his compassion for guiding and protecting all living beings. His readiness to sacrifice his own life for this cause fully demonstrates his devotion to seek scriptures despite tremendous adversities, and he was certainly no backpacker and sojourner.

3.2 Xuanzang followed no strict schedule on his westbound journey in search of Buddhism. He traveled or stayed whenever he considered it necessary to do so, and he almost traveled through all five parts of India. What is the reason behind this kind of travel?

Long before Xuanzang's westbound journey, he was already well known, had earned everyone's respect, and had a prominent reputation in the circle of Chinese Buddhism. However, Xuanzang was not concerned about his accomplishment, reputation, and status. Instead, he was firmly determined to travel west in search of truth. When we analyze all the routes he took before and after his arrival at his destination, Nālandā Monastery, we notice that Xuanzang kept no strict schedule on his westbound journey. He traveled or stayed based on his learning purposes and demands. When he had something to learn, he would stay three to five days, or temporarily live for one to two years. Xuanzang stayed longer in three places: one was Kaśmīra, where he was taught *Abhidharmakośabhāṣya* (C. *Jushe lun*), *Abhidharmanyāyâ-nusāraśāstra* (C. *Shunzhengli lun*), Buddhist logic (C. *Yinming*), and grammar and composition (C. *Shengming*) by Monk Sengcheng. Another was Parvata, where from two to three *bhadanta* he learned *Saṃmitīya* (C. *Zhengliang bu*) scriptures, such as *Genbenapitamo, Shezhengfa lun*, and *Jiaoshi lun*. The third was Zhanglinshan, where commentator Jayasena (C. *Sheng Jun*) taught him *Weishijueze lun, Yiyili lun, Chengwuwei lun, Buzhu niepan shi er yinyuan Lun, Zhuangyan jing lun, Yuqie shidi lun*, and *Yinming* (Chu, 1981). Xuanzang spent about two years in each place. However, the place that played the most significant role in Xuanzang's scripture-learning was Nālandā Monastery, where he took Venerable Śīlabhadra as his mentor. He learned the teachings of *Yuqie shidi lun* three times; *Shunzhengli lun, Xian yang shengjiao lun* (S. *Āryavācāprakaraṇaśāstra*), and *Duifa* once each; *Yinming, Shengming*, and *Jiliang* twice each; the *Treatise on the Middle-Way* (C. *Zhong lun*) and the *Treatise in One Hundred Verses* (C. *Bai lun*) each three times. He also studied *Jushe, Po sha, Liu zu*, and *Pitan*, and so on (CBETA, T50, no. 2053, pp. 244, c1–3). He traveled through five parts of India, and his journey spanned hundreds of thousands of kilometers. Carrying numerous scriptures, Xuanzang walked to visit eminent monks and laypeople for the Dharma, and he came to have a comprehensive understanding of *Yinming, Shengming*, and some scriptures of Brahmanism, as well as all doctrines of both the Great and Small Vehicles. On his journey, Xuanzang studied scriptures of different mainstream Buddhist schools, such as *Sautrāntika* (C. *Jingliang bu*), *Mahāsāṃghika* (C. *Dazhongbu*), *Sammatīya* (C. *Zengliang bu*), and *Sarvāstivāda* (C. *Sapoduo bu*). He never missed any chance to learn, so he was knowledgeable and well experienced. As he took tonsure at 13, he vowed to "inherit and carry forward the Tathagata's lineage." The long-term objective was to inherit Buddha's lineage and pass on the seed of Buddhism; the short-term objective was to take action to seek Buddhism, to obtain original scriptures and spread Buddhist teachings, and to make the true Dharma remain forever in the world to benefit all living beings. Hence, Xuanzang almost traveled through the five Indian kingdoms. He traveled through many countries, so he was familiar with the theories and thoughts of every Buddhist school in India. He helped clarify some deviant ideas in every school, promoted the teachings of the Great Vehicle to the public, and taught them how to remove wrongdoings and to promote good deeds. His audience ranged from kings and aristocrats to ordinary people and thieves, and he taught them patiently while delivering the due Dharma at its right timing.

3.3   *The routes Xuanzang took on his westbound journey were not arranged solely for finding answers to personal questions about the Dharma. A more significant purpose of his arrangement was to engage in debates with heterodox followers and other monks from different Buddhist denominations, to refute deviant ideas, and to highlight correct understanding so as to guard Mahāyāna Buddhism, and to investigate thoroughly and pass on the ultimate truth of profound tathāgatagarbha.*

After Xuanzang left Nālandā Monastery, he traveled around paying visits and giving lectures over a period of five years and thenreturned. Venerable Śīlabhadra assigned Xuanzang to deliver lectures on *She dacheng lun* and *Weishijueze lun*. At one time, the worthy Siṃhaprabha, who previously had lectured the assembly on *Chong lun* and *Bai lun*, stated that his aim was to refute the *Yogācāra* principle. While Xuanzang, himself well trained in the subtleties of the topics regarding *Zhong lun* and *Bai lun*, as well as being skilled in the *Yogācāra* doctrines, took it to be the case that the sages who established each of those teachings did so with the same intent; there were no contradictions or oppositions between them. Those who were confused and unable to understand this complementarity would talk about them as contradictory, but this was a fault with the transmitters, not with the Dharma (CBETA, T50, no. 2026, pp. 448, b16–18). He composed the *Treatise on the Complementarity of Tenets* (C. *Huizong lun*) in 3,000 verses (CBETA, T50, no. 2053, pp. 244, c1–3) in order to explain that the tenets espoused by both systems, *Madhyamika* (C. *Zhongguan*) and *Yogācāra* (C. *Yuquie*), are to be considered a harmonious complementarity. When the treatise was completed, he presented it to Venerable Śīlabhadra and the great assembly; there were none who didn't praise its value, and all shared and propagated it, finally resolving the doubts of contemporary people.

Two more intense confrontations took place between Xuanzang and the Hīnayāna believers and non-Buddhists at Nālandā Monastery. King Harsha issued a formal invitation to Nālandā, requesting the monastery send monks to defend Mahāyāna in Uḍra. Upon receiving the invitation, Śīlabhadra selected a team of four monks: Sāgamati, Jñānaprabha, Siṃhaprabha, and Xuanzang. The other three monks were worried because Prajñāgupta had extraordinary wisdom, but Xuanzang had confidence even though he was a Chinese monk, so he promised to take full responsibility for the outcome of their debates. Right before their departure, a Brahmin came to Nālandā, challenging anyone to take him on and even wrote 40 verses and hung the note at the door. Under the witness of all Nālandā monks, Xuanzang took the challenge and defeated the Brahmin and pointed out one by one the erroneous practices of the heterodox doctrines, such as covering oneself in ashes, complete nudity, wearing bones around one's neck, eating feces and putting them on one's clothes, and so on. This Brahmin finally surrendered. When Xuanzang was preparing to go to Uḍra, he obtained a treatise in 700 stanzas, the *Treatise on Refuting Mahāyāna Doctrine*, composed by Prajñāgupta in refutation of the Mahāyāna teachings. Xuanzang read through it and found out the erroneous points and refuted them with Mahāyāna teachings in a treatise he wrote in 1,600 stanzas, entitled *Zhi ejian lun* (*Treatise on the Refutation of Wrong Views*). He presented the work to Venerable Śīlabhadra, who showed it to his disciples, who all praised it.

Later, Xuanzang attended a great open assembly in Kanauj (CBETA, T50, no. 2026, pp. 453, a22–23), where he extolled the profound and significant principles of Mahāyāna to the Śramaṇas, Brahmins, and non-Buddhist followers throughout the five Indian kingdoms and subdued their intention to defame Mahāyāna. His gracious demeanor, insightful Buddhist cultivation, and excellent eloquence convinced all comers, and no one challenged Xuanzang during the entire 18 days. Xuanzang's praise of Mahāyāna and his admiration of Buddha's deeds, as well as his highlighted right understandings of the Dharma and refuted erroneous views, caused numerous people to return from the wrong to the right and to discard Hīnayāna theories and embrace Mahāyāna teachings. Xuanzang won the debate and made his name known throughout the five Indian kingdoms. He was honored as the "deity of the Great Vehicle" by the Mahāyāna practitioners" and as the "deity of liberation" by the Hīnayāna practitioners. Xuanzang's remarkable eloquence highlighted the distinction of Mahāyāna by refuting wrong ideas. He was alone while confronted with the massive powers of Hīnayāna monastic communities, Brahmins, and non-Buddhists, but he showed no fear when promoting the correct Dharma. Xuanzang stated, "It is difficult for the correct Dharma to manifest if the false concepts are not destroyed" (CBETA, T31, no. 1585, pp. 12, c18–19). That is, Xuanzang had to lead Buddhists to the right path by destroying wrong doctrines and promoting the correct Dharma so that the four communities could truly comprehend and follow Buddha's right Dharma to the path of liberation and the Bodhi Way. In addition to actualizing the truth by the scriptures he found, Xuanzang's westbound journey established the meanings of Mahāyāna Buddhism by engaging in doctrinal debates with other masters, which is in fact another feat. Xuanzang was not only a dedicated and perseverant seeker of Buddhism with readiness to sacrifice his life, but also a brave promoter of the correct Dharma who destroyed false doctrines.

4 CONCLUSION

This paper takes another perspective to explore the routes Xuanzang took on his westbound scripture-seeking journey. Xuanzang spent five years reaching Magadha, and he spent about five years learning the Dharma from Venerable Śīlabhadra. Right after he obtained the scriptures and the true meanings of *Yuqie shidi lun*, he did not return to China right away. Xuanzang traveled throughout almost the five Indian kingdoms because of his grand ambition to seek the Dharma of Buddhism and his passion for remembering past saints. Xuanzang followed no strict schedule. Instead, in order to acquire the ultimate truth of Buddhism and to develop a complete cultivation sequence of Buddhism for attaining Buddhahood, Xuanzang traveled or stayed whenever he considered it necessary to do so. He even engaged in debates with monks from different mainstream Buddhist schools and with non-Buddhists in order to refute wrong ideas, highlight correct understandings, defend Mahāyāna Buddhism, and investigate thoroughly and pass on the ultimate truth of profound *tathāgatagarbha*. Dharma Master Xuanzang was no mere westbound scripture-seeker, scripture-translator, traveler, and explorer. What made him widely recognized and admired in the five Indian kingdoms and beyond was his outstanding excellence in Buddhist principles. Xuanzang was an actualizer of the seeking of Buddhism and an advocate of the Dharma. His life is a real demonstration of "inheriting and carrying forward the Tathāgata's lineage."

REFERENCES

A Biography of the Tripitaka Master of the Great Ci'en Monastery of the Great Tang Dynasty. Vol. 1 (CBETA, T50, no. 2053, pp. 222, b23–24).
A Biography of the Tripitaka Master of the Great Ci'en Monastery of the Great Tang Dynasty. Vol. 1 (CBETA, T50, no. 2053, pp. 224, b16–18).

A Biography of the Tripitaka Master of the Great Ci'en Monastery of the Great Tang Dynasty. Vol. 4. Trans. D. Lusthaus (CBETA, T50, no. 2053, pp. 244, c1–3).

Cheng weishi lun. Vol. 3 (CBETA, T31, no. 1585, pp. 12, c18–19).

Chu, B. S. 1981. *An Introduction to the Studies of Xuanzang.* 1st ed. Taipei: Shin Wen Feng Print Company.

Continuation of the Biographies of Eminent Monks. Vol. 4. Trans. D. Lusthaus (CBETA, T50, no. 2026, pp. 448, b16–18).

Continuation of the Biographies of Eminent Monks. Vol. 4 (CBETA, T50, no. 2026, pp. 453, a22–23).

Zhang, X. 1977. *A Revised Edition of Records of Western Lands.* Shanghai: People's Publishing House.

Innovation in Design, Communication and Engineering – Lam et al. (eds)
© 2020 Taylor & Francis Group, London, ISBN 978-0-367-17777-5

Study of the Indian customs to ward off diseases based on the *Great Tang Records on the Western Regions*

Tai-Ming Huang* & Huann-Ming Chou
Mechanical and Energy Engineering, Kun Shan University, Tainan, Taiwan

ABSTRACT: The *Great Tang Records on the Western Regions* (*Records*) serves as a very significant piece of literature on the history and culture of India during the seventh century. With reference to the Indian custom of praying at Buddhist pagoda (S. *stūpa*) and before Buddha statues for healing as recorded in *Records*, this paper seeks to analyze various healing prayers and reveals that healing prayers only worked with *stūpas* and statues related to Buddha Śākyamuni, and there are no records of praying at the *stūpas* of the great *bodhisattvas* and *arhats* in India during that time.

Keywords: *stūpa*, *Great Tang Records on the Western Regions*, healing prayers, Xuanzang

1 INTRODUCTION

The *Great Tang Records on the Western Regions* (*Records*) keeps track of the countries Xuanzang travelled through on his journey from Xinjiang to India in the seventh century and features information such as geography, culture, religion, local produce, and politics, so it occupies a prominent position in the studies of Indian history and cultures. One passage about curing illness appears in *Records*: "Those who fall ill have to stop eating for seven days. Most people recover during this period; those who fail to recover then need to seek medical remedies" (CBETA, T57, no. 2087, pp. 877, c22–23). At that time, the first act a sick Indian took was to fast for seven days, and most people would recover during this time. Only those who couldn't be healed by this seven-day fast needed to take medicine. In those days, it was a custom for patients to pray to Buddha and Bodhisattva for recovery, in addition to seeking medical treatment, which was not always available. This can be confirmed in *Records*, where Dharma Master Xuanzang puts down the custom of performing healing prayers in Buddhist pagodas (*stūpas*) and before Buddha statues in many places. Based on these accounts, this paper sorts out, analyzes, and discusses ways of performing healing prayers to Buddha and Bodhisattva.

2 THE INDIAN CUSTOMS OF HEALING PRAYERS DURING THE TANG DYNASTY

2.1 *People fully recovered by circumambulating the* stūpas.

A. The Country of Kanyākubja

At a distance of six or seven li to the southeast of the great city there is a stūpa over two hundred feet high built by King Aśoka at the site where the Tathāgata once preached on the doctrine that the physical body is impermanent, sorrowful, empty, and impure. Beside it is a site

---

* Corresponding author: taiming0219@yahoo.com.tw

where the four past buddhas used to sit and walk up and down. There is also a small stūpa containing hair and nail relics of the Tathāgata. Anyone who suffers from an illness can surely be cured and benefited if he or she circumambulates the stūpa with a pious mind. (CBETA, T57, no. 2087, pp. 896, a10–12)

B. The Country of Magadha

In the stūpa to the south of the statue of Avalokiteśvara Bodhisattva are kept hair and fingernail [relic]s of the Buddha, shaven and clipped over a three-month period. Sick people are often cured of their illnesses by circumambulating the stūpa. (CBETA, T57, no. 2087, pp. 924, a19–20)

In these two regions, most ill people recovered by performing healing prayers sincerely while they were walking around the pagodas that contained relics of Buddha's hair and nails. References appear to the custom of circumambulating the *stūpas* for recovery, such as *Yourao fota gongde jing*, "The person who circumambulates the stūpas clockwise chanted the name of Buddha, and his subtle voice spread far and deep, making anyone who heard it feel delighted. The person himself is always joyful and disease-free afterwards" (CBETA, T16, no. 700, pp. 801, b22–24). In fact, circumambulating the pagodas clockwise was a way to pay homage to Buddha. Paying respect to Buddha could gain many merits and people could use these merits to pray for recovery. Therefore, there was a custom of circumambulating the Buddhist *stūpas* for recovery.

2.2   *People recuperated well by drinking the water and taking a bath in the gushing springs next to the* stūpas.

A. The Country of Udyāna

Beside the cliff at the north of the Sanirāja River there is a stūpa that often cures sick people who come to pray for the recovery of their health. When in a former life the Tathāgata was a peacock king, he came here with his flock. As it was the hot season, the peacocks were thirsty but they could not find any water to drink. The peacock king then pecked the cliff to let water flow out of the rock. Now a pond has been formed there, and its water is effective for healing illness. There are traces of the peacocks still visible on the stone. (CBETA, T57, no. 2087, pp. 883, a27–28)

B. The Country of Kapilavastu

Outside the south gate of the city, on the left side of the road, is a stūpa at the place where the prince, competing with other Śākyas in the arts of war, shot at iron drums. At a distance of more than thirty li to the southeast from here is a small stūpa, beside which is a spring flowing with clear water. While competing with other Śākyas in the skill of archery, the prince drew his bow, and, as the arrow left the bow it pierced through the surface of the drums and hit the ground, sinking into the earth up to its fletching. The spring of pure water formed at that spot, by tradition called Arrow Spring. When people are sick they drink the water or bathe in it and in most cases they are cured. People travel from distant places to collect the clay of the spring and make it into a paste, which is applied on the forehead whenever they have any ailment. As the clay is protected by spirits and deities it has a healing effect in most cases. (CBETA, T57, no. 2087, pp. 902, a13–15)

Above are the two gushing springs next to the Buddhist *stūpas*. One gushing spring was allegedly developed when the Peacock King (Mahamayuri), the reincarnation of Buddha at that time, used its beak to peck the cliff. The other came into existence when the Śākya prince's arrow penetrated the target and shot into the bottom of a pond. A spring then gushed out, accordingly called Arrow Spring. The existence of both gushing springs was said to come from Buddha. Most of the patients who drank or took a bath recovered. As for those who lived too far away to bathe in these gushing springs in person, they could apply the mud of these springs to their afflicted part, or to their forehead. Because the Guardians of Dharma offered protection, their health usually recovered well. The customs of drinking the water and taking a bath in the gushing springs next to the *stūpas* for recovery come from the disciples' belief that the gushing springs contained hidden protection from Buddha and the Guardians of Dharma, thereby they could recover through drinking the water or applying spring mud to their body.

2.3 People fully recovered by taking a bath in the hot springs near the stūpas: *The Country of Magadha*

More than ten li to the southwest of Yaṣṭi Wood there are two hot springs to the south of a great mountain. The water is very hot and after the Buddha had caused these springs he bathed in them. They are still in existence and the flow of clear water has never diminished. People come here from far and near to bathe and the springs may effect a cure for those who suffer from chronic illness. Beside the springs is a stūpa built at a place where the Tathāgata walked up and down. (CBETA, T57, no. 2087, pp. 920, b04–06)

The healing powers of hot springs have been recognized since a long time ago, just as Shuijing Zhu (*Commentary on the Water Classic*) states, "Rice can be cooked at Lushan Huangnu Tang Hot Springs. Hundreds of ailments can be cured by drinking it. A Taoist priest took a bath there three times a day for forty days, and his numerous diseases were cured." In modern days, it is still a common medical intervention for patients to take a bath in hot springs. The custom of taking a bath in the hot springs near the *stūpas* formed partly because hot springs themselves have certain healing powers, but mostly because these hot springs were transformed by Buddha himself, who also took a bath there. Hence, taking a bath in the hot springs near the *stūpas* was regarded by Buddhist believers as gaining subtle protection from Buddha, which could reinforce the healing powers of the hot springs. It is recorded that those with obstinate maladies of long standing could be cured by taking a bath in this kind of hot spring.

2.4 People perfectly recovered by offering fragrant incense and flowers to the stūpas: *The Country of Kāpiśī*

To the south of the pipal tree there is a stūpa constructed by King Kaniṣka. In the four hundredth year after the Tathāgata's demise King Kaniṣka ascended the throne and ruled over Jambudvīpa. He did not [originally] believe in the theory of retribution for good and evil deeds and contemptuously defamed the Buddha-dharma. Once he was hunting in a marsh when a white hare appeared. The king chased after the hare and it suddenly disappeared at this place. In the woods he saw a young cowherd building a small stūpa three feet high. ... These two stūpas still stand. Those who are ill and wish to pray for recovery offer incense and flowers to the stūpas with pious minds, and in most cases they are cured. (CBETA, T57, no. 2087, pp. 880, a09–11)

The *Sutra on Distinguishing Good and Evil Karmic Retribution* (C. *Fenbie shan-e baoying jing*) says:

If someone makes flower offerings to the Tathāgata pagodas, the person will be granted ten merits. What are these ten merits? The first merit is good looks, the second unparalleled beauty, the third an illness-free olfactory faculty, the fourth an odour-free body, the fifth pure and wondrous fragrance, the sixth the rebirth in the Pure Land and seeing Buddha, the seventh the dissemination of precept incense, the eighth being widely regarded and obtaining great happiness from the Dharma, the ninth the rebirth in a heaven where one can live freely, the tenth fast attainment of nirvana. These merits are conferred on the very one that makes flower offerings to the Buddhist pagodas, where sariras are stored. (CBETA, T01, no. 0081, pp. 900, a19-22)

Those who make flower offerings in the Buddhist pagodas by scattering flowers around can be presented with the 10 merits, such as an illness-free olfactory faculty and an odor-free body. Based on this reference, it is known that the merits accumulated by making flower offerings at the pagodas could be a means of praying for recovery. What's more, *Fenbie shan-e baoying jing* says:

If someone offers fragrant incense to the Tathāgata pagodas, the person will be granted ten merits. What are these ten merits? The first merit is a pure olfactory faculty, the second an odour-free body, the third good physical and mental conditions with pure fragrance, the fourth a majestic countenance, the fifth being revered by the world, the sixth being well-informed and passionate for the Dharma, the seventh being born in noble ranks with a blessing of living freely, the eighth a greatly recognized reputation, the ninth ascending to the heavens when passing away, the tenth fast attainment of nirvana. These wondrous merits are conferred on the very one that makes offerings of fragrant incense to the Buddhist pagodas, where sariras are stored. (CBETA, T01, no. 0081, pp. 900, b06–10)

Those who make offerings of fragrant incense at the Buddhist pagodas can be presented with the 10 merits, such as a pure olfactory faculty, an odor-free body, and good physical and mental condition. As a result, the merits people gather by offering fragrant incense at the Buddhist pagodas can be a means of praying for recovery.

In addition to accumulating merits by making offerings of fragrant incense and flowers at the Buddhist pagodas, invalids also had to reverently convert to Buddhism. With the merit of faith after becoming a Buddhist, they might elicit subtle protection from Buddha and then recover.

2.5 People recovered well by worshiping the stūpas earnestly and reverently: The Country of Rāma[grāma]

From the stūpa where the prince had his head shaved, going southeast for one hundred eighty or ninety li through a wilderness, I reached a banyan grove in which there is a stūpa over thirty feet high. When the Tathāgata entered nirvana and his relics were distributed, the brahmans who had not obtained a share of the relics collected the ashes and charcoal from the ground of the niṣṭapana (meaning "burning," formerly known as shewei by mistake) and brought them home; they built this holy reliquary for worship. Since then it has manifested many miracles and most of the sick people who have prayed for recovery here have received a response. (CBETA, T57, no. 2087, pp. 903, a25–29)

Fenbie shan-e baoying jing states:

At that time, Buddha told elder Soka:

If someone worships the Tathāgata pagodas with joint palms, the person will be granted ten merits. What are these ten merits? The first merit is being born in noble ranks in the future lifetimes, the second a dignified countenance in the future lifetimes, the third a great body form in the future lifetimes, the fourth rich with four sorts of resources in the future lifetimes, the fifth ample treasures and wealth in the future lifetimes, the sixth a good reputation in the future lifetimes, the seventh a built-in wholesome roots for the Dharma in the future lifetimes, the eighth a good recollection of practices in the future lifetimes that continues to cultivate a wholesome root, the ninth being intelligent enough to practice Buddhism and to teach others in the future lifetimes, the tenth being well versatile enough to cultivate insightful true wisdom in the future lifetimes. As this is the correct way, elder man, these merits are conferred on the very one that worships Tathāgata's pagodas reverently with joint palms. (CBETA, T01, no. 0081, pp. 899, c12–16)

People who deferentially paid homage at pagodas with joint palms were granted many merits, such as a dignified countenance in their future lifetimes. A reverent worship alone could also gather such subtle protection from Buddha that helped recovery even without offering flowers, incense, and other items.

2.6 People recuperated utterly by pasting gold foil onto Buddha statues: The Country of Gostana

Going eastward for more than thirty li from the battleground, I reached the city of Bhīmā, in which there is a standing image of the Buddha carved out of sandalwood, over twenty feet high, which has shown spiritual responses many times and often emits a bright light. If someone who suffers from a painful ailment pastes a piece of gold foil on the image at the part that corresponds to where his ailment is, he may be instantly relieved of the pain. If someone says prayers to the image with earnest devotion their wishes will be fulfilled in most cases. (CBETA, T57, no. 2087, pp. 945, b06–10)

To produce majestic effects, Buddha statues are created with golden bodies. Such bodies are formed either by pasting gold foil all over Buddha statues or using pure gold as the construction materials. *Flower Garland Sutra* (C. *Dafang guang fo huayan jing*), *Inconceivable State of Tathāgatas*, Vol. 1, states, "At that time, Samantabhadra Bodhisattva told Virtue Store Bodhisattva, 'Tathāgata's son! As you ask, how should one cultivate to attain prajñā wisdom? ... One should go to an ashram to view a Buddha statue, adorned in golden color or cast in pure gold, complete with a Buddha's physical marks. In its halo one should see innumerable magically manifested Buddhas, who sit in order and are in samādhi" (CBETA, T10, no. 0300, pp. 907, c09–11).

Since Buddha's time, disciples have followed a custom of offering gold foil to adorn Buddha statues. The invalids pasted gold foil onto the part of the Buddha statues that corresponded to the afflicted part of their bodies, and the merits accumulated by such offering reinforced their subsequent sincere healing prayers, which were mostly answered.

2.7 People regained health by applying scented oil to the Buddha statues of shrines: The Country of Magadha

In a wood to the east of the dragon Mucilinda's pool is a shrine with an image of the Buddha in an emaciated condition. Beside the shrine is the place where he walked up and down, more than seventy paces long, and there are two pipal trees, one at the south and the other at the north side of the promenade. In the past and at present it is the custom of the local people to anoint the image with fragrant oil when they are afflicted with a disease, and in most cases they are cured of their illness. (CBETA, T57, no. 2087, pp. 917, b18–24)

One of the common offerings made during the Buddha's time was to apply scented oil to Buddha statues. Just as stated in the *Fo shuo dazizai tianzi yindi jing*, "The next immortal being offered a metal piece to Buddha, applied some scented oil to Buddha's foot, and then took a vow, saying, 'By the merits I earn from this good deed, I will be granted Narayana Deva as the master of the Three Realms'" (CBETA, T15, no. 0594, pp. 128, c02–03). With time, this practice of applying scented oil to Buddha statues as an offering developed into the custom that invalids performed prayers by applying scented oil to emaciated Buddha statues. Most of these prayers were later granted with subtle protection from Buddha.

These are all the documentation of the customs in *Records* related to healing prayers. Two methods are presented of performing a healing prayer: praying at the *stūpa* and making offerings in front of Buddha statues. According to the custom of praying at the *stūpa*, praying with flower or incense offerings shown in *Fenbie shan-e baoying jing*, as well as reverently circumambulating the Buddhist pagodas clockwise recorded in *Yourao fota gongde jing*, are ways to accumulate merits in the hope of obtaining subtle protection from Buddha and eradicating illness. In addition to making offerings at the pagodas, invalids can also drink from or bathe in the gushing springs beside the pagodas that were brought forth by the Buddha, or bathe in the hot springs beside the *stūpas* that were transformed by Buddha himself. As for making offerings before Buddha statues, there are two customs: one is Magadha's custom of applying scented oil to Buddha statues, and the other is Gostana's custom of pasting gold foil onto Buddha statues. The objects of these prayers are all Buddha. There is no reference of praying to *bodhisattvas* or *arhats*.

3 CONCLUSION

Based on the seventh-century Indian customs of healing prayers documented in Dharma Master Xuanzang's *Records*, it is known that invalids had to reverently make offerings before they performed any healing prayers. They offered items such as flowers, incense, and fragrant essence, worshipped Buddha with joint palms, or placed gold foil on Buddha statues. These ways of making offerings are all mentioned in Buddhist scriptures (CBETA, T09, no. 0262, pp. 0030, c07–23). The recent practice of redeeming a vow in Buddhist temples, a subsequent offering made after a prayer is heard and an illness is gone was unheard of in India at that time. Buddha embodies four boundless minds – namely the immeasurable mind of kindness (S. *maitrī*), the immeasurable mind of pity (S. *karuṇā*), the immeasurable mind of joy (S. *muditā*), and the immeasurable mind of impartiality (S. *upekṣa*). His mind of *maitrī* blesses all living beings with security and happiness, and his mind of *karuṇā* blesses all living beings with exemption from all sufferings. The invalids pray for recovery, and Buddha or Bodhisattva will not ask anything in return. Thus, there is no need of their redeeming a vow to Buddha. Such a redeeming practice in modern times is supposed to be the result of influences from other religions.

The objects of believers' prayers are all the Buddha statues or the Buddhist pagodas where Buddha's *sariras* are kept, and it is never seen to invoke the statues or pagodas with relics that are connected with other great *bodhisattvas* or *arhats*. A number of pagodas with *arhats'* relics were reserved in every part of India then, so it was not inconvenient for Buddhist believers to perform prayers at such kind of pagodas. However, there was no record of this kind of practice. It can be inferred that it works to pray to only Buddha. Buddhist followers considered it truly helpful for rapid recovery to obtain subtle protection from Buddha. On the other hand, *arhats* were considered without enough merits to offer such protection. The practice of performing healing prayers to Buddha was verified over generations and developed into a custom. This custom shows the fact that *arhats* did not have as many merits as Buddha, which was already the consensus in seventh-century India.

REFERENCES

Dafang guang fo huayan jing (CBETA, T10, no. 0300, pp. 907, c09–11).
Fenbie shan-e baoying jing (CBETA, T01, no. 0081, pp. 899, c12–16).
Fenbie shan-e baoying jing (CBETA, T01, no. 0081, pp. 900, a19–22).
Fenbie shan-e baoying jing (CBETA, T01, no. 0081, pp. 900, b06–10).
Fo shuo dazizai tianzi yindi jing (CBETA, T15, no. 0594, pp. 128, c02–03).
Great Tang Dynasty Record of the Western Regions. Trans. Li Rongxi (CBETA, T57, no. 2087, pp. 877, c22–23).
Great Tang Dynasty Record of the Western Regions. Trans. Li Rongxi (CBETA, T57, no. 2087, pp. 896, a10–12).
Great Tang Dynasty Record of the Western Regions. Trans. Li Rongxi (CBETA, T57, no. 2087, pp. 924, a19–20).
Great Tang Dynasty Record of the Western Regions. Trans. Li Rongxi (CBETA, T57, no. 2087, pp. 883, a27–28).
Great Tang Dynasty Record of the Western Regions. Trans. Li Rongxi (CBETA, T57, no. 2087, pp. 902, a13–15).
Great Tang Dynasty Record of the Western Regions. Trans. Li Rongxi (CBETA, T57, no. 2087, pp. 920, b04–06).
The Great Tang Dynasty Record of the Western Regions. Trans. Li Rongxi (CBETA, T57, no. 2087, pp. 880, a09–11).
Great Tang Dynasty Record of the Western Regions. Trans. Li Rongxi (CBETA, T57, no. 2087, pp. 903, a25–29).
Great Tang Dynasty Record of the Western Regions. Trans. Li Rongxi (CBETA, T57, no. 2087, pp. 945, b06–10).
Great Tang Dynasty Record of the Western Regions. Trans. Li Rongxi (CBETA, T57, no. 2087, pp. 917, b18–24).
Lotus Sutra, "Making offerings to it in various ways with flowers, incense, necklaces, powdered incense, paste incense, incense for burning, silk canopies, flags, banners, robes, or music, as well as revering it with palms together, these people will be looked up to and honored by the whole world. Offerings should be made to them as if they were tathagatas." Vol. 4 (CBETA, T09, no. 0262, pp. 0030, c07–23).
Yourao fota gongde jing (CBETA, T16, no. 700, pp. 801, b22–24).

User interface design

Innovation in Design, Communication and Engineering – Lam et al. (eds)
© 2020 Taylor & Francis Group, London, ISBN 978-0-367-17777-5

A VR bike system designed to play on Google street view

Chien-Shun Lo
Department of Multimedia Design, National Formosa University, Hu-Wei, Yunlin, Taiwan

Bo-Syun Jhan*
Department of Multimedia Design, National Formosa University, Hu-Wei, Yunlin, Taiwan.
Graduate Institute of Digital Content and Creative Industries, National Formosa University, Hu-Wei,
Yunlin, Taiwan

ABSTRACT: In this study, a VR bike system designed to play on Google Street View was proposed. The VR Bike System combines a VR System connected to a Game-Bike on which a player can ride and a VR helmet which the player wears to facilitate immersion in the VR environment. The VR environment contains a series of VR360 pictures of street views from Google maps. The series of VR360 pictures construct a series of VR spaces which connect to form a tunnel. When the player rides the Game-Bike, the camera of the VR system starts to move along the tunnel. By wearing the VR helmet, the player can easily look around the VR envi- ronment. The player can ride the Game-Bike to control the speed of the VR system camera moving along the tunnel. In our system, it is easy to travel all over the world by riding Game-Bike in the VR environment.

1 INTRODUCTION

The exercise bike is a popular machine used for working out. It is useful indoors and in limited space. Unlike a real bike, it does not require a comfortable road and favourable weather conditions. However, working out on an exercise bike makes people feel more bored than riding a bike outdoors especially for longer periods. This is due to the monotony of seeing the same surroundings.

In order to fix this issue, game bike has been developed. Game bike [1] is a kind of exercise bike connected to a mobile device or a computer to run a game. Participating in a game is a good way to improve the user's motivation to keep working out. In such a device, the handle of the bicycle is modified as a joystick for the game. The bicycle pedal is used as a trigger in the same way as a fire button of a joystick. If we improve the game bikes connection to a VR system, then it becomes a VR bike system. This means the user could wear a VR helmet to move and be immersed in a VR environment.

However, another purpose for biking is to simulate travel to the specific target scene. Therefore, it is necessary to have the practical panoramas of a specific road. This becomes possible by using Google Street View with panoramas of stitched images. Most photography is done by car, tricycle, boat, snowmobile, underwater apparatus and on foot [2]. Most popular bicycle routes of the world are available from Google Street View.

In our study, A VR bike system designed to play on Google Street View was developed for the purpose of biking in the world with practical VR 360 views.

---

* Corresponding author: 10768113@gm.nfu.edu.tw

2 METHODS AND RESULTS

2.1 *Download panoramas*

There is a panorama downloading software offered by Google, called *street view download 360* which can be obtained from [3], shown in Figure 1.

When we press the left down link 「iStreetView.com」 shown in Figure 1, it turns to a specific website to show the positions where we want to download shown in Figure 2. There are two panels making up this web page. The left side shows the position of map and the right side shows the street view corresponding to the specific point. In the right side panel, users can use a mouse to rotate the VR view to see around. In the top-left of the right side panel, it shows the download link of this panorama.

An example is shown in Figure 3-4. We choose a street view and copy a panorama ID from the top- left hyperlink shown in Figure 3.

Then use *street view download 360* to download a sequential panorama shown in Figure 4. This may be done by copying a sequential panorama IDs for download shown in Figure 4. The sequential panoramas obtained by the viewer provide a directional button to choose the next path. The directional buttons as in Figure 3 may be the left or right direction. While users press the directional button, it jump to the next view following the path. In this way, users can obtain sequential panorama IDs and copy them to the download software shown in Figure 4. Then all of them were download into the local site.

2.2 *Build a virtual reality path*

Our system was built in Unity 3D platform [4]. As shown in Figure 5, a sphere is used to contain a panorama. The downloaded panorama was applied to the sphere. The result was shown as Figure 6.

Figure 1. The screen panel of *street view download 360*.

Figure 2. iStreetView.com is a specific website to select the position to view the street view.

Figure 3.   Choose a street view and copy the panorama ID.

Figure 4.   Choose the target local directory and copy a sequential panorama Ids for downloads.

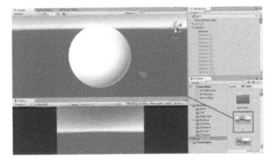

Figure 5.   The panorama was applied to the sphere.

The top window of Figure 6 is the edit window. It shows a sphere containing a panorama. There is a camera placed in the center of the sphere. The bottom window shows the view of the camera. Actually, the camera captures the panorama based on the panorama being attached to the inside of sphere. This idea uses an insideout shader applied to this sphere for this purpose shown in Figure 7.

175

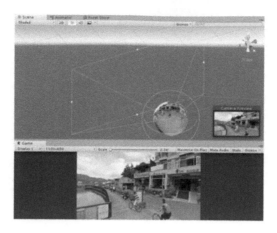

Figure 6.   The panorama was shown in the sphere.

```
5    Shader "insideout" {
6    Properties {
7        _MainTex ("Base (RGB)", 2D) = "white" {}
8    }
9
10   SubShader {
11       Tags { "RenderType"="Opaque" }
12       Cull front    // ADDED BY BERNIE, TO FLIP THE SURFACES
13       LOD 100
14
15       Pass {
16           CGPROGRAM
17               #pragma vertex vert
18               #pragma fragment frag
19
20               #include "UnityCG.cginc"
21
22               struct appdata_t {
23                   float4 vertex : POSITION;
24                   float2 texcoord : TEXCOORD0;
25               };
26
27               struct v2f {
28                   float4 vertex : SV_POSITION;
29                   half2 texcoord : TEXCOORD0;
30               };
31
32               sampler2D _MainTex;
33               float4 _MainTex_ST;
34
35               v2f vert (appdata_t v)
36               {
37                   v2f o;
38                   o.vertex = UnityObjectToClipPos(v.vertex);
39                   // ADDED BY BERNIE
40                   v.texcoord.x = 1 - v.texcoord.x;
41                   o.texcoord = TRANSFORM_TEX(v.texcoord, _MainTex);
42                   return o;
43               }
44
45               fixed4 frag (v2f i) : SV_Target
46               {
47                   fixed4 col = tex2D(_MainTex, i.texcoord);
48                   return col;
49               }
50           ENDCG
51       }
52   }
```

Figure 7.   The code of insideout shader.

Figure 8 shows how we create a path which is formed of spheres. This path is designed for the user to ride. These spheres are placed non-overlapped. Between any two spheres, there is a transparent box as a collider to detect user movement. The box shown as Figure 9. Actually,

176

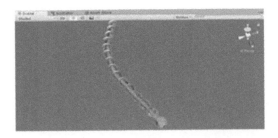

Figure 8.   The path is consisted of spheres.

a model of the bicycle is put in the front view of camera to be a virtual bicycle seen by the user. The virtual bicycle is shown in Figure 10. Figure 11 shows the result of system view, the virtual bicycle put in the front view of the main camera. The path and location are shown in the top left of the window. The distance, speed, and calories are shown in the top right of the window. The corresponding panorama is shown in the window based on the position of the map.

When a user starts to ride the game bike, it triggers the virtual bike to move. When the virtual bike moves to the transparent box collider shown in Figure 9, it triggers the virtual bike to move to the next sphere. The virtual bike moves to pass one by one through the spheres. The moving speed of the virtual bike is calculated by the ride speed. The distance of movement is counted by the passing count of the spheres. This path consists of 437 spheres. The path is 3000 meters. The diameter of a sphere is about 6.8m. In practise, if the user trigger the pedal sensor (corresponding to one crank roatation) once in 0.7 seconds a speed of 10km/hour can be achived. The system following this linear relationship rule to calculate the speed. As such the game bike is considered as a single speed bike.

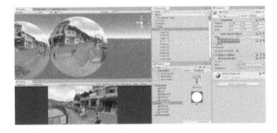

Figure 9.   A transparent box is placed between two spheres.

Figure 10.   A model of bicycle used as a virtual bicycle.

Figure 11.   The screen consists of a virtual bicycle, map and panorama.

3   CONCLUSION

Exercise bikes are a popular work out in many countries, especially in countries with cold and raining weather and especially during winter months. Riding an exercise bike for prolonged periods is boring. Therefore, most people would rather choose to ride a bike outdoors than indoors as it is more interesting. For committed cyclists real track training is important. However, this study is useful to overcome the above two issues. It helps people to ride indoors in a virtual reality environment like outdoors. The ride speed and distance are similar to the real ride. It is an effective VR application to improve the exercise bike.

REFERENCES

[1] Performance, 2019, "About Game-Bike", from http://www.game-bike.com.tw/about.php.
[2] Wikipeida, 2019, "Google Street View", from https://en.wikipedia.org/wiki/Google_Street_View.
[3] Thomas Orlita, 2019,"Street View Download 360", from https://svd360.istreetview.com.
[4] Unity Technologies, 2019, "Unity for all", from https://unity.com.

Innovation in Design, Communication and Engineering – Lam et al. (eds)
© 2020 Taylor & Francis Group, London, ISBN 978-0-367-17777-5

Comparative cultural research between Russian and Taiwanese sign interpretation

Siu-Tsen Shen*, Mariia Ominina & Po-Heng Shih
Department of Multimedia Design, National Formosa University, Huwei, Yunlin, Taiwan

ABSTRACT: This sarticle discussed the relationship between cultural background in society and cross-cultural sign perception based on the results of a survey. The survey related to this article was conducted in August and September 2019 for two cultural groups of participants from Taiwan and Russia. There were 10 signs in total, the first set of 5 were Taiwanese, and the other were Russian. The goal of the study was to find out which signs were more likely to be perceived correctly and why. In order to analyze the results, concepts such as culture and cultural dimensions, semiotics, communication theory, and cognitive psychology were included. Based on the results of the survey, possible explanations to the difference of visual signs perception were highlighted. The discussion and analysis led to creation of a set of signs that would be understandable to both residents and other nations in future.

Keywords: visual perception, cultural dimension, sign perception, cultrual difference

1   INTRODUCTION

1.1   *Culture issues–related research*

Culture is a complex of factors that affect a person's unconscious behavior. Cultural variables are categories that organize cultural data (Evers, 1997). A cultural model compares the similarities and differences of two or more cultures by using dimensions of culture (cultural variables). Societal cultures reside in (often unconscious) values, in the sense of broad tendencies to prefer certain states of affairs over others (Hofstede, 2001, p. 5). Hofstede's cultural theory includes six dimensions in which cultural values could be analyzed: power distance, uncertainty avoidance, individualism versus collectivism, masculinity versus femininity, long-term versus short-term orientation, and indulgence versus restraint. This system was used to compare the cultures of the two groups of participants in the current research. The analysis data was collected through the web page Hofstede Insights where countries can be compared online. Our goal was to compare Taiwan and Russia in selected dimensions (https://www.hofstede-insights.com/).

In Figure 1, the results of Russia were blue, and Taiwan indicated by purple. In this research, only three of the six dimensions were used, namely, uncertainty avoidance, individualism, and masculinity. These three dimensions were chosen since others relate more to the analysis of organizations and workflow than to the study of cultural aspects related to visual recognition of signs.

1.2   *Communication theory–related research*

Communication theory explains communication as the transmission of information among people and information as knowledge. This theory has many interpretations from different

---

* Corresponding author: stshen@nfu.edu.tw

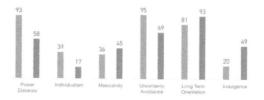

Figure 1.   Results of a comparative diagram of six cultural dimensions between Russia and Taiwan.

authors, but a lot of them include cultural and semiotic aspects. "The idea that different people, from differing social and cultural backgrounds, can actively interpret graphic designs in different ways cannot be accounted for except as a failure" (Barnard, 2013).

1.3   Semiotics–related research

Semiotics is one of the critical factors for current research because the main objects of study were groups of signs. Semiotics involves the study not only of what we refer to as signs in everyday speech, but of anything which "stands for" something else. In a semiotic sense, signs take the form of words, images, sounds, gestures, and objects (Chandler, 2007).

1.4   Cognitive psychology–related research

Cognitive perception includes many ways of receiving information, such as listening, smelling, seeing, feeling, etc. Decades of research in psychology tend to undermine the assumption that people's perception reflects what is happening in the world. There were plenty of researchers who tried to discover correlations in data analysis cultural comparison research between two or more cultures. The specific issues were different from time to time, but overall were significant for the current study. There were 13 signs to convey consumer safety information about a product which were tested by Davies, Haines, Norris, and Wilson in 1998. The signs were weak in terms of understanding by customers. "In fact, it appears to be very difficult to design highly representative symbols requiring no learning to convey public information such as consumer information, warnings, or information in public places" (Tijus et al., 2007). Fink and Laupase (2000) compared the perceptions of Web sites between Malaysian and Australian groups of participants as represents of Eastern and Western cultures. They found that "effectiveness was significantly higher, it was for sites originating in that group's country" (Fink & Laupase, 2000).

2   ONLINE PILOT STUDY AND DISCUSSION

2.1   Survey

This survey was conducted in August 2019 as an attempt to rethink the previous survey conducted in December 2018. The previous survey was not sufficient to provide the more-or-less definite conclusions needed for further work. The current survey was conducted via Google-forms.

2.2   Participants

The total number of participants was 59 people, which included two groups of participants: Taiwanese and Russian. The number of participants in the groups was unequal. There were 39 people in the Taiwanese group and 20 in the Russian group. The participants were recruited from the mailing list and the information post on the authors' Facebook. Female participants dominate in the Russian group, and the gender balance is roughly equal in the Taiwanese group. Also, in the author's opinion, an important aspect is that the Taiwanese participants were mainly up to 25 years old (65%), and the Russian group mostly included people from 25 to 35 years old (59%).

2.3 Preselected signs

There were 10 preselected signs tested with the primary meanings: "Healthcare system," "Blood donation," "Political party," "Railway," "National Electricity." The signs were divided into pairs by their meaning; each meaning has a representation in the form of Taiwanese and Russian signs as shown in Figure 2. All signatures in all languages were blurred in the questionnaire and all signs were randomly mixed.

2.4 Result

2.4.1 Healthcare signs

There were two localized signs related to healthcare. First, the National Health Insurance sign in Taiwan which is commonly used all around Taiwan. Second, the Russian sign used in one region of Russia as a sign of a healthcare system particular to this area. The Taiwanese group recognized the Taiwanese sign with 100% accuracy. Only 41% of the Russian group recognized the Taiwanese sign correctly and 52% regarded it as "Family center." The Russian sign was recognized by only 21% of Taiwanese participants and 26% of Russians. In both cases, the majority chose "Family center" as a correct answer (47% and 63%). The Taiwanese sign had a higher level of abstraction, according to visual analysis, which could be a reason for the low rate of correct answers in the Russian group. In previous research, it was mentioned that the survey (conducted in 2018) showed foreigners could not recognize the National Health Insurance sign as a healthcare service sign on the street and find the service they need (Ominina & Shen, 2019). The cause of low recognition for both signs by participants not from the original country of the sign may be the silhouettes of the people depicted there. This could cause trouble with its metaphoric association and not the transparent transmission of the information.

2.4.2 Electricity signs

The Taiwanese sign with a direct translation "use electricity efficiently" depicts an electricity plug and cable in the shape of the heart. However, there was no electricity efficient usage sign in Russia, hence the alternative sign of "usual electricity usage" was used for the survey. The Taiwanese sign was correctly identifies by 42% of the Taiwanese group, and 16% of the Russian group. Also, 47% of the Taiwanese group regarded thie sign as "Heart problems center." The positive result was for the Russian sign, which had a 97% recognition rate for the Russian group of participants, and 68% for Taiwanese. Usage of the heart shape could lead to the misunderstanding of the Taiwanese sign even for Taiwanese. At the same time, the sign represented as Russian is more well known. These factors could affect the recognition ability of participants in this research.

2.4.3 Blood donation

The sign "Taiwan Blood Services Foundation" was named as "Blood donation" in the survey. The Russian sign was "National donor center" named the same as the Taiwanese one ("Blood donation"). For the Taiwanese sign, 93% of Russian participants answered correctly, and only 79% of Taiwanese participants (the rest of Taiwanese participants chose "LGBT community" as a correct answer). The Russian sign had quite similar results (97%; 74%). The Taiwanese

Figure 2. The set of the pre-selected signs: "T" stands for Taiwanese and "R" for Russian.

sign is quite abstract with the heart shape and allusion to human silhouettes. It could lead to the disappointment of some Taiwanese participants who replied "LGDT community" (21%). Meanwhile, the Russian sign had a red cross and red (blood) drop behind the cross, which was distributed around the world, like signs related to medicine. This kind of sign decoding showed higher perception rate than a more abstract metaphor.

2.4.4 *Political party*
There were two signs related to political parties in Taiwan ("The Democratic Progressive Party" (DPP)) and Russia ("United Russia"). Both of them are the most prominent political parties in their countries. In the survey, the definition "Political party" was used. For the Taiwanese sign, the correct rate was high for the Taiwanese group (80%) and low for the Russian group (22%). The Russian group of participants regarded it as "National park" (70%). The Russian sign had high recognition rate among Russian participants (97%) and lower among Taiwanese (31%). The signs of the "Political parties" were very diverse from each other, for example, on the Taiwan sign there was nothing related to the flag of Taiwan; however, it shows the silhouette of the island of Taiwan. At the same time, the Russian sign included the outline of the Russian national flag color, and the image of a bear was depicted, apparently implying a symbol of the party. These two signs demonstrated that localized signs, even with the visual hints, could go unrecognized by people.

2.4.5 *Railway signs*
Both Railway signs were represented: Taiwan Railways Administration and Russian Railways in Russia. The Taiwanese railway sign was recognized by only 33% of Russian participants. The rest of the answers were split into two other results (Political party [38%] and Repair shop [30%]). The Taiwanese group of participants recognized the sign at 95%. The Russian railway sign had a rate of 100% for the Russian group of participants. The Russian sign for the Taiwanese group of participants was 53% as the correct rate. Railway signs were widely used and common signs in the countries to which they belong. In both cases, the signs were quite abstract and did not have any figurative image on them related to their original meaning. These signs were used in the survey, most likely not to test their visual effectiveness, but to check the difference in perception of local signs on audiences from other countries. The results proved expected results that common local sign does not work for people from other countries.

3 DISCUSSIONS

Based on the results of the signs of "Healthcare", "Electricity," and "Blood Donation," we found that excessive detailing of the signs can lead to a low level of recognition of the sign, both among the local population and among non-original countries of the sign. Balcetis and Dunning (2006) also conducted similar studies which they found "provided converging evidence to suggest that participants' desires, hopes, fears, or wishful thinking led them to perceive a representation of the visual environment they desired" (Balcetis and Dunning, 2006). Taiwan is a collectivistic type of society, according to Hofstede's cultural dimensions. People in this type of society consider themselves as a part of a group and take responsibility for other members. This cultural specific could be a reason for the use of a human image and heart shape in the Taiwanese signs. "Blood donation" and "Electricity" signs, which were close to standardized and world-common signs, have shown the best results for both groups of participants. This was important in this research, because it showed the need to design sets of universal signs.

For "Railway" and "Political party", it was apparent that local signs did not work for foreign users. Despite the widespread use of signs in the original country, the signs did not show a high recognition rate in the result. In general, it could be seen that for the Russian group, the detail signs ("Healthcare", "Blood donation", "Electricity") were more complicated, rather than for the Taiwanese group. The reason could be partly explained in the Fink and Laupase's research (2000) where Australians (Western culture) prefer an environment of low context and highly explicit communications, while Asians operate in an environment of high context that stresses implicit communications. The challenging part of the survey was the fact that all signs were used without original description captions. This was done on purpose, to keep the objectivity of the survey. There were opinions that pictures, signs, and pictograms were clearer in collaboration with a textual description of the meaning of a sign. For example, Barnard (2013) highlighted that "without language, the image would not be experienced in any meaningful or communicable way at all and could, therefore, hardly be described as an experience at all." This section showed that the efficiency of information transfer by the signs without captions of the sign meaning was low for people not from the country where the sign originated. We learned that it was hard to highlight exact correlations between cultures in sign recognition, due to the complexity of the process. Some aspects of the signs (such as abstractness, prevalence, cultural orientation, and textual part of the sign) were outlined as less, rather than more effective.

4 CONCLUSION

The difference in sign perception among international participants was evident. Conjectures such as inefficiency in the usage of the signs with lack of a text have been addressed. Further work aims to solve what was found in this and previous surveys. First, the use of standardization as a basis for signs could be helpful. Second, the usage of text captions would be useful, preferably in English language, based on David Crystal's 1997 book: language can be recognized as global if "a language can be made a priority in a country's foreign-language teaching, even though this language has no official status"; which obviously could be English for most of the countries (Crystal, 1997).

Taiwan, under the influence of globalization, is becoming a popular destination to move for a permanent place of residence and tourism; a desirable sign system shall be made universal and unambiguous in meaning for people. Future surveys should be iteratively conducted to test these designed signs for their usability and correct recognition rate.

REFERENCES

Balcetis, E., & Dunning, D. (2006) "See what you want to see: Motivational influences on visual perception. *Journal of Personality and Social Psychology, 91*(4), 612.
Barnard, M. (2013). *Graphic Design as Communication*. Routledge.
Chandler, D. (2007). *Semiotics: The Basics*. Routledge.
Crystal, D. (2012). *English as a Global Language*. Cambridge University Press.
Evers, V. (1997). "Human computer interfaces: Designing for Culture." Unpublished MS Dissertation, University of Amsterdam, Netherlands.
Fink, D., & Laupase, R. (2000). "Perceptions of web site design characteristics: A Malaysian/Australian comparison." *Internet Research, 10*(1), 44–55.
Hjelm, S. I. (2002). *Semiotics in Product Design*. CID. https://www.hofstede-insights.com/
Hofstede, G. (2001). *Culture's Consequences: Comparing Values, Behaviors, Institutions and Organizations across Nations*. Thousand Oaks, CA: Sage (co-published in the PRC as Vol. 10 in the Shanghai Foreign Language Education Press SFLEP Intercultural Communication Reference Series, 2008).
Ominina, M., & Shen, S-T. (2019, May). "Investigating non-native Chinese speakers' adoption of using localized pictograms in Taiwan." In Engineering Innovation and Design: Proceedings of the 7th International Conference on Innovation, Communication and Engineering (ICICE 2018), November 9–14, 2018, Hangzhou, China (p. 216). CRC Press.
Tijus, C., Barcenilla, J., De Lavalette, B. C., & Meunier, J. G. (2007). *The Design, Understanding and Usage of Pictograms: Written Documents in the Workplace* (pp. 17–31). Brill.

Cultural & Creative research

Innovation in Design, Communication and Engineering – Lam et al. (eds)
© 2020 Taylor & Francis Group, London, ISBN 978-0-367-17777-5

Evaluation of the attractiveness of commercial recording studios in Taiwan

Chian-Fan Liou & Hung-Yuan Chen
Department of Visual Communication Design, Southern Taiwan University of Science and Technology, Yungkang, Tainan, Taiwan

Chao-Chih Huang*
Department of Popular Music Industry, Southern Taiwan University of Science and Technology, Yungkang, Tainan, Taiwan

ABSTRACT: In recent years, the increasing processing power of personal host computers and the increasingly powerful functions of software and hardware for digital audio processing, means many previous music productions that would have required an expensive budget to rent commercial recording studios for working, can now completed by consumers doing almost the same audio production tasks to their own satisfaction, to some extent, on a budget. However, what is the attractiveness of commercial recording studios to consumers in terms of audio recording, editing, mixing, or mastering for music production? It is a research topic worthy of exploration to see why consumers are willing to spend considerably higher recording studio rental fees and recording engineers' employment fees without handling the production tasks themselves at home. Accordingly, this article performs in-depth interviews and the evaluation grid method to investigate the attractiveness of commercial recording studios in Taiwan. A subjective evaluation survey of commercial recording studio attractiveness and a linear statistical analysis are then used to infer the critical attractiveness factors that determine consumers' psychological expectations toward commercial recording studios. The findings of this research are expected to serve as a useful source of reference for commercial recording studio organization teams in preparing future studios. Moreover, the study methods and procedural concepts employed herein can be extended to other academic or practical investigations of attractiveness.

1 INTRODUCTION

With the advent of the era of digital technology, the work and consumption patterns of the digital audio recording engineering industry and its practical application process have been changed. Consumers are beginning to try to deal with some of the professional tasks that had to be processed in commercial studios in previous years, such as recording, editing, mixing, and mastering. But different from the traditional audio recording engineering processes in the past, recent consumer demands in digital audio recording engineering, with hardware and software, start paying attention to the self-mastery of home or personal studio operations, as well as gradually paying attention to the cooperative experience with other music creators or audio engineers. It is worth noting that once home or personal studios' cooperation modes are successfully developed, the overall work efficiency might be no less than the result of working in commercial studios.

The convenience provided by the development of digital audio hardware and software allows consumers to choose to work at home or personal recording studios, which not only

* Corresponding author: z7m@stust.edu.tw

affects the rental benefits of the commercial studios, but also the direction and scale of purchasing digital audio software and hardware in home or personal recording studios and commercial recording studios. Therefore, for commercial recording studio operators, the environment and condition of their audio software and hardware and replaceability and reproducibility issues are worth discussing, whilst consumers have their own audio software and hardware in certain levels, but why still willing to plan additional budgets to rent and experience the one-time service of commercial recording studios. Thus, these topics are worth exploring.

2 EXPERIMENTAL METHOD

In recent academic research, many scholars have proposed relevant research on attractiveness, such as consumer taste perceptions (Lin, Hoegg, & Aquino, 2018) and visual research (Rossit et al., 2019). However, there are not many studies focused on the attractiveness and psychological expectations (PE), of commercial recording studios to consumers. If appropriate research procedures can be developed to look at the attractiveness of commercial recording studios, and the PE to consumers, the results will help commercial recording studio operators with subsequently preparation and operation strategies. There are many research methods for the study of attractiveness. One methods, the evaluation grid method (EGM) (Onoue et al., 2017; Park et al., 2016), is a method based on in-depth interviews and a three-layer grid table consisting of the "original layer," the intuitive attractiveness evaluations and impressions; the "upper layer," the abstract attractiveness with psychological expectation associations' and the "lower layer" of the descriptions of the attractiveness types. This study is based on the EGM and descriptive statistics (Libman, 2010; Marshall & Jonker, 2010) to explore the attractiveness of commercial recording studios to consumers, and the psychological expectations of consumers.

This study selected 30 commercial studios active in Taiwan for the last five years. The information is shown in Table 1, and an A4 format paper card is produced for each commercial studio, with one main representative photo of the studio on the paper card, and two photos of the scene environment or equipment, as shown in Figure 1, as the visualization card for each studio, to provide the respondents for recall purposes in the EGM in-depth interviews.

In this study, 15 consumers, who had participated in music production work more than three times in commercial studios, and with the age between 20 and 35 years old, were invited as respondents to be guided through the following procedures to express their evaluations to the attractiveness of commercial studios. The interview procedures were as follows:

(1) The respondents were asked to check the 30 commercial studio visualization cards, and select one to three cards of commercial studios where they had participated in any kind of music production tasks.
(2) With the aforementioned selected cards, respondents were asked to rank the cards from high to low according to their preferences.
(3) For the selected preferred cases, starting from the case with the highest preference evaluation, respondents were asked to provide one to three main intuitive reasons or impressions (intuitive attractiveness evaluations) of the commercial studios (original-layer).
(4) The respondents were further asked to explain the most one to three specific details of the intuitive reasons (lower-layer), and one to three of the most important abstract feelings to the commercial studios (upper-layer).
(5) Every respondent followed the aforementioned interview procedures, and records were made with all respondent's responses to construct the individual associated three-layer hierarchical diagrams for final statistics.

Table 1.   Studio lists.

No	Studios
1	MuScene Studio
2	Beautiful Recording Studio
3	ff studio
4	Mega Force Studio
5	Yu Chen Recording Studio
6	B-mic Studio
7	Music People Studio
8	Skawa Music Studio
9	Wei Studio
10	Huang Shuai Studio
11	Sense Sound Studio
12	Cozi Sound
13	DIO Studio
14	Golden Apple Studio
15	Shihor Studio
16	Nil-Ravine Sound Assembly Studio
17	Under Ground Studio
18	Eli Recording Studio
19	Dynasty Recording Studio
20	Music Dictionary Studio
21	Listen Studio
22	DAP Studio
23	In-Joy Studio
24	PM Music Studio
25	M.y Studio
26	Platinum Studio
27	168ch Studio
28	Singfon Studio
29	Nice923 Boss Studio
30	Heart Music Studio

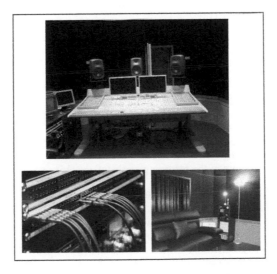

Figure 1.   Sample of visualization card for in-depth interviews.

3 RESULTS AND DISCUSSION

According to the EGM-based in-depth interviews of the 15 respondents, the cumulative numbers of responses were obtained in each form through the three-layer hierarchical diagrams. The results are as shown in Table 2, Table 3, and Table 4: the "upper-layer" abstract consumer psychological expectations were divided into four groups: "Faith of trust (24)," "Participation and self-awareness (12)," "Sense of professional and technological (22)," and "Sense of mission inheritance (15)." The numbers in the brackets of each psychological expectation indicate the numbers of times the same and similar associated responses are consolidated. The lower-part contents are the integrated psychological expectation descriptive responses. Evaluations of the intuitive attractiveness of the "original-layer" were divided into five parts: "A1: The novelty of studio equipment's hardware and software (28)," "A2: Engineers' (team) trait charm (33)," A3: Interactive feedback (15)" "A4: Studio images and environmental conditions (17)," and "A5: Trust and task stability (13)". Each intuitive attractiveness factor has two or three "lower-layer" concrete attractiveness types of descriptions.

Figure 2 shows that for the original-layer: intuitive attractiveness factors, studio customers first focused on factor A2, engineers' (team) trait charm (33), which includes considerations of the operator's (team) style, and recording, editing and mixing engineers' professionalism. Secondly, studio customers focused on factor A1, the novelty of studio equipment's hardware

Table 2 . Three-layer hierarchical diagram, original-layer.

Original-layer: intuitive attractiveness factors
A1: The novelty of studio equipment's hardware and software (28)
Studio equipment, studio software, studio hardware, sound quality, overall novelty…
A2: Engineers' (team) trait charm (33) Operator (team) style; recording, editing and mixing engineers' professionalism…
A3: Interactive feedback (15) Interaction with customers, response to commissions, on-site work atmosphere handling…
A4: Studio images and environmental conditions (17) Studio images, workspace design styles, publicity, and traffic convenience…
A5: Trust and task stability (13) Trustworthy, safe, stable, passionate…

Table 3. Three-layer hierarchical diagram, lower-layer.

Lower-layer: concrete studio attractiveness types
A11: Windows-based recording system (17) A12: Mac-based recording system (23) A13: Selections of the core of the recording workstations (15) A21: Mature and steady route (12) A22: Theoretical and professional route (30) A23: Thinking innovation route (18) A31: Direct music production advices provided (17) A32: Indirect music production opinions provided (13) A41: Cultural retro style (9) A42: Modern fashion style (26) A43: Community neighborhood style (17) A51: Reasonable charge (13) A52: Technical professional orientation (18)

Table 4.   Three-layer hierarchical diagram, upper-layer.

Upper-layer: abstract consumer psychological expectations

Faith of trust (24)
Stableness, audio-visual satisfaction, realism, experienced, feelings of enthusiasm, convenience with close distance...

Participation and self-awareness (12)
Sense of belonging, recognition with the operator, commonality of software and hardware using habits, resonance, self-realization and satisfaction...

Sense of professional and technological (22)
Professional equipment, novel equipment, well-maintained equipment, sufficient quality and quantity in software and hardware...

Sense of mission inheritance (15)
Familiar processing techniques new and old crossed techniques, technological innovation, retro style, classical sense...

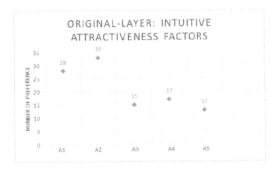

Figure 2.   Statistic data, original-layer: intuitive attractiveness factors.

and software (28), which include the considerations of checking the studios' equipment, software, hardware, sound quality, as well as the overall novelty.

From Figure 3, it can be seen that the number one concrete studio attractiveness type for studio customers was A22, the theoretical and professional route (30), which is part of the attractiveness factor of A2, the engineers' (team) trait charm (33). The number two concrete studio attractiveness type for studio customers was A42, modern fashion style (26), which belonged to the attractiveness factor of A4, the studio images and environmental conditions (17), which include the customers' considerations with studio images, workspace design styles, publicity, and traffic convenience.

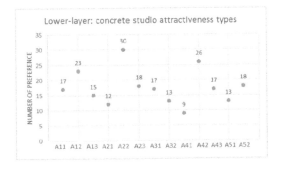

Figure 3.   Statistical data, lower-layer: concrete studio attractiveness types.

191

Figure 4 shows that for abstract consumer psychological expectations, customers first focused on the topic of faith of trust (24), which includes considerations of stableness, audio-visual satisfaction, realism, experienced, feelings of enthusiasm, and convenience of proximity. The second abstract consumer psychological expectation was the sense of the professional and technological (22), which includes considerations of professional equipment, novel equipment, well-maintained equipment, sufficient quality and quantity in software and hardware.

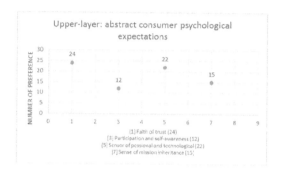

Figure 4.   Statistic data, upper-layer: abstract consumer psychological expectations.

4   CONCLUSION

In this article, the study introduced EGM-based in-depth interviews to survey consumers to express their evaluations of the attractiveness of commercial studios by using the collected information of 30 commercial studios active in Taiwan for the last five years, then statistically induce the results of consumers' views with commercial studios in Taiwan: (1) intuitive attractiveness evaluations of the commercial studios (original-layer), (2) specific concrete details of the intuitive reasons (lower-layer), and (3) abstract feelings to the commercial studios (upper-layer). The research findings and future suggestions are as follows:

(1) There are five intuitive attractiveness factors (original-layer), the novelty of studio equipment's hardware and software, engineers' (team) trait charm, interactive feedback, studio images and environmental conditions, and trust and task stability. Each factor accompanys two or three concrete studio attractiveness types (lower-layer).
(2) For the abstract consumer psychological expectations (upper-layer), faith of trust, participation and self-awareness, sense of professional and technological, and sense of mission inheritance are most considered expectations by studio consumers.
(3) This study introduced EGM-based in-depth interviews as the main research method. For future research, it is suggested that maybe researchers can apply with some other survey methods, such as VIKOR (VlseKriterijumska Optimizacija I Kompromisno Resenje) or SAW (Simple Additive Weighting) methods to check out more research results.

REFERENCES

Libman, Z. (2010) "Alternative assessment in higher education: An experience in descriptive statistics." *Studies in Educational Evaluation, 36*(1–2), 62–68.
Lin, L., Hoegg, J., & Aquino, K. (2018) "When beauty backfires: The effects of server attractiveness on consumer taste perceptions. *Journal of Retailing, 94*(3), 296–311.

Marshall, G., & Jonker, L. (2010) "An introduction to descriptive statistics: A review and practical guide." *Radiography*, *16*(4), e1–e7.

Onoue, Y., Kukimoto, N., Sakamoto, N., Misue, K., & Koyamada, K. (2017) "Layered graph drawing for visualizing evaluation structures." *IEEE Computer Graphics and Applications*, *37*(2), 20–30.

Park, J., Kim, J.-H., Park, E.-J., & Ham, S. M. (2016) "Analyzing user experience design of mobile hospital applications using the evaluation grid method." *Wireless Personal Communications*, *91*(4), 1591–1602.

Rossit, D. G., Vigo, D., Tohmé, F., & Frutos, M. (2019) "Visual attractiveness in routing problems: A review." *Computers & Operations Research*, *103*, 13–34.

Innovation in Design, Communication and Engineering – Lam et al. (eds)
© 2020 Taylor & Francis Group, London, ISBN 978-0-367-17777-5

What personality traits drive distinguished craft artists' career success: Those devoting to crafts over 50 years

Jeng-Chung Woo
Department of Industrial Design, FuJian University of Technology, China

Chun-Ho Lu
Department of Arts and Plastic Design, Taipei University of Education, Taiwan

Artde Donald Kin-Tak Lam
Department of Industrial Design, FuJian University of Technology, China

ABSTRACT: This study examined the effects of personality traits on distinguished craft artists' career development. Previous winners of Taiwan's Crafts Achievement Award (mean age = 75.67, professional experience > 50 years) and 363 general artists were participants in the study. The Big-Five Mini-Markers was employed to evaluate the participants' personality traits. To analyze highly recognized personality traits among the participants and related factors, this study adapted the five stages of career development proposed by Super and divided the participants' creative arts careers into five stages, namely growth, exploration, establishment, maintenance, and disengagement. Quantitative and qualitative research methods were combined to summarize the research findings, which could serve as a reference for policy implementation and as guidance for prospective craft artists.

Keywords: Artist, Big five, Narrative research, Career development, Chinese intangible cultural heritage

1 INTRODUCTION

Mainland China and Taiwan have diligently preserved and promoted Chinese craft arts for decades. In addition, the Taiwanese Government has been continuously concerned with craft arts in its efforts to implement policies to preserve cultural heritage. Moreover, related incentives such as selections for National Important Folk Artists, National Heritage Award, and Crafts Achievement Award have provided skilled craft artists with an understanding of the importance of their long-term dedication to craft arts. Craft arts are no longer merely a context of daily life and are regarded as a form of the intangible cultural heritage of the nation and society. Many aspects of Chinese intangible cultural heritage such as preservation, inheritance, and promotion are facing unprecedented challenges. Efforts to preserve and pass on intangible folklore arts before their extinction are imperative for the promotion of relevant cultural measures.

Previous winners of Crafts Achievement Award (sponsored by the National Taiwan Craft Research and Development Institute, an agency under the Taiwan Ministry of Culture) were recruited as participants in this study. Through a theoretical research design and literature review, this study conducted an in-depth investigation of the effects of distinguished craft artists' personality traits on their career development before deducing quantitative and qualitative analytical findings as the research outcomes. These distinguished craft artists' personality traits and the factors

---

* Corresponding author: wwwjc2000@yahoo.com.tw, yoyolu_@outlook.com, artde.lam@qq.com

of their acknowledgment may serve as a reference for policymakers and those pursuing a career in craft arts.

2 LITERATURE REVIEW

The Taiwanese Government has implemented numerous plans for national cultural policies. The Cultural Heritage Preservation Act promulgated in 1982 designates traditional craft arts as folklore arts to facilitate the development of preservation policies to promote cultural heritage coherence in Taiwan. In addition, the Taiwanese Government has adapted Japan's "living national treasure" system to honor outstanding folklore artists as *yishi* (masters). The Crafts Achievement Award is a national-level prize. A winner has been selected annually since its establishment in 2007. It is the highest-ranking craft award in Taiwan. The emphasis in judging is on artists with more than 30 years' experience who have demonstrated outstanding achievements and contributions in the craft field. Overall, eight sessions were held during the implementation of this study in 2015. Figure 1 shows the first 8 Crafts Achievement Award winners since the award's inception.

Figure 1.   First 8 crafts achievement award winners (2007-2014).

Allport defined personality as "the dynamic organization within the individual of those psychophysical systems that determine his unique adjustments to his environment" and argued that personality should encompass an individual's physiological characteristics. Individuals' behaviors reflect their personality characteristics [1]. When these characteristics continue to surface under specific circumstances, they are regarded as personality traits. Therefore, personality traits are consistent and crucial components throughout an individual's entire life [2]. The five-factor model of personality proposed by McCrae and Costa [3] has made remarkable progress in recent years. Over the past half century, the most crucial individual differences in personality traits have been consistently divided into five traits. The Big Five personality traits, one of the most common methods of classifying personality traits [4-5], are extraversion, openness to experience, emotional stability, conscientiousness, and agreeableness. Super divided career development into the following five stages according to age: growth, exploration, establishment, maintenance, and disengagement [6]. Each stage contains various developmental tasks, each of which exhibits the cycle of the aforementioned five stages. Understanding trajectories of vocational interest development is important for understanding how to maximize job satisfaction, job performance, and individual well-being [7].

3 PARTICIPANTS AND METHODOLOGY

3.1 *Participants*

This study recruited two groups of participants. The Group A is the distinguished craft artists who are previous winners of the Crafts Achievement Award, and the Group B is the general artists, including college students, teachers of the art design department and artists. The award has honored one winner each year since 2007, and the first 8 winners were recruited as participants in this study.The two eldest award winners declined to be interviewed in this study, citing health reasons. The six participants in Group A had a mean age of 75.67 ± 4.84 years (Table 1), and the 363 participants in Group B had a mean age of 28.84 ± 7.73 years (Table 2).

Table 1. Group A participants' background.

Subject	A	B	C	D	E	F	
Age	70	80	71	73	80	80	75.67±4.84 (Mean age ± SD)
Age of beginning practicing craft arts	14	25	22	14	18	16	

Table 2. Group B participants' background.

Group B	n	Mean age ± SD
Group B1 (age of 18-30)	248	24.45 ± 3.32
Group B2 (age of 31-40)	80	35.24 ± 2.91
Group B3 (age of 41-52)	35	45.31 ± 4.47
Total	363	28.84 ± 7.73

3.2 Methodology

Although Super classified the five stages in relation to age, the interview outline devised by for the current study did not comply with age restrictions and instead classified the five stages according to the participants' growth, exploration, establishment, maintenance, and disengagement periods from the beginning of their first encounter with craft arts [6]. After the career development literature [8-9] review had been conducted, the interview outline was developed. In-depth interviews were conducted based on this interview outline, and the Traditional Chinese version of the International English Big-Five Mini-Markers [10] was employed to assess the interview content. For the formal questionnaire survey, participants were required to rate each question on a 5-point scale (from 1 = strongly disagree to 5 = strongly agree). The second author conducted interviews and questionnaire sampling in Group A; the participants in Group B only received the questionnaire. The strategy of narrative research is to emphasize events experienced by the subjects, such as their development processes and histories of individual or group lives. Storytelling is the primary method for understanding and presenting the progress of events, individuals, or group history. The holistic–content analysis [11] in this study focused on the content of the participants' complete life stories and was supplemented by other data such as the participants' background information, media reports, and related documents for further induction. Finally, the texts were compared with and verified against one another to achieve triangular validity.

4 RESULT AND DISCUSSION

4.1 Narrative analysis of distinguished craft artists' personality traits

Table 3 shows an indexing example and the meaning of each code. For example, A-I-06 denotes Participant A's speech recorded on page 6 of the interview transcript. This study analyzed the factors of distinguished craft artists' personality traits through narrative research. Table 4 shows an example of the indexing process for extraversion.

This study summarized the characteristics of distinguished craft artists' personality traits as follows:

(a) Introversion and a quiet nature; focused and dedicated: Because the participants were generally introverted and inactive, they were more likely to remain focused when learning

Table 3. Definitions of indexed symbols.

Symbol	1	2	3
Meaning	Participant code	Data category	Serial number Page number on transcript
Content	A-F	I: Interview M: Biography or memoir O: Other document (e.g., media report or documentary) P: Research log	Serial numbers starting with 1

P.S. Em dashes are included between symbols.

Table 4. Indexing process for extraversion.

Topic	Concept	Content
Extraversion	Introverted and quiet	"I was very shy before and did not talk much (What period does 'before' refer to?). 'Before' means when I was young. Because of stress in daily life, and by stress I do not mean a sense of inferiority, I always felt I was different from everyone else. During my school days, everyone enjoyed college life, played with peers, exercised, or engaged in romantic relationships. However, I had to work every day, and naturally I was more introverted than everyone else..." (B-I-08) "I have been very introverted since I was young. I guess it is not easy to change your personality!" (C-I-15) "In fact, I am an introverted person who does not talk much. Now I have changed and talk more, but I was relatively reserved before. (Do you talk more now because you are forced to?) Yes, sometimes I have to give speeches. I feel I am somewhat forced into gradually developing my courage to talk." (D-I-05) "Actually, I think most artists are relatively introverted. Like you said, extroverted people are impatient in learning our techniques and are easily distracted..." (F-I-04)
	Focused and dedicated	However, I have spent a considerable amount of time and number of trials testing aspects such as thickness and the order of each layer of glaze. You can only seek desired results such as delicate, simplistic, bright, and vivid colors through continual failure. This is also how many artists present their unique styles." (C-M-02) "...If you are extroverted and wander around every day, you work few hours a day. Second, you cannot focus or enhance your skills." (F-I-04) "Previously, it has been said that it takes 3 years and 4 months to learn a craft art skill. Our zhuangfo skill takes at least 10 years to master to a degree where the final products will not receive criticism." (F-I-05)

craft art skills. In addition, all the participants were dedicated to one particular craft art throughout their lives. This phenomenon reflects the following quotation from famous French educator Jean-Jacques Rousseau: "He who wills the end wills the means also,"

(b) Innovativeness: Those who engage in creative arts are usually imaginative, excellent at multifaceted thinking, and courageous and ambitious in terms of new experiences. This generalization also applies to craft artists. These personality traits can be observed among the participants in this study.

(c) Open-mindedness: The participants credited their accomplishments to their mentors sharing the skills that they possessed. In addition, the participants expressed willingness to pass on the skills they had learned throughout their lives to younger artists to extend craft art practices to future generations.

(d) Optimism and tolerance: The participants had all been attracted to artistic and aesthetic objects since childhood, at which time, creative arts were regarded as a profession without future or fortune. Nevertheless, the participants listened to their hearts and advanced in their journeys of creative works without hesitation. Despite low income, they enjoyed producing creative works and did not complain of financial hardship.

(e) Calmness and steadiness: Slumps and severe stress inevitably occurred in their seemingly satisfactory lives; however, despite countless such setbacks, the participants remained calm and steady to gradually overcome every challenge.

(f) Confidence: Because of their love of art, they have diligently devoted themselves to their fields of specialty. Some participants believed that from an early stage, they had been destined to engage in craft arts as a profession. Their diligence and effort yielded sound fundamental skills and professionalism alongside positive acknowledgment, which boosted their confidence to continue moving forward.

(g) Self-demand: Rigorous demand on the self is a decisive factor in acknowledging a distinguished craft artist. The participants have all maintained self-discipline and highly rigorous attitudes throughout every step from learning craft art skills to generating artworks; they have been challenging themselves and others for decades. Furthermore, they have continually tested their craft art levels and have overcome challenges to progress by listening to others' advice and accumulating experience.

(h) Perseverance: The participants attributed their winning of the Craft Achievement Award primarily to their perseverance. Because of their perseverance in art, creation, and perfection in relation to every step, method, and material, they remained diligent in refining their professional skills and artistic literacy throughout the relatively satisfying establishment stage despite failures during the exploration stage.

(i) Humility: Although the participants are regarded as leaders in their respective fields of specialty, they exhibited no arrogance regarding their accolades and accomplishments. This phenomenon reflects the following quotation from Isaac Newton: "If I have seen further it is by standing on the shoulders of Giants."

(j) Grace and friendliness: According to their experiences, the participants encouraged youngsters who intended to practice craft arts to be persistent and optimistic in addition to cultivating the required professional skills. The participants graciously and warmly reminded upcoming artists that a positive attitude and confidence are crucial to overcoming adversity. They answered every question in detail and without reservation, and their gracious and friendly attitudes demonstrated the so-called "master style."

4.2 Statistical analysis of personality traits of distinguished craft artists and general artists

Table 5 shows the descriptive statistics of the personality trait scale. The average scores for the distinguished craft artists' responses to the five dimensions (extraversion, openness to experience, emotional stability, conscientiousness, and agreeableness) were 22.5, 29, 29.5, 30.5, and 32.83, respectively, whereas the average scores for the general artists' responses were 28.21, 30.10, 26.87, 30.35, and 32.31, respectively. The distinguished craft artists had the lowest

Table 5. The participants' descriptive statistics of personality traits.

Group	Subgroup	n	Means				
			Extraversion	Openness to experience	Emotional stability	Conscientiousness	Agreeableness
The distinguished craft artists		6	22.5	29	29.5	30.5	32.83
The general artists	G1	248	28.17	29.81	26.81	30.14	32.09
	G2	80	28.61	30.81	27.09	30.73	32.46
	G3	35	27.66	30.57	26.83	31.00	33.51
	Total	363	28.21	30.10	26.87	30.35	32.31

average score in extraversion and the highest in agreeableness. By contrast, the general artists had the lowest average scores in emotional stability and extraversion and the highest average score in agreeableness. Moreover, the distinguished craft artists and the general artists have a greater amount of difference in extraversion and agreeableness. The average scores for the distinguished craft artists's responses to extraversion and emotional stability were 22.5 and 29.5, whereas the average scores for the general artists's responses were 28.21 and 26.87. Therefore, the distinguished craft artists were more introverted and emotionally stable than the general artists were.

The general artists were divided into three subgroups, namely G1 (aged 18–30 years), G2 (aged 31–40 years), and G3 (aged 41–52 years). As the analysis of variance results were shown in Table 6, the personality traits of extraversion (F = .924, p > .05), openness to experience (F = 2.120, p > .05), emotional stability (F = .415, p > .05), conscientiousness (F = 1.359, p > .05), and agreeableness (F = 2.312, p > .05) were nonsignificant in the general artists. This revealed that there were no significant differences among the subgroups of the general artists in average scores for the Big Five personality traits. This means that the general artists of all ages have consistent personality traits.

4.3 Discussion

This study found that the general artists of different age groups did not exhibit significant differences in the Big Five personality traits, indicating that general artists of all ages have consistent personality traits [12]. Among these, emotional stability and extraversion were relatively low, which is consistent with the literature. Additionally, the general artists possessed high agreeableness.

"As a young chap he (Van Gogh) had been slightly morose and had avoided companionship...." [13]. In addition, although the internationally renowned Japanese artist Yoshitomo Nara is known for depicting large-eyed evil dolls in his works, Nara is an extremely introverted and shy person [14]. This study found differences between the distinguished craft artists and general artists in extraversion and emotional stability. The distinguished craft artists were more introverted than the general artists were, and they were more emotionally stable. This finding is consistent with the personalities of distinguished artists such as Vincent van Gogh and Yoshitomo Nara, both known for their introvertedness. In addition, past studies show that people's happiness and emotional regulation increase with age [15-16]. Gibson showed that age was positively correlated with positive moods (vigor, friendliness) and negatively correlated with negative emotions (depression, anger etc.)[17]. The

Table 6.   Summary of one-way analysis of variance results regarding the effects of the general artists on personality traits.

Personality traits		Sum of squares	df	Mean square	F	p
Extraversion	Between Groups	24.145	2	12.072	.924	.398
	Within Groups	4703.095	360	13.064		
	Total	4727.240	362			
Openness to experience	Between Groups	69.760	2	34.880	2.120	.122
	Within Groups	5923.469	360	16.454		
	Total	5993.229	362			
Emotional stability	Between Groups	4.846	2	2.423	.415	.661
	Within Groups	2102.069	360	5.839		
	Total	2106.915	362			
Conscientiousness	Between Groups	36.855	2	18.427	1.359	.258
	Within Groups	4882.010	360	13.561		
	Total	4918.865	362			
Agreeableness	Between Groups	64.765	2	32.382	2.312	.101
	Within Groups	5042.679	360	14.007		
	Total	5107.444	362			

Table 7. Analysis of distinguished craft artists' five personality traits.

Dimension	Means	Concept of personality traits generalized in narrative research
Extraversion	22.5	Introversion and a quiet nature/focus and dedication
Openness to experience	29	Innovativeness/open-mindedness/optimism and tolerance
Emotional stability	29.5	Calmness and steadiness/confidence
Conscientiousness	30.5	Self-demand/perseverance
Agreeableness	32.83	Humility/grace and friendliness

Source: Compiled by this study.

aforementioned arguments may explain why the distinguished craft artists' exhibited higher emotional stability in comparison to the general artists in this study.

This study combined qualitative and quantitative research methods to analyze the five dimensions of extraversion, openness to experience, emotional stability, conscientiousness, and agreeableness. Extraversion had the lowest average score (22.5), whereas agreeableness had the highest (32.83). Moreover, openness to experience (29), emotional stability (29.5), and conscientiousness (30.5) were all positive. Table 7 demonstrates the similar results of the qualitative and quantitative approaches. This study provides further support for the effectiveness of Big-Five Mini-Markers scales.

5 CONCLUSION AND SUGGESTIONS

The results of this study are as follows:

(a) Introversion and a quiet nature foster focus and dedication; openness to experience cultivates the courage to innovate; and personality traits such as emotional stability, confidence, conscientiousness, perseverance, agreeableness, grace, and friendliness enable distinguished craft artists to reach notable career achievements.
(b) The distinguished craft artists had the highest score in agreeableness (32.83) and lowest in extraversion (22.5). Emotional stability, openness to experience (29), and conscientiousness (30.5) were positive, and the distinguished craft artists were more introverted and emotionally stable than the general artists were.
(c) General artists in different age groups exhibited no significant differences in the Big Five personality traits, indicating that they had consistent personality traits, among which agreeableness was the highest (32.31), whereas emotional stability (26.87) and extraversion (28.21) were relatively low.

The numbers of teaching artists at school are increased because of the stimulation by social demands for innovative educational practices. These artists contribute to the development of students' creative imaginations beyond conventional classroom practices [18]. In addition, successful people are able to inspire others regardless of context [19]. Therefore, this study recommends that government agencies continue to expand promotional policies similar to the Crafts Achievement Award as an intangible power to facilitate the longevity of intangible cultural heritage. This study is based on the analysis of the personality traits of Chinese artists, and the results cannot be inferred to other cross-cultural groups.

ACKNOWLEDGMENT

This work was supported by Research Center for Design Innovation of Humanities and Social Sciences, Fujian Province, and Fujian University of Technology [grant numbers GY-S18084, 2018].

REFERENCES

[1] Allport, G. W. (1937) *Personality: A psychological interpretation*. New York: Holt, Rinehart, & Winston.

[2] Costa, P. T. & McCrae, R. R. (1992) Four ways five factors are basic, *Personality and Individual Differences*, Vol. 13, No. 6, pp. 653–65.

[3] McCrae, R. R. & Costa, P. T. (1987) Validation of the five-factor model of personality across instruments and observers, *Journal of Personality and Social Psychology*, Vol. 52, No. 1, pp. 81–90.

[4] Dale, L. & Harrison, D. (2017) How the Big Five personality traits in CPSQ increase its potential to predict academic and work outcomes (online). Available at: http://www.admissionstestingservice.org/images/419493-the-big-five-personality-traits-in-cpsq.pdf (accessed 18 August 2014).

[5] Soldz, S. & Vaillant, G. E. (1999) The Big Five personality traits and the life course: A 45-year longitudinal study, *Journal of Research in Personality*, Vol. 33, pp. 208–32.

[6] Super, D. E. (1984) Career and life development, in D. Brown & L. Brooks [Eds] *Career choice and development*. San Francisco: Jossey-Bass, pp. 192–234.

[7] Schultz, L. H., Connolly, J. J., Garrison, S. M., Leveille, M. M. & Jackson, J. J. (2017) Vocational interests across 20 years of adulthood: Stability, change, and the role of work experiences, *Journal of Research in Personality*, Vol. 71, pp. 46–56.

[8] Chang, C. C. (2014) *A study of the career development of excellent aboriginal professional baseball player* (Master thesis). Southern Taiwan University Science and Technology, Tainan.

[9] Hsu, L. S. (2014) *The study of the career development from a senior teacher in elementary school and the related factors for experience narration* (Master thesis). National University of Tainan, Tainan.

[10] Teng, C.-I., Tseng, H.-M., Li, I.-C. & Yu, C.-S. (2011) International English Big-Five mini-markers: Development of the traditional Chinese version, *Journal of Management*, Vol. 28, No. 6, pp. 579–600.

[11] Lieblich, A., Tuval-Mashiach, R. & Zilber, T. (1998) *Narrative research: Reading, analysis, and interpretation* (Vol. 47). Thousand Oaks: Sage.

[12] Costa, P. T. & McCrae, R. R. (1994) Set like plaster? Evidence for the stability of adult personality, in T. F. Heatherton & J. L. Weinberger [Eds] *Can personality change?*. Washington: American Psychological Association Books, pp. 21–40.

[13] Stone, I. (1984) *Lust for life*. New York: New American Library.

[14] Lin, H.-T. (1992) *How to plan your career*. Taipei: China Productivity Center.

[15] Gross, J. J., Carstensen, L. L., Pasupathi, M., Tsai, J., Skorpen, C. G. & Hsu, A. Y. (1997) Emotion and aging: Experience, expression, and control, *Psychology and Aging*, Vol. 12, No. 4, pp. 590–9.

[16] Inglehart, R. (1990) *Culture shift in advanced industrial society*. Princeton: Princeton University Press.

[17] Gibson, S. J. (1997) The measurement of mood states in older adults, *Journals of Gerontology Series B: Psychological Sciences and Social Sciences*, Vol. 52, No. 4, pp. 167–74.

[18] Paek, K. M. (2018) Social Expectations and Workplace Challenges: Teaching Artists in Korean Schools, *International Journal of Art & Design Education*, Vol. 37, No. 3, pp. 507–18.

[19] Klein, J. W., Case, T. I. & Fitness, J. (2018) Can the positive effects of inspiration be extended to different domains? *Journal of Applied Social Psychology*, Vol. 48, No. 1, pp. 28–34.

Management science

Innovation in Design, Communication and Engineering – Lam et al. (eds)
© 2020 Taylor & Francis Group, London, ISBN 978-0-367-17777-5

Comparative analysis on technical efficiency of different operation models in natural rubber production in Hainan state farms

Su-Fang Zheng
School of Internet Economics and Business, Fujian University of Technology, Fuzhou, Fujian, China

ABSTRACT: Three operation models existed in national rubber production in Hainan State Farms: national farms, shareholding farms, and self-support farms. Based on the comparison of cost benefits, distribution situation of input-output efficiency and calculation results of input-output efficiency, optimum proposals were put forward. The conclusion was the national farms had higher technical efficiency and scale efficiency, the other two models' returns to scale were rising and had great growth rate. So existing factor combinations need to be optimized and adjusted to raise the input-output efficiency.

1 INTRODUCTION

There were three operation models in natural rubber production in Hainan State Farms. Quantitative analysis of technical efficiency was based on comparing the revenue products of three different operation models. Due to the great differences among the three models on cost structure, common cost components were chosen to conduct the comparative analysis. Furthermore, different farms had great differences in cost structure, so the average data was chosen to analyze the technical efficiency.

2 MODEL AND METHODS

2.1 Data sources

The data resources of family long-term contracts were from Hainan Rubber Group (headquarters), and data under the shareholding cooperative system and self-support economy was from Hainan Agriculture Reclamation Administration and subsidiary farms.

2.2 Comparison of technical efficiency of different operation models

2.2.1 Comparative analysis of the cost–benefit of different operation models
Cost accounting was an important link of rubber production and management, providing information on expenses associated with fertilization, tools, labor, and other costs. We also learn about the processing charges of dry rubber, productivity, the influence of risk factors, if the labor organization was reasonable, and if the production technical measures conformed to the economic benefits. Cost accounting also offered important references for price decisions, which objectively reflected the value of natural rubber and was helpful to promoting the development of the rubber industry (Howard, 2010).

Generally, the costs of rubber products could be divided into two parts, reflecting the production costs and nonproductive-period compensation costs. According to the standard definition of cost accounting at home and abroad, direct costs meant direct material and direct labor costs; indirect costs meant the manufacturing expenses indirectly allocated in the production costs. These two parts of costs were called production costs and period costs.

By using the manufacturing cost method, direct costs include production worker salary and welfare, tress depreciation, pesticides, fertilization, fuel and energy, transportation, and protection expenses. Indirect costs included executive salary and welfare, repair charges, utilities, worker insurance expenses, depreciation of plant equipment, and office allowances.

After analyzing the cost structure of the three models, the national rubber model had the highest comprehensive cost, shareholding had the second highest cost, and self-supporting rubber had the least cost. The first one was 37.58% higher than the second one and 39.75% higher than the third one. The main reason for the high national cost was it had two extra costs:

One was processing expenses, including fuel and energy, chemical materials, process equipment maintenance, and other direct expenses such as entertainment expenses and purchasing office supplies. The rubber products were processed and sold in 13 processing factories, so process expenses should be included when cost accounting. However, the workers of shareholding and self-support systems only needed to turn in 30–40% administrative expenses to the farms. And the process expense had been included in the administrative expenses, so there was no need to separately list it.

The other high cost was period cost or indirect cost, accounting for 15.85% of the total costs. It included administrative expenses, financial expenses, sales expenses, and taxes. The workers were official employees of state-own enterprises, so the salary, welfare expenses, and endowment medical insurance expenses should be apportioned into the cost. By contrast, the shareholding rubber had strong cost advantages and its total costs were mainly direct production costs. Because the shareholding rubber was household management, there was no indirect expense, no welfare, no insurance expense, so its cost was lower. However, the costs of the self-support group were lowest: besides the direct production costs, the workers just needed to turn in certain land expenses and training expenses, and union expenses, and what was left over all belonged to them. And the members of family were responsible for protecting and managing the rubber plantation and selling products, so these expenses were included in the labor costs.

As a result, in order to improve the operation performance and efficiency, national rubber should reduce its administrative expenses. Although the self-support rubber had the lowest costs, it was a model with low efficiency: its efficiency should not be restricted by low cost. The shareholding rubber had medium costs and it was a model with relatively high overall efficiency.

With the same average price per ton, total profits = operating income minus operating costs/expenses, national rubber had the lowest profit, shareholding rubber was the medium and self-support rubber had the highest profit. In terms of profit ratio of sales, profit ratio of sales per ton = total profit/operating income × 100%, national rubber was the lowest, shareholding the medium and self-support rubber the highest.

In terms of the average profit, profit ratio of sales and cost–profit ratio, although self-support rubber had the highest of them, it cannot be illustrated that it was the most efficient. It should be measured from product quality, production efficiency, and workers' income. Firstly, in terms of product quality, due to the decentralized operation of self-support rubber, the workers had independent property rights, but they lacked unified quality standards and technical management. So the product quality was uneven and this reduced the total quality of Hainan State Farm. Secondly, in terms of production efficiency, loose management and lack of unified technical regulations caused low efficiency. Thirdly, in terms of workers' incomes, low product quality could not bring more income to workers so incomes were lower than before. Lastly, in terms of constraints, due to the limited land suitable for planting rubber, scale effects could not be reached. So this model was subsidiary and its scale could not be expanded.

2.2.2 *Distribution situation of input-output efficiency*

Every rubber planting farm of every operation model was used as DMU of efficiency evaluation, input-oriented CRS, and VRS of DEA method was chosen and DEAP2.1 software was used to calculate and analyze the technical efficiency of every rubber planting farm

(Huelsbeck, Merchant, & Sandino, 2011). The calculation results of overall efficiency, pure technical efficiency, and scale efficiency were as follows:

Table 1 showed the input-output efficiency of sample national rubber planting farms. Generally, overall efficiency was the lowest, with an average of 0.671. The average of pure technical efficiency was 0.733 and scale efficiency was the highest, with an average of 0.919. This illustrated that the national rubber planting farms gained returns of scale, but average efficiency of resource allocation using existing technical conditions was relatively low. Under the circumstance of unchanged environmental factors, technical level, and existing inputs, input-output efficiency could be improved 32%. So, in the condition of unchanged production factor market price, the improvement of input-output efficiency could increase income and profits of rubber planting farms.

Individually, there were great differences among different farms and their efficiency was from 0.276 to 1.0, among which only 9% of all farms had overall efficiency which was totally effective. The overall efficiency of most farms was medium level which has considerable room for improvement, and the usage of existing technical and production resources was in average level. Only 18% of all farms have totally effective in pure technical efficiency which was from 0.8 to 1.0. Its reason was the risks existing in adopting new technology.

Table 2 showed the distribution condition of input-output efficiency under the shareholding cooperative system. Generally, overall efficiency was the lowest: average was 0.685. Pure technical efficiency was 0.798 and scale efficiency was the highest, with an average of 0.868. This illustrated that the system improved production efficiency to some extent but its improvement range was limited and could be improved 31.5%.

Table 1. Distribution table of technical efficiency of the state-owned rubber.

OE		TE	SE		
1	6	10	6	Average of OE	0.67
0.9–1	3	3	28	Average of TE	0.73
0.8–0.9	5	5	9	Average of SE	0.91
0.7–0.8	4	5	1	Increased ROA	18
0.6–0.7	12	12	1	Constant ROA	4
0.5–0.6	8	7	0	Decreased ROA	22
<0.5	8	4	1		

Data resources are from the model results.
OE = Overall efficiency, TE = Pure technical efficiency, SE = Scale efficiency, ROA = returns of scale

Table 2. Distribution table of technical efficiency of the rubber under stock cooperation system.

OE		TE	SE		
1	3	5	5	Average of OE	0.685
0.9–1	0	0	3	Average of TE	0.798
0.8–0.9	1	0	1	Average of SE	0.868
0.7–0.8	1	2	0	Increased ROA	5 farms
0.6–0.7	1	1	0	Constant ROA	6 farms
0.5–0.6	1	2	0	Decreased ROA	0 farm
<0.5	4	1	2		

Data resources are from the model results.
OE = Overall efficiency, TE = Pure technical efficiency, SE = Scale efficiency, ROA = returns of scale

Table 3 showed the distribution condition of input-output efficiency under the self-support economy. Generally, overall efficiency was lowest, only 0.303. Scale efficiency was 0.497, and pure technical efficiency was the low, only 0.605. This illustrated that under the self-support economy the efficiency was low and it could be improved 70%.

2.2.3 *Calculation of input-output efficiency of different models*

Table 4 showed there were large differences in input-output efficiency of different models. The average efficiency in 2010 was: shareholding rubber > national rubber > self-support rubber, primarily because workers under the shareholding system had 60–70% property rights, which is clear. All product revenues of rubber were shared by the farms and workers as 3:7 or 4:6, so the workers took the farms as their own and the input-output efficiency was greatly improved. However, self-support rubber had the lowest average input-output efficiency, largely due to big differences betwen the different planting families. Another reason was the low technical level and management level.

The average pure technical efficiency of different model was shareholding rubber > national rubber > self-support rubber, too. The scale efficiency was national rubber > shareholding rubber > self-support rubber. This illustrated that the planting families of shareholding rubber would rather to use new technology, but the technical efficiency of national rubber was lower than shareholding rubber due to unclear property rights. In terms of economies of scale, national and shareholding rubber almost reached optimal size and the effect was limited. According to the above comprehensive analysis, management levels under different models

Table 3. Distribution table of technical efficiency of self-supporting rubber.

OE	TE	SE			
1	2	5	2	Average of OE	0.303
0.9–1	0	0	0	Average of TE	0.605
0.8–0.9	0	1	0	Average of SE	0.497
0.7–0.8	0	1	1	Increased ROA	8farms
0.6–0.7	0	0	1	Constant ROA	2farms
0.5–0.6	0	1	1	Decreased ROA	4farms
<0.5	12	6	9		

Data resources are from the model results.

Table 4. The calculation results of input-output efficiency in different pattern.

National Rubber		Shareholding Rubber	Self-support Rubber
Number of Farms	44	11	14
Increased ROA	18	5	8
Constant ROA	4	6	2
Decreased ROA	22	0	4
Overall efficiency			
Average	0.671	0.685	0.303
Maximum	1	1	1
Minimum	0.276	0.289	0.043
Pure technical efficiency			
Average	0.733	0.798	0.605
Maximum	1	1	1
Minimum	0.451	0.498	0.066
Scale efficiency			
Average	0.919	0.868	0.497
Maximum	1	1	1
Minimum	0.276	0.385	0.127

Data resources are from the model results.

differed due to different property rights, so there were big differences in input-output efficiency. Due to all factors and combining the opinions of planting families during the field research, the conclusion was that the shareholding cooperative system had the highest efficiency, so it was the best model from the point of view of both farms and workers.

2.2.4 *Optimum proposal for efficiency of different models*

The improvement of input-output efficiency could be achieved by improving pure technical efficiency and scale efficiency. It could be realized in two ways: one was redundancy adjustment, that is, if some input factor had redundancy, which meant excess investment, optimal input could be gained by the original value minus the redundancy quantity. If some output factor had redundancy, which meant it still could be improved, the target value could be gained by output plus the redundancy. The other was slack adjustment. If some input factor had slack, which meant there was excess investment, the optimal input could be gained by the original value minus the slack quantity. If some output factor had slack, which meant it still could be improved, the target value could be gained by output plus the slack. According to the calculation results of input-output efficiency based on an output-oriented DEA model, that is, under the conditions of not changing the input, how much could output increase by proportion. Based the data of 2010, the output of three models was adjusted as follows:

Table 5. Projection analysis of input-output efficiency based on output orientated in different pattern.

Operation Model	Original		Redundancy	Slack	Target	Improvement Proportion
National	Output	3672.591	988.118	30.582	4691.291	27.74%
	Output Value	6509.159	1,738.917	134.057	8382.133	28.77%
Shareholding	Output	774.799	181.498	0.124	956.421	23.44%
	Output Value	1972.483	459.263	0.315	2432.061	23.30%
Self-support	Output	262.090	161.277	35.619	423.367	75.13%
	Output Value	667.230	401.151	90.680	1159.061	73.71%

Data resources are from the model results.

After adjustment, the target output under the national system could be increased to 4691 tons, an improvement proportion of 27.74%. The output value could be increased to 8382 yuan, an improvement proportion of 28.77%. Under the shareholding system, the former could be increased to 956 tons, a proportion of 23.44% and the latter could be increased to 2432 yuan, a proportion of 23.3%. Under the self-support system, the former could be increased to 423 tons, or 75.13% and the latter could be increased to 1159 yuan, or 73.71%. This illustrated that the output in all three systems could be improved. The output in the national system had the highest target output and the self-support system had the lowest. According to the above analysis, under the fixed input all three models needed to raise the management level of rubber planting and improve the use of technology to maximize output (Moore, 2011).

Based on Input Orientated DEA model, that is, under the condition of not decreasing output, how much, proportionally, should input be decreased. According to the calculation results, the input in three models was adjusted based on 2018data. The average of original value, adjustment value, and target value was used to calculate and the results were as as in Table 6.

After adjustment, the target tapping area of national rubber could be decreased to 50,903, an improvement proportion of 25.58%. The number of workers could be decreased to 1,190, an improvement proportion of 22.58%. The target tapping area of shareholding rubber could be decreased to 7,523, an improvement proportion of 19.52%. The number of workers could be decreased to 176, an improvement proportion of 23.48%. The target tapping area of self-support rubber could be decreased to 1,617, an improvement proportion of 49.69%. The number of workers could be decreased to 172, an improvement proportion of 33.33%. These data showed that the tapping area and number of workers had redundancy of some certain extent. In terms of tapping area, national rubber had the most redundancy, shareholding was

Table 6. Projection analysis of input-output efficiency based on input orientated dea in different pattern.

Operation Model		Original Value	Redundancy	Slack	Target Value	Improvement Proportion
National	TA	68,400.43	-15,559.45	-1,937.73	50,903.22	25.58%
	NOW	1,537	-338	-9	1190	22.58%
Share	TA	9,349.09	-1,825.15	0	7,523.93	19.52%
holding	NOW	230	-45	-9	176	23.48%
Self-support	TA	3215.13	-1,397.42	-200.28	1,617.43	49.69%
	NOW	258	-86	0	172	33.33%

Data resources are from the model results.
TA= Tapping area, NOW=Number of Workers

the second and self-support had the least (Krishnamoorthy & Rajasekharan, 1999). This illustrated that in terms of tapping area bigger was not necessarily better and it should remain at a moderate scale to reach the highest efficiency. In the case of number of workers, national rubber had the most redundancy, shareholding was second and self-support had the least. This illustrated that labor input under the three models had redundancy of some certain extent and numbers of workers should be reduced. In terms of room for improvement, the tapping area of the national system had the most and that of shareholding system had the least. The results showed that the shareholding cooperative system had the relatively best efficiency.

3   RESULTS AND DISCUSSION

Compared to the other models, the household long-term contract system had the largest advantage with family workers having management rights and ownership and management rights were separated. Workers families were trained in self-management, self-financing, self-development, and self-restraint so that the business entity of the rubber plantation could achieve sustainable development and increase workers' income. But the course of implementing the system was different from what was expected due to all kinds of reasons. This model was only applicable to family decentralized management, not to scale and intensive management. At present, the rubber industry needs industrialization, scale, and intensification to maximize capacity, so this model wasn't applicable to rubber production.

Self-support plantation was only applicable to family decentralized operation. Although the workers had complete property rights, which allowed for the lowest costs and highest profits, efficiency was also the lowest (Spector, 2011). The reason was self-support plantations lack coordinated management and technical training. So although the cost-profit ratio and profit ratio of sales was highest, it still was not mostly applicable for rubber production.

Compared to the other two systems, the shareholding cooperative system had the most important feature of application to moderate scale operation. This model also conformed to the agricultural development direction from decentralized to scale operation. It conformed to the requirement for forming community interests.

It was imperative for Hainan State Farm to implement shareholding reform for workers' participation in stock and to strengthen business management according to modern enterprise systems. Most people worked hard to improve their quality of life by pursuing material gain and a shareholding system directly and closely combined enterprise and workers. The system made the workers feel they not only worked for the enterprise but also worked for themselves, so it was a win–win system for both enterprise and workers.

In conclusion, the shareholding cooperative system was the most suitable for producing a level of productivity development compared to historical and existing operational models. It was the inevitable choice for operation system reform and it might be the best choice.

4 CONCLUSION

There are great differences among the different business model of natural rubber production in production efficiency. This study calculated and compared the overall efficiency, pure technical efficiency, and scale technical efficiency of three different business models by using DEA method. The conclusions are follows:

(1) In general, production efficiency was highest under the stock cooperative system, the national rubber was the second and self-support was the lowest. This conclusion was a little different from the past judgment. The main reason why the efficiency of stock cooperative system was the highest was the complete property rights, which greatly aroused workers enthusiasm. The efficiency of national rubber was a little lower, primarily because of unclear property rights and unstable policy, which greatly affected production efficiency.

(2) There are great differences among the various business modela of natural rubber production in technical efficiency. Different farms also had different technical efficiency. National rubber had high technical and scale efficiency, but the output had room for improvement and input had some redundancy. However, returns of scale of stock cooperation and self-support were both increasing and growth space was great. The output had considerable room for improvement and the input had surplus. So the existing factors needed to be optimization and adjustment. By using advanced technology, optimal allocation of resources was realized to raise the input-output efficiency.

(3) By comparing the cost and profits of three operational models, we found that the costs of the national system were higher than for stock cooperation and self-support. The reason was that the indirect expenditures of the national system were far greater than for the other two models, including administration expenses, financial costs, and sales expenses. Regarding the unified purchase and sales of natural rubber products by Hainan Natural Rubber Group, there were no selling expenses in the stock cooperation and self-support systems. To some extent, these costs were inevitable but could be effectively reduced by downsizing the staff.

REFERENCES

Howard, F A. "Synthetic rubber." (2010) *Harvard Business Review*, 1, 1–9.

Huelsbeck, D. P., Merchant, K. A., & Sandino, T. (2011) "On testing business models." *Accounting Review*, 86, 1631–1654.

Moore M. (2011) "NR prices on upward spiral" *[J]. Tire Business*, 28(23), 4.

Krishnamoorthy, S., & Rajasekharan, P. (1999) "Technical efficiency of natural rubber production in Kerala: A panel data anaylsis[J]." *Indian Journal of Agriculutral Economics*, 12, 54.

Spector, Y. (2011) "Theory of constraint methodology where the constraint is the business model[Z]." *International Journal of Production Research*, 49(11), 3387–3394.

Innovation in Design, Communication and Engineering – Lam et al. (eds)
© 2020 Taylor & Francis Group, London, ISBN 978-0-367-17777-5

A study of the service quality for real estate industry based on the SERVQUAL model

Tsung Xian Lin & Hsiu-Yueh Lin*

Huashang College Guangdong University of Finance and Economics, Guangzho, Guangdong, China

ABSTRACT: Development of the real estate industry promotes development of the local economy. In order to survive in fierce competition, real estate enterprises have changed from product and price competition to service competition. More and more enterprises are improving their potential by means of improving customer value, honesty, and service innovation to guarantee long-term development of the enterprises in the market. The present study investigated customers' perceived quality of service provided by real estate industries based on the SERVQUAL model, by adopting 5 aspects and 21 criteria as dependent variables. Through comparison of the weight of the factors by the analytic hierarchy process, the results of data analysis indicated what customers cared about and also suggested how real estate industries could improve management of the enterprise.

Keywords: service quality evaluation method, analytic hierarchy process, service quality

1   INTRODUCTION

According to statistics of the National Economic and Social Development Statistics Bulletin issued by the National Bureau of Statistics in China in 2017, the investment of real estate in 2017 is 10,979.9 billion, an increase of 7% compared with the previous year; the total GDP of the year is 82,712.2 billion, increased 6.9% compared with the previous year. Real estate investment increased 1.01% of economic growth in 2017. Compared with an increase of 0.68% in 2016, it increased 48.5% more than that in 2017. Therefore, investment in real estate is very important to facilitate economic growth.

Service management is a new management mode arising from service competition. Service management is developed from the research of marketing service. "Marketing" is a kind of business activity involving business management, production operation, organization theory, human resource management, quality management, and so on. After analyzing extensive business management principles and investigating business activities, we arrive at conclusions about business management rules and improve the discipline of management systems. Admittedly, cost reduction and product quality are still important for enterprises because they are decisive factors determining how customers evaluate products. But in order to realize customer satisfaction and competitive advantages under the market economy, business must create additional values besides the value of products in order to attract customers and expand market share. In a competitive market, rapidly improving service quality and helping enterprises transform from "production-orientation" to "customer-orientation" is a must for managers.

Having observed the rapid growth of real estate groups (e.g., Hengda, Wanke, Zhonghai, Bi Gui Garden), the authors in the present study investigate the service quality of real estate industries. The present study used the SERVQUAL model to investigate customers' perceived service quality of real estate businesses. The authors compared the weight of the factors by means of the analytic hierarchy process (AHP) to understand the customers' attitudes towards

---

*Corresponding author: rainbox123@hotmail.com

the service quality so as to suggest managers provide good services to win the customer favor, rather than merely emphasizing marketing strategies.

2 SERVQUAL MODEL

2.1 *The characteristics of customer perceived service quality*

Service is critically important to an enterprise's development. High quality services play a vital role in gaining social recognition and customer support. If an enterprise only cares about profits and ignores services, they will not develop sustainably. Therefore, the enterprise should pay more attention to improving service quality in order to be competitive in the market.

The characteristics of service quality can be explained as follows: (1) service quality is different from quality of products; (2) service quality is not a specific concept of service, but a relatively abstract concept; (3) service quality and attitude are how customers evaluate the enterprises; and (4) evaluation of quality is mostly based on comparison. That is, according to the customers' attitudes in the survey, the data analysis concluded the relative superiority or excellence of services provided among different real estate enterprises.

The measurement of service quality in this study is based on the SERVQUAL (Service Quality) model, proposed by PZB in "Servqual: a method of multivariable customer perceived service quality measurement" in 1988. The questionnaire investigating customer attitudes is based on the theory of "customer perceived service quality."

Quality of service is based on customer perception and expectations towards the services provided by the enterprises. The service quality system is a complex system involving many factors. In this study, AHP was used to break down the major system into interrelated subsystems and form a framework of three subsystems: target, aspect, and criteria. Combined with the model of SERVQUAL, the present study adopted five aspects of quality of service: facility, immediacy, efficiency, reliability, and humanity. The present study constructed a model of business service quality by considering the status of domestic real estate and relevant studies. Specific aspects and criteria of the SERVQUAL model in the present study is presented in Table 1.

3 DATA COLLECTION INSTRUMENT

The model of this study was based on SERVQUAL (Service Quality). Each item had a five-point Likert scale (very agree = 5, very disagree = 1). A total of 100 questionnaires were distributed, of which 85 questionnaires were valid. KMO = 0.733 > 0.7. This meant the effect of the factorial analysis on the extraction of common factors was good, while the Bartlett sphere value was 724.692, significant = 0.000 < alpha = 0.01. This showed that the items in the questionnaire were suitable for factorial analysis: 20 factors were selected, and results are shown in Figure 1.

3.1 *Data analysis method: Analytic Hierarchy Process (AHP)*

Analytic hierarchy process (AHP) is mainly used in analyzing uncertainty or multiple variables. It integrates both qualitative and quantitative methods. According to psychology theory, multiple variables were classified into different categories. Through statistical analysis, all factors will be reorganized and restructured hierarchically. The results of statistical analysis will be valuable for managers in making policies.

The AHP method stratified related factors from top to bottom to establish a hierarchical structure. After item analysis, if the consistency ratio C.I. is less than 0.1, this indicated that the items were internally consistent. When C.I. > 0.1, the items were not internally inconsistent, and reliability of the questionnaire was weak. The study used geometric methods to calculate the average of all reliable items, got the relative weight of all variables, then carried out the test again to examine internal consistency and weight of all test items. On evaluating relative weight of all items, Super Decisions software was used to analyze the relative weights of

Table 1. SERVQUAL model.

Major factors	Criteria	Connotation
Humanity Humanity refers to the extent of personalized customer services provided by the enterprise in order to manifest the management philosophy of the business, and humanized the overall customers service.	C11 Consideration C12 Coordination C13 Consultation C14 Courtesy C15 Customer rights C16 Exclusive services	When customers encounter problems, customers' interest would be the top priority to be considered. Refers to effective communication between colleagues in the work place. Customers can get an immediate response when confronted with a problem. Refers to etiquette in the process of providing services, to ensure unmistakable friendly service. When providing welfare, enterprises will do the best to win the rights and interests for customers. To provide customers with one-to-one and personalized services to meet individual customer requirements.
Efficiency Efficiency means whether the enterprise can immediately respond to customers' needs by providing corresponding services, which reveals the efficiency of a business.	C21 Service patience C22 Proper arrangement C23 Professional competence C24 Rapid reaction	Regardless of customers' purchase intention, enterprises can maintain consistent service attitude and patiently answer customers' doubts. When customer service staff faces a large number of customers, customers can still be tended, so that customers will not be overlooked. Means that service personnel have sufficient knowledge and professional skills to provide services. Refers to the ability to promptly respond to and solve customers' requests.
Immediacy Immediacy refers to whether an enterprise can immediately provide timely and correct services in accordance with commitments.	C31 Complete information C32 Timely service C33 Accurate service C34 Service flow	When providing customers with information, the information is accurate and without mistakes. Refers to the ability to solve customer problems in a timely way. Refers to whether the enterprise can provide timely and appropriate responses to customer requests. Means that there is a clear procedure in providing services to customers and completing transactions.
Reliability Reliability refers to whether the enterprise has professionalism, confidence, and trustworthiness.	C41 Courtesy C42 Service strongholds C43 Customer commitment C44 Customer assistance	Means service staff are always courteous when serving customers every time. There are more service strongholds and parking space. Something that promises to be done will have to be done. When customers need services, the enterprises can help in a timely fashion.
Facility Facility means that the enterprise has perfect hardware, facilities, and equipment to satisfy customers' needs.	C51 Reception environment C52 Recognition and publicity C53 Operation procedure	Refers to the facilities that are comfortable, perfect, and humanized to serve customers. Clear and attractive corporate identification and publicity campaign to help customers to identify clearly. Refers to the overall environmental design from the entrance of the building, the district map, and the sand table to the construction site.

Table 2. Demographic information of the participants.

Category	Distribution
Gender	Male: 8; female: 3
Age	<30 years old: 4; 31–40: 5; 41–50: 2; 51–60: 0;>60: 0
Education	Senior high school: 1; college: 3; university: 5; postgraduate: 2
Experiences in real estate industry	<1 year: 1; 1–3 years: 5; 3_5 years: 4; 5–10 years: 1; <10 years:;0
Interest in buying houses	No: 0; Yes: 11
Experiences of buying houses	No: 4; Yes: 7
Frequency of buying houses	1 time: 3; 2 times: 4; 3 times: 3; >3 times: 1

TOPIC	Major factors	Criteria
The SERVQUAL model of real estate service quality	Humanity	C11 Consideration
		C12 Coordination
		C13 Consultation
		C14 Courtesy
		C15 Customer rights
		C16 Exclusive services
	Efficiency	C21 Service patience
		C22 Proper arrangement
		C23 Professional competence
		C24 Rapid reaction
	Immediacy	C31 Complete information
		C32 Timely service
		C33 Accurate service
		C34 Service flow
	Reliability	C41 Courtesy
		C42 Service strongholds
		C43 Customer commitment
		C44 Customer assistance
	Facility	C51 Reception environment
		C52 Recognition and publicity
		C53 Operation procedure

Figure 1.   Model of real estate service quality.

the selected variables in each category so as to calculate the relative importance of the factors, and obtain the appropriate weight ratio of the selected items.

After analyzing the factors, the present study used AHP to analyze the 21 criteria to explore the factors affecting the quality of customer service, and put forward suggestions to the real estate industry to help them improve the quality of customer service.

4 THE RESULT OF DATA ANALYSIS

4.1 Participants

The study was divided into two stages. The participants in the first survey were those who had no experience buying houses and those who had experience of buying house once. After factorial analysis, some critical factors influencing the quality of customer service were left and used in the formal experiment. In terms of service providers and customers, the formal survey was expertise oriented: the participants in the second survey were those who had several experiences buying houses and also senior real estate agents. After item analysis, some factors were selected and left. Of the 20 questionnaires distributed, 11 valid were left.

Table 3.　Relative weight of all aspects.

Aspect	Weight	Ranking
B1 Humanity	0.232	2
B2 Efficiency	0.154	4
B3 Immediacy	0.146	5
B4 Reliability	0.250	1
B5 Facility	0.219	3

Data sources: This study

4.2　The survey of real estate industry

The participants in the formal survey were real estate owners or the senior real estate brokers in the real estate industryy. Through item analysis, we could understand how the service providers felt about the services they provided. Super Decision software was used to analyze all factors to obtain the relative weight of each variable.

A. *Analysis of the relative weight of all aspects is presented in Table 3.*
Geometric mean was used to get the relative weight of all variables.

According to the results of data analysis, reliability ranks the top. It means that customer purchase intention is greatly influenced by the professional ability and credibility of the sales staff. Humanity ranks second, meaning that if enterprises take the initiative to care for customers and provide personalized customer service, it will humanize the service. Facility ranks third, meaning that the real estate industry is different from other industries. Comfortable facilities and equipment provided to the customers will affect customer decision-making. These three factors all ranked 20% above without a large discrepancy between them.

Efficiency ranks fourth and immediacy ranks the fifth. Real estate is a durable property and sales must provide efficient service while ensuring quality of service. Doing this can not only improve the quality of personalized customer service, but also improve customer satisfaction. The services provided by the real estate industry are complex, and include not simply being polite to customers, but often also involve sharing of mortgage, design, interior decoration, and other resources. Because of unexpected factors in the actual service process, when conflicts between customers and service providers take place, the enterprises may not be able to solve the conflict immediately, which may affect customer satisfaction.

B. *Analysis of relative weight between criteria*
According to the data in the Table 4, identification and publicity, reception environment, courtesy, customer commitment, and service strongholds are among the top in the overall weight. The purchase of real estate is absolutely not induced by unreasonable impulse. Consumers must make a decision after careful consideration of various factors. Brand, professional ability, construction quality, and customer commitment are customer concerns.

The top five factors still seem to focus on physical factors, perhaps because Chinese consumption habits are different from those in other countries, or most of the consumers are blinded by the promotional strategies, or the participants in the study are changed from the consumption type to the investment type.

In the book *Monetary Philosophy*, Simmel mentioned that formation of society begins with interaction between people, and interaction lies in exchanges, especially in the exchange of money, which cannot be accomplished without trust. In fact, the operation of the whole society is inseparable from trust, and trust is critical to long-term development of social relations. A real estate purchase is the exchange between customers and sales. While dealing with real estate, there might be some miscommunication and conflicts, and mutual trust plays an important role in business transactions. Trust is based on customers' recognition of the company's corporate image, commodity value, quality assurance, and so on. Accurate customer commitment and customer assistance will also

Table 4. Relative weight of different criteria.

Major factors	Criteria	Weight	Ranking	level	Ranking
Humanity 0.232	C11 Consideration	0.150	5	0.035	16
	C12 Consideration	0.156	4	0.036	15
	C13 Consideration	0.166	2	0.039	13
	C14 Courtesy	0.232	1	0.054	7
	C15 Customer rights	0.160	3	0.037	14
	C16 Exclusive services	0.135	6	0.031	19
Efficiency 0.154	C21 Service patience	0.180	4	0.028	21
	C22 Proper arrangement	0.300	2	0.046	10
	C23 Professional competence	0.211	3	0.032	18
	C24 Rapid reaction	0.309	1	0.048	9
Immediaty 0.146	C31 Complete information	0.279	2	0.041	12
	C32 Timely service	0.295	1	0.043	11
	C33 Accurate service	0.203	4	0.030	20
	C34 Service flow	0.223	3	0.033	17
Reliability 0.250	C41 Courtesty	0.299	1	0.075	2
	C42 Service Strongholds	0.233	3	0.058	5
	C43 Custmor commitment	0.243	2	0.061	4
	C44 Customer assistance	0.225	4	0.056	6
Facility 0.219	C51 Reception environment	0.343	2	0.075	2
	C52 Recognition and publicity	0.434	1	0.095	1
	C53 Operation procedure	0.223	3	0.049	8

Data sources: This study

enhance customer trust in the business to a certain extent. This may explain why reliability ranks in first place in the survey.

In overall weight, humanity ranks second, while other factors, such as providing advice, understanding customers, mutual coordination, customer rights, and exclusive services rank less important among all factors. According to the survey of international authoritative data, poor service will cause 94% of customers to leave, and the importance of creating a comfortable experience is self-evident. Observing market trends, competition in today's market has changed from product competition and price competition to service competition. The concept of consumption gradually develops towards the externalization and individuation of quality. Service competition in accordance with the principle of "customer first" tends to create personalized services. While providing services to customers, sales will consider the personal situation of customers to make them feel respected and satisfied. It all depends on building friendly relationships with customers. From Table 4, it is found that domestic consumers have not yet truly enjoyed exclusive services, which is probably because domestic customers often buy undecorated or unfinished houses and then design and decorate the interior of the houses by themselves. If this is the case, they may not enjoy services provided by service providers.

It is most surprising that accurate service and service tolerance rank lowest in the overall weight. The real estate industry is already a mature industry, and all enterprises have already developed the SOP standard service flow. However, in the eyes of the experts, these two factors are still low. In addition, the survey found that the weight of professional ability and service process are also important, which shows that our local industry pays too much attention to physical appearances, such as the sales' appearance, clothing, etiquette, the reception hall, and the publicity of public relations. On the contrary, professional knowledge and literacy, professional competence, and service processes still are ignored by service providers and need to be improved.

5 CONCLUSIONS AND SUGGESTIONS

5.1 *Conclusion*

The real estate industry has long been a motivating industry, driving employment population, increasing people's income, and also promoting the development of local economy. However, the real estate industry is not only a construction industry, but also an important part of the service industry economy, so the improvement of its service quality is important to the sustainable development of the enterprise. This study, starting with quality management of real estate services, focused on the comprehensive impact of the five aspects of customer service experience through the SERVQUAL: humanity, efficiency, immediacy, reliability, and facility. The analysis of the relative weights of these factors through the AHP show what the customer concerns are, and encourage enterprises to implement service management.

5.2 *Suggestion*

(A) *As a means to enhance humanized services, polish brands*

In fact, most domestic consumers still superstitiously pursue brands, and most enterprises work hard to create reputations to win customer favour. Establishing a brand is not about publicizing the brand, but about improving the quality of the brand. To polish up the brand means sales should start with solving problems to improve service quality. Judging from the overall weight, the humanity criteria is still low, the enterprise should be courteous and considerate in caring for customers and even their relatives by providing personalized service. Chinese are conservative in expressing true feelings. It is necessary for enterprises to encourage staff to enhance their service quality so as to make customers feel friendly and comfortable.

(B) *Emphasizing corporate commitment and creating a corporate image*

Whether the real estate industry can accurately provide good services will be critical to winning customers' trust. Enterprises must pay attention to their commitment to customers, which is particularly important for the real estate industry. Whether in advertisements, briefings, or contracts, enterprises should keep their promises and be honest in order to win customers' favour. In addition, service personnel must have professional competence and provide maximum assistance to customers. The enterprises should actively maintain a reliable corporate image by keeping the philosophy of "customer interest first" and "customer first." Establishing a brand by reputation, and maintaining reputation by being honest, will bring benefits to the enterprise.

(C) *Simplifying the service process and improving the service quality*

From the overall weight analysis, it is found that the service process ranks 17th among the 21 criteria. It means customers are dissatisfied with the service quality. Simplifing the service flow and improving the quality of service is important to solve this problem. Mature enterprises have their systematized rules and regulations. The larger the scale of an enterprise, the longer it takes to establish systematized regulations. The more that is needed to deal with problems, the worse the flexibility and contingency will be. If everything is to be approved through several managers, it takes a long time to solve customer problems in time, which in turn results in customer dissatisfaction. To solve this problem, the enterprise should start with two aspects. On the one hand, the enterprise should learn to authorize subordinates properly, deal with the privileges of the service personnel and supervisors through the authority to avoid the delay in the processing time. On the other hand, the enterprises should shorten the decision-making time and create an effective service team to solve customer problems.

(D) *Doing a good job and improving brand loyalty*

From the data available, the overall weight of immediacy is relatively low, 14.6% of total weight, ranking fourth. Corporate image is established in the process of providing good customer service. The perception of customer service quality is based on two-way communication. If you can grasp the customer's needs clearly, you can make the service right at a time, which can increase the customer's sense of identity and establish loyalty to the enterprise brand. Therefore, the enterprise must hold more professional training for service staff, enriching the professional knowledge and performance of service personnel, so that they can accurately judge customer needs, and serve customers efficiently. At the same time, when developing various cultural promotion activities, we should actively communicate with customers, actively strive for rights and interests for customers, enhance customer's sense of belonging and improve customer's loyalty.

REFERENCES

"2015 China real estate service research report." (2015) *Chinese Residential Facilities*, (Z3), 8–12.
Chen Haibo. (2015) "Brand management research of Y group real estate company [D]." Donghua University.
Li Fei. (2013) "Image building of harmonious interpersonal relationship among College Students: A sociological analysis based on impression management. *Journal of Beihua University (SOCIAL SCIENCES)*, 14(04), 120–125.
Wang Lijun. (2013) "Relationship between service quality and customer satisfaction in real estate enterprises [D]." Wuhan University of Technology.
Yan Jin Juan. (2016) "Study on the evaluation system of rural tourism service quality [D]." Northeast Forestry University.
Yang Wenkai. (2015) "Research on the evaluation of telecommunications service quality based on the modification of SERVQUAL index." *Shanghai Management Science*, 37(03), 51–54.

Others

Innovation in Design, Communication and Engineering – Lam et al. (eds)
© *2020 Taylor & Francis Group, London, ISBN 978-0-367-17777-5*

A(n) (Anti-) neo-Malthusian relationship in the Romer's model

Min-Liang Hsieh
Department of International Economics and Trade, Fujian University of Technology, Fuzhou, Fujian, China

ABSTRACT: This note analyzes the necessary and sufficient conditions concerning the principles of modeling for the emergence of a(n) (anti-) neo-Malthusian relationship between economic growth and population growth on a Romer's endogenous growth model [P.M. Romer 1986. "Increasing Returns and Long Run Growth," Journal of Political Economy 94, 1002-1037.] with endogenous fertility, by employing a relative wealth-as-status motive. In contrast to the existing numerical studies, this note is an analytical one wherein the comparative static conditions in response to all exogenous shocks including the demand side (e.g. desire of status-seeking) and the supply side (e.g. technological progress) shocks are characterized.

1 INTRODUCTION

The opinions in the empirical literature with regard to the relationship between population growth and economic growth usually lack unanimity (T.N. Srinivasan 1988, J. Thornton 2001, J.R. Faria et al. 2006, M. Eris 2010). In order to bring forth a compromise and reconciliation among these conflicting opinions, C.K. Yip et al. (1996, 1997) consider endogenous fertility choice in the P.M. Romer (1986) endogenous growth model. They claim that a neo-Malthusian (inverse) relation between population growth and economic growth emerges only when *all* exogenous variables *are controlled for*, but may not arise when some exogenous factors change. Despite much effort which has been made, they can only illustrate their results by numerical examples or attribute conflicting findings to heterogeneity in the unobserved variable in cross-country panel data sets. We still know little from the theoretical model. Unlike their numerical studies, the main aim of this note is to provide further analytical proof and some precise mechanisms for generating a(n) (anti-) neo-Malthusian relationship, so that the aforementioned claim that all exogenous variables are controlled for is corrected.

Based on the model of C.K. Yip et al. (1996), in order to obtain further analytical conditions for the emergence of a(n) (anti-) neo-Malthusian relationship, my modeling strategy is to introduce the motive of the relative wealth-as-status seeking into their model. Recently, in addition to the endogenization of fertility choice, the alternative interesting subject concerning endogenous growth in development economics is the social status-seeking modeling. As documented by P.H. Werhane (1991, pp. 99-100), "In the *WN* (*The Wealth of Nations*), part of the motive for property and capital accumulation is the desire for approval or respect, as Smith recognizes that we tend to admire the rich and avoid the poor." Accordingly, a rat-race resulted from individuals' relative wealth-as-status seeking behavior unintentionally leads to the growth prospects of the whole society.

While the concept of 'wealth-as-a-status seeking motive facilitating wealth accumulation' can be traced back to the '*The Wealth of Nations*' by A. Smith (1776), the concept of endogenous fertility modeling can be traced back to the '*An Essay on the Principle of Population*' by T.R. Malthus (1798). The nature and causes of wealth are discussed in the former while the analogues of poverty are discussed in the latter's essay. Consequently, these two subjects can

---

* Corresponding author: leomin437@fjut.edu.cn; 2772805354@qq.com

be integrated reasonably in an endogenous growth model to analyze the relationship between population growth and economic growth.

In line with a continuous-time, infinite-horizon representative agent framework without government, G. Corneo et al. (1997, 2001) show that there exists a degree of relative wealth-enhanced social status motive such that the spillover distortions from P.M. Romer's (1986) endogenous growth model are corrected. In this note, I first address how does an increase in the motives of the relative wealth seeking affect population growth and economic growth. I find that the stronger the desire for relative wealth seeking is, the *higher* the steady state economic growth rate and the *lower* the population growth rate of economy will be, which thereby generates a neo-Malthusian relationship. Finally, I explicitly point out the necessary and sufficient conditions concerning the principles of modeling for the emergence of a(n) (anti-) neo-Malthusian relationship in response to *all* exogenous shocks including demand side (e.g. desire of social status) and supply side (e.g. technological progress) shocks in a P.M. Romer (1986)-type model.

It is noteworthy that what I focus on, in this note, is the essence of the model itself rather than the magnitude of shock parameters in the model that is capable of generating a(n) (anti-) neo-Malthusian relationship. This clarification helps us figure out the mechanism behind the model and thereby let us run the model precisely for the related studies in the future.

2   THE MODEL

Our modeling strategy entails the combining of G. Corneo et al. (1997) model with C.K. Yip et al. (1996) model. Assume that the economy consists of a continuum of infinitely-lived representative households who care not only about their consumption c and fertility rate n, but also about their relative wealth, which is the determinant of social status. Specifically, I define relative wealth as physical capital goods k, relative to the average $\bar{k}$, i.e. $k/\bar{k}$. The identical households are endowed with one unit of productive time in each instant of time $t > 0$ and with the same positive amount of wealth $k_0 > 0$ at the initial date $t = 0$. They all share the technology of production that is commonly available and seek to maximize the following lifetime utility:

$$\int_0^\infty \left[\ln c + \frac{n^{1-\varepsilon} - 1}{1 - \varepsilon} + \beta v\left(\frac{k}{\bar{k}}\right) \right] e^{-\rho t} \, dt \tag{1}$$

subject to:

$$\dot{k} = A k^\alpha l^{1-\alpha} \bar{k}^{1-\alpha} - nk - c, \tag{2}$$

$$1 + \phi(n) = 1, \tag{3}$$

where c = consumption, n = the fertility rate or the population growth rate, ε = the elasticity of the marginal utility of fertility, β = a nonnegative parameter reflecting the desire for social status, k = capital stock, $\bar{k}$ = the average level of capital stock in the economy, ρ = a constant rate of time preference, A = a scale parameter, $\alpha \in (0, 1)$, and l = labor. Following G. Corneo et al. (1997), the instantaneous status utility $v(\cdot)$ is increasing, differentiable and concave. Following C.K. Yip et al. (1996), the child-rearing cost function $\phi(\cdot)$ is strictly increasing in n, but the second-order derivative can be of either sign.

Substituting equation (3) into equation (2) and letting λ be the co-state variable of the current value Hamiltonian associated with equation (2), the optimum conditions necessary for the representative household are

$$\frac{1}{c} = \lambda, \tag{4}$$

$$n^{-\epsilon} = \lambda k + \lambda(1-\alpha)Ak^{\alpha}(1-\phi(n))^{-\alpha}\phi'(n)\bar{k}^{1-\alpha,} \tag{5}$$

$$\frac{\eta}{k}v'\left(\frac{k}{\bar{k}}\right) + \lambda\alpha Ak^{\alpha-1}(1-\phi(n))^{1-\alpha}\bar{k}^{1-\alpha} - \lambda n = -\dot{\lambda} + \lambda\rho, \tag{6}$$

together with equations (2) and (3), and the transversality condition of k,

$$\lim_{t\to\infty} \lambda k e^{-\rho t} = 0.$$

Since the households are assumed to be identical, $\bar{k} = k$ is true in a symmetric equilibrium (P.M. Romer 1986). Accordingly, by using equation (4), equation (5) is rewritten as:

$$n^{-\epsilon} = \frac{1}{Z}[1 + (1-\alpha)A(1-\phi(n))^{-\alpha}\phi'(n)], \tag{7}$$

where we define the transformed variable $Z \equiv c/k$.
From equation (7), we can then derive the following instantaneous relationship:

$$n = \Psi(Z; \epsilon, \alpha, A), \tag{8}$$

where $\psi_Z = n^{-\epsilon}/(Z\epsilon n^{-\epsilon-1} + \Delta) > 0, \Delta \equiv (1-\alpha)A(1-\phi(n))^{-\alpha}\left[\frac{\alpha(\phi'(n))^2}{1-\phi(n)} + \phi''(n)\right].$

Combining equations (2), (3), and $\bar{k} = k$ together yields the goods market equilibrium condition as:

$$\dot{k} = Ak(1-\phi(n))^{1-\alpha} - nk - c. \tag{9}$$

Next, differentiating equation (4) with respect to time and using equations (4) and (6) with $\bar{k} = k$, we obtain

$$\frac{\dot{c}}{c} = \eta Z v'(1) + \alpha A(1-\phi(n))^{1-\alpha} - n - \rho, \tag{10}$$

where $v'(1)$ is the marginal utility of relative wealth-induced social status in equilibrium.
By taking the logarithmic of $Z \equiv c/k$ on both sides and differentiating it with respect to time together with equations (8), (9) and (10), we obtain:

$$\frac{\dot{z}}{z} = \left[1 + \eta v'(1)\right]Z - (1-\alpha)A[1 - \phi(n(z))]^{1-\alpha} - \rho \equiv h(Z). \tag{11}$$

Since the conditions of $\lim_{Z\to 0}\phi(n(Z)) = 0$, $\lim_{Z\to 0}h(Z) = -(1-\alpha)A - \rho < 0$ hold, and $h'(Z) = 1 + \beta v'(1) + (1-\alpha)^2 A(1-\phi)^{-\alpha}\phi'\Psi_Z > 0$, there exists a unique but determinate balanced growth path (BGP) equilibrium. This characterization is consistent to the results reported by C.K. Yip et al. (1996).

3 THE STEADY STATE AND THE EMERGENCE OF A(N) (ANTI-)NEO-MALTHUSIAN RELATION

Following C.K. Yip et al. (1996), along the BGP equilibrium, c, k grow at a constant rate $\bar{\gamma}$ (hereafter the upper bar, $\bar{}$, denotes the BGP equilibrium value of the corresponding variables). In other words, at steady-growth equilibrium, the economy is characterized by $\dot{Z} = 0$, and there exists $\bar{Z}$ such that the BGP equilibrium sustains. Hence, from equations (7), (9), and (10), we have

$$\bar{Z} = (\bar{n})^\varepsilon [1 + (1 - \alpha)A(1 - \phi(\bar{n}))^{-\alpha}\phi'(\bar{n})], \tag{7-a}$$

$$\bar{\gamma} = \frac{\dot{k}}{k} = A(1 - \phi(\bar{n}))^{1-\alpha} - \bar{n} - \bar{Z} = A(1 - \phi(n))^{1-\alpha} - \bar{n} - \overline{\Psi}^{-1}(\bar{n}; \varepsilon, \alpha, A), \tag{9-a}$$

$$\bar{\gamma} = \frac{\dot{c}}{c} = \eta\bar{Z}v'(1) + \alpha A(1 - \phi(\bar{n}))^{1-\alpha} - \bar{n} - \rho = \eta v'(1)\overline{\Psi}^{-1}(\bar{n}; \varepsilon, \alpha, A) + \alpha A(1 - \phi(\bar{n}))^{1-\alpha} - \bar{n} - \rho. \tag{10-a}$$

Accordingly, equations (7-a), (9-a), and (10-a) yield a 3×3 system for $(\bar{Z}, \bar{n}, \bar{\gamma})$. Taking the total differential of these equations, we solve them simultaneously. Comparative static conditions in response to demand shocks (increase in β, ρ, ε) and supply shocks (increase in α, A) will be established, respectively, as the following:

$$\begin{cases} \frac{d\bar{n}}{d\beta} = \frac{-Zv'(1)}{\Omega} < 0, \\ \frac{d\bar{\gamma}}{d\beta} = \frac{\left[(1-\alpha)A(1-\phi(\bar{n}))^{-\alpha}\phi'(\bar{n}) + 1 + \frac{1}{\Psi_Z}\right]\bar{Z}v'(1)}{\Omega} > 0, \end{cases} \tag{12}$$

$$\begin{cases} \frac{d\bar{n}}{dA} = \frac{-(1-\alpha)(1-\phi(\bar{n}))^{1-\alpha}\Lambda}{\Omega} \gtrless 0, \\ \frac{d\bar{\gamma}}{dA} = \frac{(1-\phi(\bar{n}))^{1-\alpha}[(1-\alpha)\Lambda + (\beta v'(1) + \alpha)\Theta]}{\Omega} \gtrless 0, \end{cases} \tag{13}$$

$$\frac{d\bar{n}}{d\rho} > 0, \quad \frac{d\bar{\gamma}}{d\rho} < 0; \quad \frac{d\bar{n}}{d\varepsilon} > 0, \quad \frac{d\bar{\gamma}}{d\rho} < 0; \quad \frac{d\bar{n}}{d\alpha} \gtrless 0, \quad \frac{d\bar{\gamma}}{d\alpha} \gtrless 0. \tag{14}$$

where $\Omega = \frac{1 + \beta v'(1)}{\Psi_Z} + (1-\alpha)^2 A(1 - \phi(\bar{n}))^{-\alpha}\phi'(\bar{n}) > 0, \Theta = \frac{\bar{n}^\varepsilon \phi'(\bar{n})}{1 - \phi(\bar{n})}(1-\alpha)^2 A(1 - \phi(\bar{n}))^{-\alpha}\phi'(\bar{n})$

$+ \frac{1}{\Psi_Z} > 0, \Lambda = \frac{(1 + \beta v'(1))\bar{n}^\varepsilon \phi'(\bar{n})}{1 - \phi(\bar{n})} - 1.$

First, from equation (12), we can notice that the stronger the desire for relative wealth-as-status seeking is, the higher the steady-state growth rate and the lower the population growth rate of the economy will be. Second, from equations (12)-(14), a neo-Malthusian relation always emerges in response to all demand shocks, but to supply shocks only on occasion. Additionally, it should be noted that due to their omission of exogenous shock parameters in $\overline{\Psi}^{-1}(\bar{n}; \varepsilon, \alpha, A)$, the results in C.K. Yip et al. (1996) are inconsistent with the results, if $\beta = 0$, in equation (13).

From equation (12), proposition 1 is established:

Proposition 1. *In a P.M. Romer (1986)-type endogenous growth model with endogenous fertility, the stronger the desire for relative wealth-as-status seeking is, the higher the steady-state growth rate and the lower the population growth rate of the economy will be.*

The intuition of proposition 1 is straightforward. Stronger motives of capital-induced-status encourage more capital accumulation. Further, since capital and labor are technical complements under a P.M. Romer (1986)-type (Cobb-Douglas) production technology, more capital accumulation will lead the representative household to change his time allocation, substituting his labor time for child-rearing time, which in turn brings down fertility (population growth rate) and speeds up economic growth eventually. As a result, a neo-Malthusian relation holds.

Analytically, how a(n) (anti-) neo-Malthusian relation, a specific set of comparative static results in response to all exogenous shocks, is obtained? For the sake of better comprehension, we substitute equation (7-a) into equations (9-a) and (10-a), respectively, as shown in the third equality in each of them, and then obtain the reduced-form equations for the rate of economic growth, i.e. equation (9-a) as $\bar{\gamma} = \bar{\gamma}(\bar{n}(\beta, \rho, \varepsilon, \alpha, A); \varepsilon, \alpha, A)$ and equation (10-a) as

$\bar{\gamma} = \bar{\gamma}(\bar{n}(\beta, \rho, \varepsilon, \alpha, A); \beta, \rho, \varepsilon, \alpha, A)$. From these two reduced equations, the comparative static conditions can be derived; meanwhile, each of them can be decomposed into two ingredients, the direct effect and the indirect effect, as follows:

$$\frac{d\bar{\gamma}}{dj} = \left.\frac{d\bar{\gamma}}{dj}\right|_{\text{given } \bar{n}} + \left(\left.\frac{d\bar{\gamma}}{d\bar{n}}\right|_{\frac{k}{k}}\right)\left(\frac{d\bar{n}}{dj}\right), \tag{15}$$

$$\frac{d\bar{\gamma}}{dj} = \left.\frac{d\bar{\gamma}}{dj}\right|_{\text{given } \bar{n}} + \left(\left.\frac{d\bar{\gamma}}{d\bar{n}}\right|_{\frac{c}{c}}\right)\left(\frac{d\bar{n}}{dj}\right), \tag{16}$$

where $j = \beta, \rho, \varepsilon, \alpha, A$. The first term on the right hand side in equations (15) and (16), respectively, is the direct effect of shock j on equilibrium steady-growth rate, and the remainder is the indirect one. It should be noted that no matter which equation we adopt, the final results of the comparative static analysis should be all equal to the ones that equations (12)-(14) reveal. To proceed in the reverse direction, the results of equations (12)-(14) revealed also must simultaneously satisfy both equations (15) and (16). It is important for us to have an understanding of the above-mentioned character to find the messages behind the essence of the model.

Similar with the treatment of endogenizing the labor supply decision in the equilibrium structure of the AK growth model proposed by S.J. Turnovsky (2000), we can find that the reduced form of both equations (9-a) and (10-a) can be described as a pair of tradeoff loci relating the equilibrium growth rate and the population growth rate. We call them "locus $\dot{k}/k$" and "locus $\dot{c}/c$" respectively. In other words, the slopes of these two loci in the neighborhood of the equilibrium point are exactly what these two terms

$$\left.\frac{d\bar{\gamma}}{d\bar{n}}\right|_{\frac{k}{k}} \text{ and } \left.\frac{d\bar{\gamma}}{d\bar{n}}\right|_{\frac{c}{c}}$$

in equations (15) and (16) exhibit.

Accordingly, the occurrence of a neo-Malthusian inverse relation indicates the opposite sign between terms

$$\frac{d\bar{\gamma}}{dj} \text{ and } \frac{d\bar{n}}{dj} \text{ for all } j,$$

which obviously implies that the claim of C.K. Yip et al. (1996), "when all exogenous variables are controlled for, there exists an inverse relation between population growth and economic growth," is incorrect. Further, from the viewpoint of equation (15), we can infer that

$$\left.\frac{d\bar{\gamma}}{d\bar{n}}\right|_{\frac{k}{k}} < 0$$

is the necessary condition of inverse relation between population growth and economic growth by the fact that the direct effect on the equilibrium steady-growth rate is zero when the shock j is exactly ρ or β. Moreover, in fact, because the term

$$\left.\frac{d\bar{\gamma}}{d\bar{n}}\right|_{\frac{k}{k}} = -\left[(1-\alpha)A(1-\phi(\bar{n}))^{-\alpha}\phi'(\bar{n}) + 1 + \frac{1}{\psi_z}\right]$$

is always negative, we infer that a neo-Malthusian relation will emerge only when the indirect effect is greater than the direct one which possesses the opposite sign to the former, if any.

Therefore, the sufficient condition of an inverse relation between population and economic growth is established. Additionally, we can obtain that the necessary and sufficient emergence conditions of an anti-neo-Malthusian relation are on the occasion when the indirect effect is smaller than the direct one which certainly possesses the opposite sign to the former.

From equation (16), since the sign of the term

$$\frac{d\bar{\gamma}}{d\bar{n}}\bigg|_{\frac{\dot{c}}{c}} = \frac{\beta v'(1)}{\bar{\psi}z} - \left[\alpha(1-\alpha)A(1-\phi(\bar{n}))^{-\alpha}\phi'(\bar{n}) + 1\right]$$

is influenced by the magnitude of $\beta v'(1)$ and thereby could be either positive or negative, we could conclude that the slope of "locus $\dot{c}/c$" in the neighborhood of the equilibrium point does not matter in affecting the emergence of a(n) (anti-)neo-Malthusian relation. This is exactly contrary to the situation where the negative slope of "locus $\dot{k}/k$" in the neighborhood of the equilibrium point is the necessary condition for the emergence of a neo-Malthusian relation. Therefore, we obtain:

Proposition 2. *In the neighborhood of the P.M. Romer (1986)-type balanced growth equilibrium point,* **the slope of the tradeoff locus between economic growth and population growth necessary to maintain the goods market equilibrium is always negative,** *which is a* **necessary condition** *for the emergence of a neo-Malthusian relation. Additionally, the* **sufficient condition** *is that the* **indirect effect** *(of an exogenous shock on the equilibrium steady-growth rate through the channel of an equilibrium population growth rate)* **is greater than the direct one which possesses the opposite sign to the former, if any.** *The necessary and sufficient condition for the emergence of an anti-neo-Malthusian relation is that the indirect effect must be smaller than the direct one.*

4   CONCLUSION

This note analyzes the principles of modeling for the emergence of a(n) (anti-) neo-Malthusian relation between economic growth and population growth on a P.M. Romer (1986)-type endogenous growth model with endogenous fertility. It shows that the emergence of these relations depends on: (1) the slope of the tradeoff locus between economic growth and population growth necessary to maintain goods market equilibrium, and (2) the relative magnitude of the indirect effect (which results from the exogenous shock on an equilibrium steady-growth rate through the channel of an equilibrium fertility rate) and the direct one (caused by the exogenous shock on the equilibrium steady-growth rate).

REFERENCES

T.N. Srinivasan, Population growth and economic development, Journal of Policy Modeling 10 (1988) 7–28.

J. Thornton, Population growth and economic growth: Long-run evidence from Latin America, Southern Economic Journal 68 (2001) 464–468.

J.R. Faria, M.A. León-Ledesma, A. Sachsida, Population and income: Is there a Puzzle? Journal of Development Studies 42 (2006) 909–917.

M. Eris, Population heterogeneity and growth, Economic Modelling 27 (2010) 1211–1222.

C.K. Yip, J. Zhang, Population growth and economic growth: A reconsideration, Economics Letters 52 (1996) 319–324.

C.K. Yip, J. Zhang, A simple endogenous growth model with endogenous fertility: Indeterminacy and uniqueness, Journal of Population Economics 10 (1997) 97–110.

P.M. Romer, Increasing returns and long run growth, Journal of Political Economy 94 (1986) 1002–1037.

P.H. Werhane, Adam Smith and his legacy for modern capitalism, Oxford University Press, New York, 1991.

A. Smith, An inquiry into the nature and causes of the wealth of nations, Clarendon Press, Oxford, 1776, 1979 reprint.

T.R. Malthus, An essay on the principle of population, Cambridge University Press, Cambridge, 1798, 1992 reprint.

G. Corneo, O. Jeanne, On relative wealth effects and the optimality of growth, Economics Letters 54 (1997) 87–92.

G. Corneo, O. Jeanne, Status, the distribution of wealth, and growth, Scandinavian Journal of Economics 103 (2001) 283–293.

F. Tournemaine, Social aspirations and choice of fertility: why can status motive reduce per-capita growth? Journal of Population Economics 21 (2008) 49–66.

S.J. Turnovsky, Fiscal policy, elastic labor supply, and endogenous growth, Journal of Monetary Economics 45 (2000) 185–210.

Innovation in Design, Communication and Engineering – Lam et al. (eds)
© 2020 Taylor & Francis Group, London, ISBN 978-0-367-17777-5

Effectiveness and usability assessments of clinical upper limb rehabilitation equipment: A case study of an incline board

Lan-Ling Huang*
School of Design, Fujian University of Technology, Fuzhou, Fujian Province, PR China

Mei-Hsiang Chen
Department of Occupational Therapy, Chung Shan Medical University, Taichung, Taiwan

ABSTRACT: The purpose of this study is to assess the effectiveness and usability of conventional and digital incline boards. The results of the assessment can be summarized as follows: (1) In effectiveness, regarding within-group changes, the groups showed improvements in upper extremity function on three assessments (the Fugl-Meyer Assessment of Physical Performance, the Box and Block Test of Manual Dexterity, and upper extremity range of motion). (2) In usability, patients were positive when using the digital incline board in rehabilitation. They thought that the information provided from the digital incline board was useful to them and so they can understand the status of their treatment.

1 INTRODUCTION

Stroke is one type of cerebrovascular disease threatening people in many countries. Upper extremity motor deficit is one of the main symptoms for stroke patients and elderly (Gowland et al., 1992). In order to restore upper extremity motor function, stroke patients must be treated with functional equipment. Most existing clinical upper extremity rehabilitation products provide no feedback to patients. The main reasons may be summarized as follows: (a) the devices are expensive; (b) the gaming interfaces are complicated for patients to independently operate the games without help from the therapists; (c) the games' contents are designed for normal people and their leisure, not for stroke patients.

Therapists have thought that the traditional rehabilitation equipment is still important and irreplaceable. Rehabilitation equipment that meets both human and user needs will help reduce the workload of the occupational therapist and enhance the patient's motivation and effectiveness in treatment. In this study, the most commonly used upper limb rehabilitation equipment (incline board) is the clinical priority.

The purpose of this study is to assess the effectiveness and usability of traditional and digital incline boards. The research contents are: (1) comparison of the effectiveness of conventional and digital incline boards; (2) research of the usability of conventional and digital incline boards.

2 METHOD

2.1 *Participants*

Stroke patients were recruited from the occupational therapy department of Chung Shan Medical University Hospital. This study used a pretest–posttest control group design. Stroke

---

* Corresponding author: anilhuang@163.com

patients who met certain conditions were invited to participate. Each patient gave informed consent.

The inclusion criteria were the following: (a) hemi-paretic with upper extremity dysfunction following a single unilateral stroke, (b) a history of a first-time stroke (during 3–24 months post stroke, the stroke patient is in a stabilizing neuro-rehabilitation status and has some recovery of hand functions), (c) upper extremity rehabilitation convalescent levels of Brunn-strom stages III to V – i.e., having basic upper extremity synergies to perform joint movement voluntarily, (d) the ability to communicate, (e) the ability to understand and follow instructions, and (f) no serious problems with balance.

The exclusion criteria were the following: (a) engaged in any other rehabilitation program during the study and (b) serious aphasia or cognitive impairment. Each patient gave informed consent. This study was approved by the human research ethics board of a local hospital.

2.2   Design and procedure

This study used a pretest–posttest control group design. The two assessors were blind to the assignment. Subjects who accepted were asked to sign an informed consent form. The functional ability of each subject's affected upper extremity was assessed by one of the assessors in two stages: (1) prior to the interventions and (2) immediately after completing the total of 10 sessions. All subjects were asked to complete a total of 10 training sessions, three sessions per week.

Three assessments were used as follows: the Fugl-Meyer Assessment of Motor Function (FMA) (Fugl-Meyer et al., 1975), the Box and Block Test of Manual Dexterity (BBT) (Mathiowetz et al., 1985), and the range-of-motion measurement of the upper extremity (ROM) (Gajdosik et al., 1987) were used at pretest and posttest.

During the test, researchers observed the patients' usage of the equipment from the side. Also, interviewer questions were asked in order to evaluate the patients' satisfaction with a traditional incline board or an incline board with digital functions. The patients were asked, "Do you feel that the conventional or digital incline board devices were useful for treatment?" and "Do you feel that the digital functions of the digital incline board devices were useful for you?"

2.3   Interventions

This study included two groups for treatment intervention: (1) a conventional group, and (2) a digital group. In addition to the regular treatment, each group was assigned specific devices for use in additional treatment sessions. Specifically, the conventional group used an incline board (Figure 1), and the digital group used a digital incline board (Figure 2) that provided the extension length to patients.

Figure 1.   The conventional incline board.

Figure 2. The digital incline board.

2.4 *Data analyses*

All data were analyzed with SPSS for Windows version 13.0. The characteristics of the three groups were analyzed with descriptive statistics. The Wilcoxon signed ranks analysis was used for within-group analyses. Differences were considered significant when $p < 0.05$.

3 RESULTS AND DISCUSSION

A total of 10 post-stroke patients were recruited from the occupational therapy department of Chung Shan Medical University Hospital. The characteristics of stroke patients are shown in Table 1.

Regarding within-group changes, the digital group showed significant improvements in upper limb functions on all three assessments (FMA, BBT, and ROM), while the conventional group had no significant differences on all three assessments (FMA, BBT, and ROM).

Regarding usability, patients in the digital group reported using the digital incline board in treatment helped them understand the status of the treatment, for example, progression in the number of stretches and stretching distance. Therapists reported the digital incline board was better than the conventional incline board in the following ways: (1) it is easy to move to

Table 1. Characteristics of stroke patients.

Characteristics	Groups		p-value[a]
	Conventional Group	Digital Group	
Gender, male/female (n)	2/3	¼	
Age in years (mean, SD)	74.8 (10.1)	81.0 (5.3)	
Paretic side, left/right (n)	4/1	4/1	
FMA (pretest) (mean, SD)	44.4 (21.1)	55.6 (1.4)	0.336
FMA (posttest)	43.4 (21.3)	60.8 (3.5)	0.102
BBT (pretest)	18.8 (13.0)	28.4 (4.7)	0.308
BBT (posttest)	23.8 (18.8)	35.2 (6.7)	0.234
ROM-proximal[b] (pretest)	73.7 (31.9)	76.6 (39.1)	0.043*
ROM-proximal[b] (posttest)	79.1 (34.2)	84.3 (38.3)	0.042*

* Significant at $\leqq 0.05$ level
[a] p for differences between pretest and posttest
[b] ROM-proximal was used to assess the range of active joint motion for the shoulder and elbow joints.
[c] ROM-distal was used to assess the range of active joint motion for the forearm and wrist joints.
  FMA: Fugl-Meyer Assessment of motor function; BBT: Box and Block Test of Manual Dexterity; ROM: Range-of-motion measurement of the upper extremity

different spaces to meet the treatment space needs, (2) it helps to calculate the time, (3) the interface is easy to operate, and (4) the size of the irons were easy to pick up.

4 CONCLUSION

The results of the assessment can be summarized as follows: (1) for effectiveness, regarding within-group changes, the groups showed improvements in upper extremity function on all three assessments; (2) for usability, patients were positive when using the digital incline board in rehabilitation. In terms of performance outcomes, the potential efficacy of the digital incline board improved upper extremity function after stroke. Significant improvements appeared on the ROM.

ACKNOWLEDGMENT

This article was partly supported by No. FJ2018B150.

REFERENCES

Fugl-Meyer, A. R., et al. 1975. *Scandinavian Journal of Rehabilitation Medicine*, 7, 13–31.
Gajdosik, R., et al. 1987. *Physical Therapy*, 67, 1867–1872.
Gowland, C., et al. 1992. *Physical Therapy*, 72, 624–633.
Mathiowetz, V., et al. 1985. *American Journal of Occupational Therapy*, 39, 386–391.

Innovation in Design, Communication and Engineering – Lam et al. (eds)
© 2020 Taylor & Francis Group, London, ISBN 978-0-367-17777-5

A study of the fractal graph generation of higher-order Julia sets based on a complex plane

Artde Donald Kin-Tak Lam*
Fujian University of Technology, Fuzhou, Fujian, PR China

ABSTRACT: In this paper, a method of fractal graph generation is presented. Based on a complex plane, a higher-order Julia set is applied so as to generate fractal artistic graphics. First of all, the relationship between the fractal graphics and mathematical morphology parameters has been studied, especially in complex planes; an escape time algorithm was used in order to generate fractal artistic graphics. Then, a computer program was developed so as to generate fractal artistic graphics. Finally, the artistic fractal graphics generated by the developed program were applied to the artistic graphics design.

Keywords: fractal graph generation, higher-order Julia set, complex plane, fractal artistic graphics

1 INTRODUCTION

In comparisons of fractal artistic graphics and traditional artistic graphics, the fractal artistic graphic is a kind of art with scientific characteristics. A fractal artistic graphic comes from a fractal graph, and the fractal graph comes from the calculation of fractal geometry. Thus fractal geometry has become one of the main scientific researches in modern graphic design.

Fractal geometry originated in the 19th century. Famous mathematicians have studied the continuous non-differentiable curves and discovered a kind of "sick" curve different from the traditional geometric curves, such as the Cantor set, Koch curve, Peano curve, Sierpinski set, etc. (Mandelbrot, 1977). Due to the complexity of calculation, these geometric curves are often expressed by mathematical functions. With the help of powerful computing and graphic display functions of modern computers, fractal graphs can be transformed from mathematical expression to visual expression, which makes the application of fractal geometry more extensive (Valor et al., 2003).

Based on fractal geometry and computer generation theory, this paper develops a method for generating a fractal graph. Also, the relationship between the morphology and the mathematical parameters of this method are discussed. In particular, on the basis of the complex plane high-order function, an actual method of generating a fractal graph is established, taking into account the conditions of the generation, the morphology, and the corresponding generation results of various kinds of fractal graphs. Finally, a system for generating a fractal artistic graphic is developed by integrating the method of generating a fractal graph and computer graphics.

2 FRACTAL GEOMETRY

The word *fractal* originated in the 1970s and refers to irregular and fragmented objects. The English *fractal* is derived from the Latin *fractus*.

---

*Corresponding author: lam@fjut.edu.cn

Mandelbrot, an American mathematician, found that many natural phenomena do not conform to Euclidean geometry and so an effective mathematical concept is needed in order to describe them. He proposed fractal geometry so as to explain these natural phenomena. However, since Mandelbrot published the concept of fractal geometry in 1975, it still lacks a clear mathematical definition. In 1977, Mandelbrot described the fractal phenomena of nature in detail, which include three elements: form, chance, and dimension. In 1984, Mandelbrot reviewed the insufficiency of fractal geometry and proposed several important concepts, including irregularity, self-similarity, non-integer dimension, etc. (Mandelbrot, 1977, 1983). These important concepts are near to the description of natural things. In 1991, Falconer applied the method of defining life in biology to defining fractals, and sought features from set theory in order to explain fractal sets: (1) a fractal set is an infinitely fine structure; (2) fractal sets are irregular; (3) fractal sets have some degree of self-similarity; (4) the dimension of a fractal set is larger than its topological dimension; (5) fractal sets can be generated by a simple recursive method. Different fractal sets may have all of these properties at the same time (Falconer, 1991).

According to Euclidean geometry, if segment n of length 1 is divided equally and each segment length is r, as shown in Figure 1, then:

$$n \cdot r = 1 \tag{1}$$

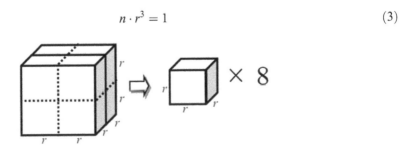

Figure 1.   Diagram of line dimension.

If the square with an area of 1 is divided into n equal parts and the side length of each small square is r, as shown in Figure 2, then:

$$n \cdot r^2 = 1 \tag{2}$$

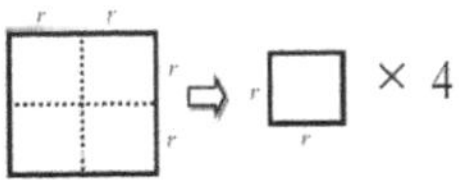

Figure 2.   Diagram of square dimension.

If the normal cube with a volume of 1 is divided into n equal parts and the side length of each small normal cube is r, as shown in Figure 3, then:

$$n \cdot r^3 = 1 \tag{3}$$

Figure 3.   Diagram of cube dimension.

In these three equations, the power of r is actually the space dimension of the geometric form that can obtain the constant measurement, so we have:

235

$$n \cdot r^D = 1 \tag{4}$$

where D is the space (fractal) dimension. Taking the logarithm of this equation to get the expression of space dimension, we have (Falconer, 1991):

$$D = -\frac{\ln(n)}{\ln(r)} \tag{5}$$

3 FRACTAL GRAPH GENERATION

3.1 *Complex plane*

The complex plane is represented by the real number on the X-axis of the Euclidean plane and the virtual number on the Y-axis of the Euclidean plane. The mathematical expression of a complex number z is:

$$z = z_x + z_y i \tag{6}$$

where z_x is the real part in the complex number ($\text{Re}z$); z_y is the virtual part in the complex number ($\text{Im}z$); i is the virtual constant in the complex number, and is defined as $i = \sqrt{-1}$, that is, $i^2 = -1$.

3.2 *Julia set on a complex plane*

The Julia set is obtained by iterative computation using a complex function. Obviously, this set is a kind of fractal set. The mathematical expression of the quadratic complex function of a Julia set is as follows:

$$F(z) = z^2 + c \tag{7}$$

where variables z and c are complex forms, $z = z_x + z_y i$, and $c = c_x + c_y i$, respectively. The iterative process of this equation can be written as follows:

$$z_{n+1} = z_n{}^2 + c \tag{8}$$

There are:

$$\text{Re}\, z_{n+1} = z_{n+1,x} = z_{n,x}^2 - z_{n,y}^2 + c_x \tag{9}$$

$$\text{Im}\, z_{n+1} = z_{n+1,y} = 2z_{n,x} z_{n,y} + c_y \tag{10}$$

Obviously, in the Julia set, the value of complex c is the constant value that controls the complex function to be iterated on the complex plane. Therefore, the generation of a Julia set is the iterative process of Equation (7). After the constant value c is given and the real and virtual parts of complex number z_0 are taken in the complex plane, we have:

$$z_1 = z_{1,x} + z_{1,y}\, i = (z_{0,x}^2 - z_{0,y}^2 + c_x) + (2z_{0,x}\, z_{0,y} + c_y)i \tag{11}$$

The Julia set graph can be obtained by matching these sets with different colors to the points of different types of sets. This process is called an *escape time algorithm*. For higher-order Julia sets, the following complex functions can be shown as:

$$F_n(z) = z^n + c = z^{n-1} \cdot z + c = \overbrace{z \cdot z \cdot z \cdots z}^{n} + c \qquad (12)$$

Obviously, this equation can also be substituted for the escape time algorithm. The algorithm virtual code is shown in Table 1.

4 FRACTAL GRAPH GENERATION

According to the process of higher-order Julia set generation, we have developed a computer program for an escape time algorithm for higher-order Julia set generation. Table 2 shows the fractal graph generation of a third-order Julia set on the complex plane; Table 3 shows the fractal graph generation of a fourth-order Julia set on the complex plane; Table 4 shows the fractal graph generation of a fifth-order Julia set on the complex plane.

Table 1. The virtual code of escape time algorithm for higher-order Julia sets.

Initialization	Procedure Julia();
	double cx,cy;
	double zxmin,zxmax,zymin,zymax;
	int wa,wb;
	double dx = (zxmax-zxmin)/(wa-1);
	double dy = (zymax-zymin)/(wb-1);
	double L = 100;
	int N = 255;
Iterative procedure	for int i = 0 to wa
	for int j = 0 to wb
	double zx = zxmin+i*dx;
	double zy = zymin+j*dx;
	int d = 0;
	for int k = 0 to N
	double zxk = zx*zx-zy*zy+cx;
	double zyk = 2*zx*zy+cy;
	double r = zxk*zxk+zyk*zyk;
	if r>L then
	int d = k;
	exit for k;
	end if
	zx = zxk; zy = zyk;
	end for k
	call Set_Draw_Color(d);
	call Draw_Point(i,j);
	end for j
	end for i

Table 2.   The fractal graph generation of a third-order
Julia set.

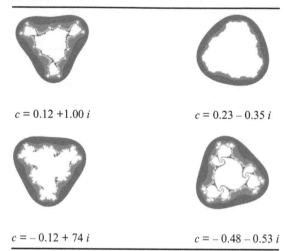

$c = 0.12 + 1.00\,i$	$c = 0.23 - 0.35\,i$
$c = -0.12 + 74\,i$	$c = -0.48 - 0.53\,i$

Table 3.   The fractal graph generation of a
fourth-order Julia set.

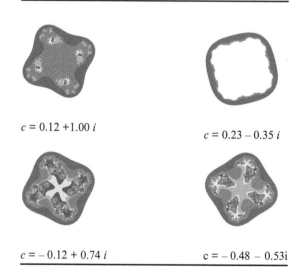

$c = 0.12 + 1.00\,i$	$c = 0.23 - 0.35\,i$
$c = -0.12 + 0.74\,i$	$c = -0.48 - 0.53i$

Table 4. The fractal graph generation of a fifth-order Julia set.

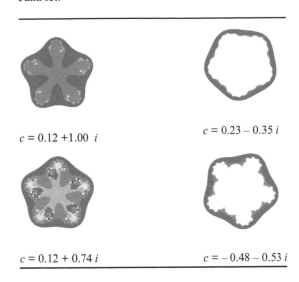

$c = 0.12 + 1.00\ i$	$c = 0.23 - 0.35\ i$
$c = 0.12 + 0.74\ i$	$c = -0.48 - 0.53\ i$

5 CONCLUSION

In this paper, a method is presented for generating fractal graphs for higher-order Julia sets. According to the complex plane generation method, fractal graphs are generated and designed. Fractal graph generation not only has artistic aesthetic properties but also has self-similar visual features and performance, which can be used in the design of actual artistic graphics and patterns (Lam, 2017; She & Lam, 2016). Thus, the process can also be used to develop a program for generating fractal artistic graphics.

ACKNOWLEDGMENT

This paper was made possible thanks to the Social Science Foundation of Fujian Province under item No. GY-S19045, and the Foundation of Fujian University of Technology under item No. GY-S18014.

REFERENCES

Falconer, K. J. 1991. *Fractal geometry: Mathematical foundations and application*. New York: Wiley.
Lam, A. D. K. T. 2017. A study on fractal patterns for the textile design of the fashion design. *Applied System Innovation for Modern Technology*, 676–678.
Mandelbrot, B. B. 1977. *Fractal: Form, chance, and dimension*. San Francisco: W. H. Freeman.
Mandelbrot, B. B. 1983. *The fractal geometry of nature*. 3rd ed. New York: W. H. Freeman.
She, G. H., & Lam, A. D. K. T. 2016. A mathematical model for the visual complexity of lacquer paintings artistic style. *Journal of Internet Technology*, 17(1),109–115.
Valor, M., Albert, F., Gomis, J. M., & Contero, M. 2003. Textile and tile pattern design automatic cataloguing using detection of the plane symmetry group. *Proceedings of Computer Graphics International*, 112–119.

Material science & Engineering

Innovation in Design, Communication and Engineering – Lam et al. (eds)
© 2020 Taylor & Francis Group, London, ISBN 978-0-367-17777-5

Research on the detection of toxic gases in the environment using APP and the IoTs

Jui-Chang Lin
Department of Mechanical Design Engineering, National Formosa University, Yunlin, Taiwan
Smart Machine and Intelligent Manufacturing Research Center, National Formosa University, Huwei, Yunlin, Taiwan

Cheng-Jen Lin
Institute of Mechanical and Electro-Mechanical Engineering, National Formosa University, Yunlin, Taiwan

ABSTRACT: This project is the detection of environmentally toxic substances such as CO, gas, NO, etc. When the sensor detects toxic gases such as CO and excessive gas concentration, it will automatically send a message to the designated customer's mobile phone, which can handle gas poisoning events in a timely manner. In this study, the device toxic substance sensor and WIFI were applied to control the high-quality environment, and the toxic substances were monitored through the Internet of Things. When the human body load is exceeded, the information is transmitted to the APP mobile phone or the management unit. This research and development is aimed at CO, gas, NO and APP online monitoring and management services to achieve friendly environmental management.

The design of the system can be divided into three parts: (1) It can be combined with a APP program to detect whether there is excessive concentration of CO, gas and other toxic substances in the air, and to transmit a message to block the risk factor. (2) It can be installed on toxic equipment such as gas stoves, and can detect the concentration of CO, gas, NO and other toxic substances at any time, and activate the solenoid valve to block the gas. (3) It can be installed independently in the required environment to detect the concentration of CO at any time. These three can be linked to the APP and send information to the monitoring unit via WIFI or Bluetooth. This study applies APP and IoT systems to effectively monitor the access of CO and toxic substances. The accuracy of detection of toxic gases by experimental statistical analysis is over 99%. The data access time error value is 3 sec. The data access error is less than 1%.

Keywords: Toxic Gases, APP program, Detection, CO, NO

1 INTRODUCTION

This study is to install CO and GAS sensing devices in the environment. When detecting excessive CO and GAS, it will automatically transmit notification to the designated customer's mobile phone to timely handle CO and GAS poisoning events.This study will combine the device Bluetooth or WIFI Internet access mechanism with quality environmental management. The user can also use it on the general household gas water heater, and can check the gas content of the gas water heater at any time, and intelligently manage the safety protection of the household gas hot water mechanism. Because of the general gas water heater, the first safety protection.

The general gas water heaters in the market lack safety sensing and intelligent design management.

This research and development is aimed at the COX security part of the APP online service, in order to achieve the product security and use the social security responsibility. The design of the system can be divided into three parts: (1) It can be combined with gas water heaters to manage environmental poisons to detect whether there is excessive concentration of CO or GAS in the air. (2) It can be installed in equipment such as gas stoves, etc. It can also detect the concentration of CO and GAS and activate the solenoid valve to block the gas. (3) It can be installed independently in the required space to detect the concentration of CO at any time.

These three can be linked to the APP to transmit information to the user via WIFI or Bluetooth. This system will first incorporate a gas water heater. The CO safety protection system to be developed by this project is mainly composed of gas water heater, CO sensing device and APP software and hardware. The higher and safer configuration of products is always the development trend of high-end products [1-5]

2 RESEARCH METHODS AND BASIC STRUCTURE

This study is to install CO and GAS sensing devices in the environment. When detecting excessive CO and GAS, it will automatically transmit notification to the designated customer's mobile phone to timely handle CO and GAS poisoning events. The Institute developed the software and hardware construction of the toxic substance detection Internet of Things system for the detection of toxic substances in the environment CO and gas. The equipment designed by this system is mainly composed of software and hardware such as toxic substance sensing system, Internet of Things system and data management system, and combined with app and big data management: (1) basic sensing system: it can also be used in other organisms in the future, such as bio-sensing, chemical sensing, and human body sensing systems. (2) Installation design and data acquisition of the sensing system: concentration status records of various substances, environmental changes, status change records, and cloud big data records. (3) Cloud monitoring and APP software design: including monitoring software design, APP software calculation and data statistical analysis records. (4) Intelligent software: The basis for environmental monitoring changes will be provided based on the aforementioned records and usage data.

(Development of a large data collection and sensing system for environmental detection, using a gas-sensing device to link the APP and the server, as shown in Figure 1. The digital signal is transmitted to the computer, and the program written by the compiling software is combined with the signal of the micro switch selected by the Arduino hardware dedicated platform, converted into the corresponding density value, and can be directly output as an Excel file, so as to facilitate the subsequent establishment of the database. And data storage, specifications are shown in Table 1. The Arduino specification used is Arduino Uno, as shown in Figure 2. Arduino Uno is suitable for developing a wide range of sensors or IoT applications.

The data extractor temporarily uses the instruNet Model 100 as a product of the US GW Instruments company or other data extractor such as the 8050, which has 8 channels, and its specifications are shown in Table 2.

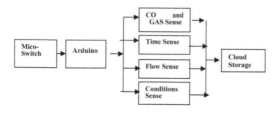

Figure 1.   Basic structure.

Table 1. Arduino Uno specifications.

Name	Current	Specifications	Unit
Mico-Switch	ATmega328P	Analog input Pins	6Sets
Digital I/O Pins	14 Sets	PWM digital I/O Pins	6 Sets
Maximum input voltage	6-20V	Memory	32 KB
Recommended input voltage	7-12V	3.3V Pin DC Current	50 mA
Operating Voltage	5V	Each pin DC current	20 mA

Figure 2. Arduino chip.

Table 2. InstruNet Model 100 specifications.

Current	Specifications
A/D Conversion Time	4μs m in
Voltage Input Range	±5V, ±.6V, ± 78mV max (80 typ), ± 8mV max (10 typ)
Number of Channels	8di
Input Resistance	10Mohm±1%

3 EXPERIMENTAL DESIGN ANALYSIS AND RESULTS

The APP software is connected to the cloud network layer, the sensor component of the sensing layer, and the application layer after the data is collected. Based on the IoT architecture, it is designed for automatic CO, GAS concentration detection and remote truncated gas systems. The system is divided into three parts, as explained below:

(1) Perceptual layer:
  (A) Embedded chip built-in Bluetooth wireless communication technology will instantly detect the CO content and gas flow. The detector will return the CO content and gas flow to the host server. And when the CO and GAS concentrations are too high, the app will notify the monitoring unit of the message.
  (B) Data back-transmission technology adopts Bluetooth wireless communication protocol, which fully complies with the design of IoT architecture, has low power consumption, low power, and meets the rapid growth trend of the wireless connection market in the world in the future.
  (C) Embedded wafers, in addition to installed CO, GAS content sensors and gas flow sensing, at least 10 sensor components can be expanded. If the system needs to increase the sensing components at any time in the future, it can be customized according to the needs.

(2) Network layer:
  (A) The embedded chip collects the CO and GAS concentration data and returns it to the host server. The host server schedules the application to periodically upload data to the cloud server database.

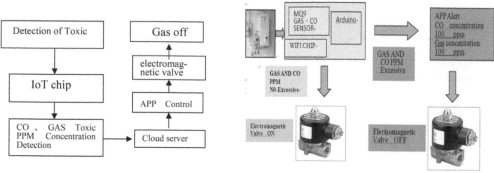

Figure 3. Operation architecture.

Figure 4. APP operation system.

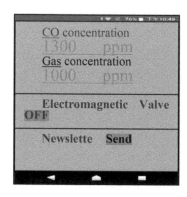

Figure 5. APP mobile phone contro.

(B) The cloud database collects concentration data at various time points, mainly for establishing a database of big data. It can be used to integrate and analyze intelligent systems in the future.

(C) Consider the importance of the cloud database, the cloud server automatically schedules backup of all data.

(D) Cloud data transmission is based on https transmission security agreement

(3) Application layer

(A) Smart Cloud Interface Design; (B) User Management System: (C) Limited Management System; (D) CO、GAS Content Detection. The system software and hardware operation architecture is shown in Figure 3, the APP and Equipment architecture diagram is shown in Figure 4.

The APP control program contains: (1) GAS electromagnetic valve control function; (2) CO, GAS and other toxic substance sensing data; (3) GAS flow display and other functions. Its mobile phone screen is shown in Figure 5.

4 RESULTS AND DISCUSSION

In order to verify the accuracy of the signal reading of the system, set the appropriate CO, GAS concentration and read the data and upload the data to the cloud database. This study then took out the cloud data for statistical analysis of its accuracy. The test was conducted with CO and

GAS concentration tests at 100PPM, 200PPM, 300PPM, 400PP M, and 500PPM, and the solenoid valve was closed and the newsletter was transmitted. Calculate the accuracy of the combination of software and hardware in the IoT system.Temperature measurement analysis.

5 CO AND GAS CONCENTRATION MEASUREMENT ANALYSIS

First, this study focused on project experiments to detect CO concentration and turn off the valve switch. The CO concentration experimental detection amounts were 100 PPM, 200 PPM, 300 PPM, 400 PP M, and 500 PPM, respectively. And take 10 experimental values as an average and calculate the success rate. As shown in Table 3, the experimental statistics.

According to the statistics in Table 3 the accuracy of the IoT sensing linkage valve is 100%. The GAS concentration experimental detection amounts were 100 PPM, 200 PPM, 300 PPM, 400 PP M, and 500 PPM, respectively. And take 10 experimental values as an average and calculate the success rate. As shown in Table 4 the experimental statistics.

According to the statistics in Table 4 the accuracy of the IoT sensing linkage valve is 100%. The actual online operation is shown in Figure 6.

Table 3. CO concentration experimental analysis.

Exp.NO.	CO concentration	Number of experiments	valve turn off (Success rate)
1	100PPM	10	100%
2	200PPM	10	100%
3	300PPM	10	100%
4	400PPM	10	100%
5	500PPM	10	100%

Table 4. GAS concentration experimental analysis.

Exp.NO.	GAS concentration	Number of experiments	valve turn off (Success rate)
1	100PPM	10	100%
2	200PPM	10	100%
3	300PPM	10	100%
4	400PPM	10	100%
5	500PPM	10	100%

Figure 6. Online operation.

Table 5. GAS cloud data transmission.

EXP. NO.	GAS concentration	Storage amount/3 MIN	Error	Average
1	100PPM	58 DATA	3.37%	
2	200PPM	59 DATA	1.67%	
3	300PPM	59 DATA	1.67%	2.69%
4	400PPM	58 DATA	3.37%	
5	500PPM	58 DATA	3.37%	

Table 6. CO cloud data transmission.

EXP. NO.	CO concentration	Storage amount/3 MIN	Error	Average
1	100	59 DATA	1.67%	
2	200	59 DATA	1.67%	
3	300	59 DATA	1.67%	2.35%
4	400	58 DATA	3.37%	
5	500	58 DATA	3.37%	

6 CO AND GAS CLOUD DATA TRANSMISSION ANALYSIS

The CO and GAS concentration test experiments in this study were five kinds of 100PPM, 200PPM, 300PPM, 400PP M, and 500PPM. The PPM value obtained by the measurement is uploaded to the cloud database, and an indicator of accuracy statistics is performed. The experiment uploads a piece of data every 3 seconds, and a 3-minute experiment to extract data analysis. The results of the GAS measurement are shown in Table 5. The results of the CO measurement are shown in Table 6.

7 CONCLUSION

This research is applied to the research and development of the Internet of Things for the detection of toxic substances in environmental CO and gas. It is based on intelligent management and will use equipment to control toxic substances and use APP to control environmental safety. Record and monitor environmental toxic substances through the Internet of Things. When the human body load exceeds the human body load, the information will be transmitted to the APP mobile phone or management unit. . The conclusions are as follows:

(1) Using the concept of Internet of Things for remote monitoring of environmentally toxic substances, saving manpower and optimizing the management of environmental management.
(2) The environmental poison concentration record management part can be directly managed by the sensing system and recorded by the cloud. Its concentration error is within 3%.
(3) The error of cloud data storage is less than 1%, which can be used for data calculation and preservation. As a comprehensive data collection of toxic substances in various regions in the future, it can be used as a system for regional environmental safety control. Provide a safe living environment for the people of the country and improve the quality of life.

REFERENCES

[1] Android SDK, http://developer.android.com/.
[2] LeJOS, "Java for LEGO Mindstorms", http://lejos.sourceforge.net/, 2009.
[3] VNC, http://www.cl.cam.ac.uk/research/dtg/ttarchive/vnc/index.html, 1999.
[4] Linux on Android project, http://forum.xda- developers.com/showthread.php?t=1585009.
[5] Android Terminal Emulator, https://play. google. com/store/apps/details?id=jackpal. androidterm&feature=search_result.

Material & Automation engineering

Innovation in Design, Communication and Engineering – Lam et al. (eds)
© 2020 Taylor & Francis Group, London, ISBN 978-0-367-17777-5

Optimization of distribution vehicle routing about chain supermarkets

Rongxia Zhou & Shih-Ming Ou
School of Business Administration, Fujian Jiangxia university, Fuzhou, Fujian, China

ABSTRACT: In the operation cost of chain supermarkets, logistics distribution cost accounts for the main part. Strengthening the planning and design of logistics distribution in supermarket chains and rationally reducing the cost of logistics distribution are of great significance for improving the economic, social and environmental benefits of enterprises. This paper establishes a decision-making model of fleet scale and distribution routes for supermarket chains, and carries out an empirical analysis with large supermarket chains in Fujian-(China) as an example. The results show that distribution time, vehicle, fleet scale and number of delivery trips are all within a reasonable range, which conforms to the actual operation of logistics distribution and verifies the validity and feasibility of the model. Related research can provide decision-making reference for peer enterprises to reduce logistics distribution cost.

1 INTRODUCTION

As a modern business model and organizational form, chain operation has become the general trend of global retail development. Supermarket chain is a typical business form of chain operation with bright development prospect. Logistics distribution is the end of the chain supermarket logistics process, but it also directly determines the cost of chain operation, affecting the profitability and competitiveness of the chain supermarket. The traditional method of direct delivery from suppliers to each supermarket store will cause a lot of waste of energy and increase of transportation cost.At present,the logistics distribution of large supermarket chains is generally completed by distribution centers throughout the country. In order to improve work efficiency, reduce logistics cost and alleviate urban traffic congestion, it is necessary to make reasonable planning of distribution fleet routes.

In recent years, the issue of fleet size and vehicle routing problem (VRP) has been widely concerned,and has accumulated rich research results. Klincewicz and Luss [1], Coto and Potvin [2], Kopfer [3] and others constructed the fleet size decision model to provide a method for reducing distribution transportation costs. Since 1959, the VRP problem was first proposed by Dantzig and Ramsar. Since then, due to a large number of practical applications, it has gradually attracted the attention of many scholars, and its theory and application have been further developed, making the VRP model perfect. Through sorting out the existing literature, it is found that the shortcomings of the establishment and solution methods of vehicle path problem model are as follows: Scholars are often limited in modeling, generally only considering time window constraints, vehicle capacity constraints, path openness constraints,etc. [4-5] Second, most models are basically single-objective programming. Third, most scholars only consider the VRP of a single uncertain factor and fail to integrate various uncertain factors, which is not practical enough [6-8]. Fourth, the size of the currently treated VRP problem is not large enough and computing capacity is limited. In addition, the existing research is

*Corresponding author: zhourongxia@fjjxu.edu.cn

a unilateral study of fleet size or Vehicle routing problem, and lack of empirical analysis. [9-10]

In view of this, this paper explores fleet size and Vehicle routing problem as a whole. Through field visits and investigations, comprehensive consideration of various influencing factors, then builds a chain supermarket fleet delivery route model, and carries out empirical analysis with the actual data of a local large chain supermarket in Fujian(China). Finally, the best fleet size and distribution route are solved to achieve the goal of reducing the logistics and distribution cost of the chain supermarket. The research conclusions can provide decision-making reference for the logistics distribution of the supermarket chain groups in the industry.

2　RESEARCH METHODS

2.1　Construction and analysis of fleet scale model

2.1.1　Influence factors of fleet scale

(1) Distribution factors

Distribution area: Including the size of the distribution area, the traffic congestion in the distribution area, the density of distribution points in the distribution area, relevant traffic regulations, geographical environment and so on.

Demand characteristics: It can be divided into periodic orders, normal orders and sudden orders.

Type of transport: It can be divided into vehicle transportation and part-load transportation. It needs to be reasonably arranged according to the characteristics of cargo source, transportation time and transportation cost.

Product features: It can be divided into unit products and multiple products. The former generally does not require separate distribution and processing, the latter needs to be distributed separately and treated differently, such as fresh products and liquid products.

(2) Vehicle restrictions

Restrictions on vehicle load and volume: Service time limit of drivers (mainly limited by working hours and delivery quality of drivers),imitation of delivery times.

(3) Fleet planning factors

Team type: It can be divided into a single Vehicle team and multi-vehicle team.

Business model: There are three types, namely private fleet, outsourcing fleet and hybrid fleet.

(4) Cost factors

Fixed cost: It includes fixed salary, bonus, pension, license tax and management fee in equipment purchase cost and maintenance cost etc.

Variable cost: It includes mileage bonus, vehicle bonus, overtime pay, operation cost of the vehicle, vehicle repair cost, parts cost, fuel cost, etc.

2.1.2　Model and procedure of fleet scale decision

On the basis of the relevant achievements of Klincewicz and Luss, combined with the logistics distribution of chain supermarkets, a mathematical programming model is proposed to assist in the selection of distribution strategies and the determination of the optimal size of private fleet, and agree to allow the use of outsourced fleets for delivery services. The algorithm steps of the entire fleet size are as follows:

Step1: Cut the distribution area into similar size areas (marked as j) $(j = 1, 2, \cdots, m)$, and each region contains a different number of requirements points (marked as λ_j), then roughly calculate the time (marked as μ_j) required to deliver to the area j. The calculation of μ_j is shown below:

$$\mu_j = \lambda_j S_j + \frac{\sqrt{\lambda_j} r_j}{V_j} + \frac{2r_j}{(2\lambda_j + 1) V_j} \tag{1}$$

λ_j represents the average distribution points of region j ; S_j express as the average service time for each customer in region j ; r_j represents the distribution radius of region j ; V_j represents the average speed of the vehicle in region j.

Step2: Determine the number of species point, and each kind of point represents the need for a vehicle for service, species point (marked as i, i = 1, 2, · · · , n).

Estimation method of the number of points (marked as p^*):

$$p^* \geq 4 \sum_{j=1}^{m} \frac{\mu_j}{T} \tag{2}$$

T represents the maximum working hour limit of the vehicle.

Step3: Determine variable cost functions C_{ij}, that is, variable cost incurred when species point (vehicle) i delivered to each region j.

$$C_{ij} = \rho * d_{ij} / f_c \tag{3}$$

In the formula above, ρ indicates the price of diesel ; d_{ij} represent the distance traveled by species point (vehicle) i to distribution area j ; f_c indicates travel distance of per liter of diesel.

Step4: Mathematical model of fleet size planning

f_i : the fixed cost of vehicle i

$$y_i = \begin{cases} 1 & \text{Vehicle is used} \\ 0 & \text{others} \end{cases} \tag{4}$$

$$x_{ij} = \begin{cases} 1 & \text{Vehicle i delivers area j} \\ 0 & \text{others} \end{cases} \tag{5}$$

$$Min \ Z = \sum_{i=1}^{n} f_i y_i + \sum_{i=1}^{n} \sum_{j=1}^{m} c_{ij} x_{ij} \tag{6}$$

Subject to

$$\sum_{i=1}^{n} x_{ij} = 1; \ j = 1, 2, \cdots, m \tag{7}$$

$$x_{ij} \leq y_i ; \ j = 1, 2, \cdots, m \tag{8}$$

$$x_{ij}, y_i \in \{0, 1\} \tag{9}$$

This model takes into account the delivery radius in each cut area, the estimated delivery time, and the estimated delivery distance, and in calculating its variable costs, the distance between regions is also taken into account, and it is assumed that customers in all regions are served by a distribution center, and the fleet scale of a single vehicle type and a single trip is taken into account.

2.2 Construction and analysis of VRP model

This paper uses lingo software and 0-1 integer programming mathematical model to construct VRP model.

2.2.1 Parameter setting of VRP model

Sets x_{ijk} as the vehicle drives from the distribution point i (customer point) to the distribution point j :

$$X_{jki} = \begin{cases} 1 & \text{if } do \\ 0 & otherwise \end{cases} \tag{10}$$

q_i : the demand for distribution point
Q : the maximum number of vehicles
M: the total number of vehicles in the distribution center
C_{ij}: the unit cost of the vehicle from the delivery point i to the delivery point j
d_{ij}: the distance from delivery point i to delivery point j
D : delivery center

2.2.2 VRP decision model

Objective function:

$$MinZ = \sum_{k=1}^{m}\sum_{i=1}^{n}\sum_{j=1}^{n} d_{ij}c_{ij}x_{ijk} \tag{11}$$

Subject to

$$\sum_{k=1}^{m}\sum_{i=1}^{n} x_{ijk} = \sum_{k=1}^{m}\sum_{j=1}^{n} x_{ijk} = 1; \tag{12}$$

$$\sum_{k=1}^{m}\sum_{j=1}^{n} x_{Djk} - \sum_{k=1}^{m}\sum_{j=1}^{n} x_{iDk} = 0; \tag{13}$$

$$\sum_{k=1}^{m}\sum_{i=1}^{n}\sum_{j=1}^{n} x_{ijk} \leq M \leq Q; \tag{14}$$

$$k = 1, 2, \ldots, m; \; i, \; j = 1, 2, \ldots, n$$

Equation (12) means that there is only one car is available at each distribution point; Equation (13) means that each vehicle departs from the distribution center and returns to the distribution center after completing all tasks; Equation (14) indicates that the total number of vehicles carrying out the distribution service cannot exceed the total number of vehicles in the distribution center.

3 EMPIRICAL ANALYSIS

3.1 Background of Fuzhou Yonghui supermarket

In 1995, Yonghui Supermarket started in Fuzhou. At the beginning, it was just a small profit supermarket. Today, Yonghui Supermarket has developed over 830 chain supermarkets across the country, with a business area of more than 6 million square meters. It ranks among the top 6 Chinese chain companies in 2017 and the top four fast-moving consumer goods chain in China. As the first to introduce the modern supermarket fresh agricultural products circulation enterprise, Yonghui Supermarket has been praised as the promotion model of

China's "agricultural reform super". People call it "people's livelihood Supermarket, common people's Yonghui". Yonghui Supermarket has developed into a large-scale enterprise group based on the retail industry and supported by modern logistics and based on industrial development. Distribution is the link of the chain supermarket supply chain, which plays a vital role in the development of the chain supermarket. Therefore, it is of great significance to explore the vehicle distribution problem of the supermarket.

The delivery vehicle of Fuzhou Yonghui Supermarket starts from Fuzhou Nanyu Logistics Park, via designated route, the goods will be delivered to various stores in Fuzhou City. This paper divides Fuzhou into four major regions for case analysis. The background of the case is as follows:

(1) Distribution area: Fuzhou
(2) Delivery points: 42
(3) Fuzhou is divided into four major districts: Gulou district, Taijiang distric, Jin'an district and Cangshan district
(4) Distribution center location: Fuzhou Nanyu Logistics Park
(5) Distribution vehicle type: this article uses 5-ton vehicles
    Demand data of each store in the four regions is shown in Table 1:

Table 1.   Demand data of each store in the four regions (unit: t).

Gulou district				Jin'an district			
Jintai store	0.07	Tongpan store	0.38	Fuxin east road store	0.42	Gushanyuan store	0.63
Pingshan store	0.66	Ximen store	0.46	Fuma road store	0.28	Residential park theme store	0.6
Shengfu store	0.89	Daru Shijia store	0.78	Chahui store	0.9	Xiangyuan Store	0.28
Guomian store	0.59	Tianquan store	0.75	Fufei store	0.17	Hualin store	0.63
Dawn store	0.2	Xihong store	0.45	Jinhui store	0.38	Xiyuan store	0.09
Jingda store	0.11	Pingxi store	0.31	Wusibei store	0.97	Helin store	0.53
Yangqiao store	0.62	Rongqiao store	0.3	Yuefeng store	0.8	Sanmu store	0.48
Taijiang distric				Cangshan district			
Hongxing yuan Aofeng square store	0.32	-	-	Fuwan Store	0.91	Puxin store	0.84
Mass road store	0.75	-	-	New Tianyu square Store	0.96	Shoushan store	0.64
Bishui Fangzhou store	0.62	-	-	Shangdu store	0.84	Jiangnan Shuidu store	0.66
Aofeng store	0.94	-	-	Century Jinyuan Puxia store	0.23	Huida store	0.69
＂	-	-	-	Jinxiang store	0.74	Aegean store	0.13

3.2.1 *Time parameter data*

(1) The average service time for each customer (marked as S_j) is shown in Table 2:
(2) Maximum working hour limit (marked as T) for vehicles:
    The working hours include the loading time, transportation time, waiting time, and unloading time in the logistics center, which is about 20 hours per day.

3.2.2 *Distance data*

(1) According to the address information provided by all Yonghui supermarkets in Fuzhou city, We can enter the location of the distribution center and the address data of all branches in each district into the Baidu map, meanwhile according to the actual road data in Fuzhou city, the actual distance between the distribution center and each branch in each area, and each branch in each area can be obtained. Due to space limitations, only the data materials of Gulou area are attached. The other three areas are not repeated here, as shown in Table 3:

Table 2.   Average service time of each customer in four regions (unit: h).

Gulou district				Jin'an district			
Jintai store	0.06	Tongpan store	0.32	Fuxin east road store	2.09	Gushanyuan store	3.16
Pingshan store	0.55	Ximen store	0.39	Fuma road store	1.38	Residential park theme store	3.01
Shengfu store	0.74	Daru Shijia store	0.65	Chahui store	4.52	Xiangyuan Store	1.41
Guomian store	0.49	Tianquan store	0.63	Fufei store	0.83	Hualin store	3.16
Dawn store	0.17	Xihong store	0.37	Jinhui store	1.91	Xiyuan store	0.45
Jingda store	0.09	Pingxi store	0.26	Wusibei store	4.87	Helin store	2.66
Yangqiao store	0.52	Rongqiao store	0.25	Yuefeng store	3.99	Sanmu store	2.38
Taijiang distric				Cangshan district			
Hongxing yuan Aofeng square store	0.27	-	-	Fuwan Store	0.76	Puxin store	0.7
Mass road store	0.62	-	-	New Tianyu square Store	0.8	Shoushan store	0.54
Bishui Fangzhou store	0.52	-	-	Shangdu store	0.7	Jiangnan Shuidu store	0.55
Aofeng store	0.79	-	-	Century Jinyuan Puxia store	0.2	Huida store	0.58
-	-	-	-	Jinxiang store	0.62	Aegean store	0.11

Table 3.   Distance between branches and distribution centers and branches in Gulou District(unit: km).

Nanyu logistics park	Jintai	Pingshan	Shengfu	Guomian	Dawn	Jingda	Yangqiao	Tongpan	Ximen	Daru Shijia	Tianquan	Xihong	Pingxi	Rongqiao
18.3	Jintai													
20.6	2.7	Pingshan												
18.2	1.4	1.7	Shengfu											
22.1	4.4	3.7	4.2	Guomian										
15.1	3	4.3	3.6	5.5	Dawn									
19.6	2.4	1.3	1.5	2.8	5.1	Jingda								
16.8	2.8	2.9	2.3	4.1	2.4	5.6	Yangqiao							
19.1	4.2	2.1	3.4	1.9	4.7	3.3	3.2	Tongpan						
17.7	2.2	1.9	1.6	3.2	3.2	1.8	0.9	3.1	Ximen					
18.2	6.3	5.8	5.8	5.7	5	6.4	4.4	4.1	4.6	Daru Shijia				
20.4	4.7	2.6	3.7	1.7	6	3.9	4.3	1.8	3.9	5.7	Tianquan			
17.1	3.1	2.9	2.6	4	2.7	2.8	1.2	2.4	1	3.9	3.8	Xihong		
19.5	3.9	1.8	2.8	1.2	5.1	3	3.4	0.9	3.1	4.9	1.1	4.7	Pingxi	
15	4.4	5.6	5.1	6.7	2.1	5.6	3.6	5.2	3.8	3.8	6.6	3.1	5.7	Rongqiao

(2) According to the method of saving mileage, the shortest distance (marked as d_{ij}) from the vehicle to each area can be obtained, as shown in Table 4:

3.2.3 *Vehicle costs information*

Through interviewing drivers and field visits, the following parameters were obtained:

(1) Fixed costs (f_i): The price of a delivery vehicle for a 5-ton model is 100,000 yuan;

(2) Variable costs (C_{ij}): The costs incurred by the distribution of the point i (vehicle) to each area *j* are as shown in Table 5:

(3) Distribution radius of the area: It can be calculated from a barrel of oil that can run 747km, so the fuel consumption f_c is 8.3 km per liter, distribution radius(r) is 118.95km;

Table 4.   The shortest distance from the vehicle to each area (unit:km).

District	Shortest distance
Gulou distric	90.8
Taijiang distric	45.7
Jin'an district	116.8
Cangshan district	85.2

Table 5.   Cost incurred by vehicle i to delivery area j (unit: yuan).

District	Costs
Gulou distric	60.17
Taijiang distric	30.28
Jin'an district	77.4
Cangshan district	56.46

(4) Average speed of the vehicle(V) is 35km per hour;
(5) current price of diesel(ρ) is 7.5yuan per liter;
(6) It takes about 10 minutes to unload 1 ton of goods at each distribution point.

3.3   Fleet size decision for distribution to each region

Bring the above relevant parameter values into Equation (1), the time of delivery to each branch of the four major regions can be separately obtained, as shown in Table 6:

Table 6.   Distance between branches and distribution centers and branches in Gulou District(unit: km).

Gulou district				Jin'an district			
Jintai store	13.77	Tongpan store	17.36	Fuxin east road store	17.81	Gushanyuan store	20.32
Pingshan store	20.65	Ximen store	18.37	Fuma road store	16.18	Residential park theme store	19.96
Shengfu store	23.36	Daru Shijia store	22.01	Chahui store	23.5	Xiangyuan Store	16.24
Guomian store	19.83	Tianquan store	21.7	Fufei store	14.88	Hualin store	20.32
Dawn store	15.26	Xihong store	18.14	Jinhui store	17.41	Xiyuan store	13.99
Jingda store	14.17	Pingxi store	16.56	Wusibei store	24.32	Helin store	19.15
Yangqiao store	20.19	Rongqiao store	16.45	Yuefeng store	22.25	Sanmu store	18.49
Taijiang distric				Cangshan district			
Hongxing yuan Aofeng square store	8.63	-	-	Fuwan Store	18.66	Puxin store	18.05
Mass road store	10.04	-	-	New Tianyu square Store	19.09	Shoushan store	16.42
Bishui Fangzhou store	9.63	-	-	Shangdu store	18.09	Jiangnan Shuidu store	16.56
Aofeng store	10.69	-	-	Century Jinyuan Puxia store	13.02	Huida store	16.85
-	-	-	-	Jinxiang store	17.27	Aegean store	12.16

From the tables above, the total time required for distribution to the four major areas (Gulou, TaiJiang, Jin'an and CangShan) can be obtained, as shown in Table 7:

Table 7.   Summary of relevant information about delivered to each region

district	Time required (hours)	Number of species (pcs)	Variable cost (yuan)	5 tons delivery car (vehicle)	Total fixed cost (yuan)	Total variable cost (yuan)
Gulou	257.82	52	60.17	2		
Taijiang	39	8	30.28	1	700,000	2141.16
Jin'an	247	50	77.4	2		
Cangshan	166.67	34	56.46	2		

3.4   *Route decision about delivering to each district*

According to the VRP model, the distribution route from the distribution center to each regional store can be obtained, as shown in Table 8:

district	Delivery route
Gulou	Route1:Nanyu logistics park→Xihong→Tongpan→Tianquan→ Guomian→Pingxi→Pingshan→Jingda→Shengfu→Jintai→Ximen→Nanyu logistics park
	Route2:Nanyu logistics park→Rongqiao→Daru Shijia→Yangqiao→Dawn→Nanyu logistics park
TaiJiang	Route:Nanyu logistics park→Mass road→Hongxingyuan→Aofeng→Bishui Fangzhou→Nanyu logistics park
Jin'an	Route1:Nanyu logistics park→Hualin→Fufei→Wusibei→Residential theme park→Xiyuan→Jinhui→Nanyu logistics park
	Route2:Nanyu logistics park→Yuefeng→Helin→ Fuxin east road→Chahui→Gushanyuan→Fuma→Sanmu →Xiangyuan→Nanyu logistics park
CangShan	Route1:Nanyu logistics park→Fuwan→Shangdu→Aegean→Jiangnan Shuidu→Jinxiang→Puxin→Nanyu Logistics Park
	Route2:Nanyu logistics park→Shoushan→Huida→Century Jinyuan Puxia→New Tianyu square→Nanyu Logistics Park

4   CONCLUSION

4.1   *Analysis conclusion*

Based on the actual data of supermarket chains, this study uses the constructed fleet size and route model to solve the fleet size and distribution route delivered to each region. In the example, the distribution of chain supermarket is relatively scattered, and the distance between the stores is also far away, which leads to a corresponding increase in the distance traveled and the delivery time of the distribution vehicles. Therefore, this paper uses a single vehicle fleet size to conduct research. The research results show that in the case of delivery using a single vehicle (5 tons) delivery vehicle, only one vehicle in Taijiang District is required for distribution, and the delivery time is also one time; the remaining areas require two vehicles for distribution. And the frequency of delivery per vehicle is also once. For the city in the example, the delivery time, delivery vehicles, fleet size, delivery times, etc. are all within reasonable limits, and the calculation result is in line with the actual delivery operation.

4.2   *Research suggestion*

The research results of this paper can be provided to the Yonghui Supermarket Headquarters in Fuzhou, as a reference for improving the logistics system. In the future, adjustments can be made based on actual conditions. At the same time, it can provide certain guiding significance for the domestic counterpart enterprises to improve the logistics distribution system. In the follow-up study, we can consider expanding the scope of the research to the whole province, and comprehensively consider the changes in demand, multi-vehicle types, etc., and improve the model and constraints to make it more realistic.

REFERENCES

Klincewicz, J G and Luss, H 1990 Fleet Size Planning when Outside Carrier Services Are Available (Transportation Science) 24 169–182.

Coto, J F and Potvin, J T 2009 A Tabu Search Heuristic for the Vehicle Routing Problem with Private Fleet and Common Carrier (EJOR) 198 464–469.

Kopfer, H and Krajewska, M 2007 A Approaches for Modelling and Solving the Integrated Transportation and Forwarding Problem In Corsten (Produktions-and Logistik management) 1 439–458.

Wang Kun 2015 Study on the Optimization of Urban Logistics Vehicle Distribution Path, Chongqing Jiaotong University (Chongqing, China).

Hai-xia Jing 2014 Research on Path Optimization of Two-way Transport Vehicles in Logistics Distribution, Wuhan University (Wuhan, China).

Zhi-Qing Yang 2015 Research on multi-target vehicle routing optimization under urban express delivery conditions, Harbin Institute of Technology (Harbin, China).

Bai Shi-Zhen, Ding Xu 2011 Research on Vehicle Routing Problem with Dynamic Travel Time(CMS) 10 53–55.

Zi-Qiang Tong, Li Pengxiang 2018 Study on the optimization of product distribution route considering real-time road conditions and vehicle turnover rate(IEM) 5 23–26.

Lei-Zhen Wang, Ding-Wei Wang and Su-Xin Wang 2018 A Vehicle Scheduling Model for Multiple Goods Transfer and Distribution and Its Mixed Solution of Particle Swarm and Ant Colony Algorithm (Information and Control 5 564–568.

Xian-Long Ge, Wang Xu and Le-bin Xing 2012 Multi-model vehicle scheduling problem with dynamic demand and cloud genetic algorithm (JSE) 6 825–827.

Management science

Innovation in Design, Communication and Engineering – Lam et al. (eds)
© 2020 Taylor & Francis Group, London, ISBN 978-0-367-17777-5

The relationships between strategy, knowledge management, and innovation performance

Mao-Sung Chen
Ph.D. Program in Engineering Science and Technology, College of Engineering, National Kaohsiung University of Science and Technology, Kaohsiung City, Taiwan

Ming-Tien Tsai
College of Engineering, National Cheng Kung University, Tainan City, Taiwan

ABSTRACT: Past studies of strategy, knowledge management, and innovation performance have primarily targeted the electronics or information industries. Studies of other industries are lacking in comparison. The government has listed the medical device industry as one of the top 10 emerging industries in Taiwan and as a target of focused development in the 21st century. Thus, the present study explores the correlations between strategy choices, knowledge management capabilities, and innovation performance. Findings of the study relevant to the research hypotheses are discussed in greater detail in what follows. The findings were as follows: (1) Product innovation performance: the prospector and analyzer strategies improved product innovation performance. (2) Process innovation performance: none of the three strategies significantly improved process innovation performance. (3) Research and development innovation performance: the prospector strategy effectively improved research and development innovation performance.

Keywords: strategy, knowledge management, innovation performance, medical instruments, apparatus

1   INTRODUCTION

The medical device industry integrates applied professional medical knowledge with fundamental industrial technology. It acts as an indicator of a nation's technological advancement and reflects the quality of the nation's technology research. In this industry, competitiveness is based on research and development (R&D) capability. Companies must continually introduce new products and engage in R&D in order to ensure their competitiveness. In general, competitiveness has been examined from the standpoints of business strategies, knowledge management, and innovation performance. Porter (2008) believes that businesses must adopt general competitive strategies if they wish to gain and maintain competitive advantages. Previous studies have examined the degree of influence of different business strategies on operation performance. Knowledge management has become a powerful tool for individuals and businesses to set themselves apart from the competition and maintain competitive advantages. Those that can successfully manage knowledge will be in a dominating position to succeed. Drucker (1993) defines innovation as "endowing resources with the new capacity to create wealth," turning resources into true capital. In addition, management theory views innovation as an important function of management that is directly linked to business performance.

Industrial technologies in the fields of computers, electronics, and chemical engineering, as well as machinery, are quite mature in Taiwan. Therefore, Taiwan's advantages in developing a niche industry for medical devices are its comprehensive related industrial systems and its strong manufacturing capacity. Studying how companies can improve their abilities to face this highly competitive environment is particularly worthwhile. This study focuses on exploring business strategy choices, knowledge management capabilities, and innovation performance differences among various companies; the effects of business strategy and knowledge management on innovation performance were determined through a literature review and a survey.

The remainder of this investigation is organized as follows. Section 2 describes the literature review used in this investigation. Section 3 presents the hypothesis. Section 4 presents the analysis result. Finally, Section 5 draws conclusions.

2 LITERATURE REVIEW

Business strategy has a great influence on an enterprise. Good business strategy constructs a framework for future enterprise development. Especially when an enterprise has limited resources, it is very difficult for it to gain benefits if it cannot develop appropriate strategies. Similarly, managers use knowledge management so to enhance the value of organizational knowledge in order to maintain competitive advantage. Therefore, this paper discusses the impact of business strategy and knowledge management on corporate performance.

2.1 Taiwan's medical device industry

Medical device manufacturing primarily depends on the application of mature technology from related industries (e.g., the chemical engineering, machinery, and electronics industries). Therefore, the field of medical devices in the biotechnology industry is an appropriate area of development for Taiwan. However, medical device product development is usually time-consuming. Medical devices must undergo numerous procedures and certification processes before they can be introduced to the market (Figure 1). The production and marketing process of the medical device industry can be divided into four stages: (1) R&D, (2) certification, (3) pilot production and application for market approval, and (4) mass production and introduction to the market. The details are shown as follows:

(1) R&D: This stage of one to three years includes analyzing laws and regulations, examining patents, designing products, and planning for future certification processes.
(2) Certification: This stage of nine months to three years includes laboratory, safety, function, and clinical testing.
(3) Pilot production and application for market approval: This stage of six months to three years includes planning for a pilot production, pilot production, and market approval by a governing authority.
(4) Mass production and introduction to the market: This stage of three to 10 years includes mass production, marketing, and repair and maintenance, as well as monitoring or supervising possible side effects, accidents, product recalls, and resource recycling after the product has been introduced.

Taiwan has been highly reliant on imported medical devices in the past. Achieving the ability to manufacture medical devices is necessary, whether the goal is to gain a share of the increasingly massive medical device market or to decrease the growing trade deficit for medical devices. The reason could be finding that Taiwan's healthcare market is relatively small; more than 70% of the companies in Taiwan's medical device industry are small and medium-sized enterprises with 50 or fewer employees and paid-in capital not exceeding 80 million NTD.

Consequently, Taiwan's medical device industry has long been considered an import-oriented industry. According to the Industrial Technology Research Institute's (2015) estimates, import dependency for this industry peaked at approximately 60%. The bottlenecks of development are key technologies, marketing, and product certification. Important topics in the development of Taiwan's medical device industry that require careful consideration are

Figure 1. Product certification flowchart for medical devices. (Source: Industrial technology research institute's industrial economics & knowledge center-industry and technology intelligence service, may 2005).

how to establish a foothold in this industry while competing against international corporations and medical manufacturers from emerging countries and how to create a competitive niche.

2.2 Strategies

Strategy is a method by which enterprises can create advantages and a development space by considering their own resources in a competitive environment (Andrews & Roland, 1987; Bea & Haas, 2016). A company's business strategy has a tremendous impact on the company. A good business strategy acts as a blueprint for the company's development and directs the company's future efforts. The choice of business strategy can decide the future of a corporation. American, European, and Japanese medical manufacturers must be unique in some way because they have secured their places in the global market. Taiwan's manufacturers must enhance efficiency and improve their competitiveness through their business strategies. Although, the scale and environment of Taiwan's healthcare industry differ from those of the American industry, Taiwan's companies can still observe and learn from the business strategies of foreign benchmark companies. The medical device industry requires a lower level of technology than pharmaceutical or other biotechnology industries. Coupled with the appeal of high profits, new competitors from related industries such as electronics or machinery are readily attracted to the medical device industry.

Therefore, one issue of priority in choosing a business strategy is how to improve organizational performance and raise the barrier to entry into the industry (Akter, 2016; Blackburn, Hart, & Wainwright, 2013; Khedhaouria, Gurău, & Torrès, 2015). Most of Taiwan's medical companies have already contemplated how to lower production costs. Among the participants in the study, R&D innovation performance was the main impetus behind business growth and improvement. Improving R&D innovation performance by choosing an appropriate business strategy can only benefit the business.

2.3 Knowledge management

An organization's most important asset is the knowledge possessed by its people. In the digital age characterized by constant changes, increasing the value of knowledge through effective knowledge management is a valuable and critical issue for companies. The company's competitive advantage arises from its ability to create knowledge and to continually innovate (Hooge & Le Du, 2014; Nonaka & Takeuchi, 1995). In other words, knowledge management has become a powerful tool for individuals and businesses to set themselves apart from the competition and to maintain competitive advantages. The creation and accumulation of knowledge is of utmost importance in knowledge management mechanisms (Bloodgood, 2015; Leonard, 1995; Ragab & Arisha, 2013; Ukko et al., 2016). Employees exchange knowledge with or without interference, and this exchange of know-how is the source of an organization's competitiveness (Huseman & Goodman, 1999; Jansen, 2017). Knowledge has no value unless it is shared or used in some way. The value of intellectual property increases with use (Chang, Liao, & Wu, 2017; Davenport & Prusak, 2001; García-Sánchez, García-- Morales, & Bolívar-Ramos, 2017; Hidding & Catterall, 1998; Pais & Santos). Old ideas can continuously generate new ideas. By sharing their knowledge, providers not only enrich the knowledge of receivers but also preserve their own knowledge (Sultani, 2016). Therefore, sharing knowledge preserves the original value of the knowledge. More important, it creates new value.

The medical device industry is knowledge-intensive and requires an effective knowledge management process to effectively increase corporate innovation performance. Knowledge creation is indeed a powerful weapon for Taiwan's medical device companies that seek to strengthen their corporate characteristics and promote industrial upgrading. Effective knowledge creation is conducive to improved innovation performance. Using this to establish the knowledge management environment and allowing its maturation are indeed key tasks for Taiwan's medical device industry in the future.

2.4 *Innovation performance*

Innovation is a key basis of profitability for medical device producers. Prioritizing technology R&D and innovative actions will be the impetus for growth of Taiwan's medical device industry. Therefore, Taiwan's medical device manufacturers have emphasized investment in the R&D workforce (Industrial Technology Research Institute, 2015).

Currently, the majority of Taiwan's medical companies are focused on innovation performance. Companies that adopt an aggressive business strategy and apply that business strategy to achieve more innovative R&D create an advantage that cannot easily be replaced by nations with lower production costs such as China or Southeast Asian countries. This advantage provides companies with an internationally advantageous position in the future.

3 HYPOTHESIS

This study was based on the knowledge management theory of Nonaka and Takeuchi (1995), Davenport (2001), Hooge and Le Du (2014), and Pais and Santos (2015); the business strategy theory of Miles and colleagues (1978), Burke (2017), and Bryson, Edwards, and Van Slyke (2017); and the innovation performance theory of Utterback (1994), Slater, Mohr, and Sepgupta (2014), and Carlborg, Kindström, & Kowalkowski (2014). Following the literature review, this section includes a discussion of the relationships between business strategy, knowledge management, and innovation performance. The research framework is shown in Figure 2. Details of hypothesis inferences are discussed as follows.

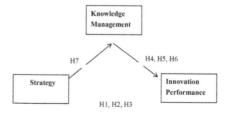

Figure 2. The research framework.

266

3.1 Strategy and knowledge management

Because acquiring and processing knowledge is expensive (Quintas, Lefere, & Jones, 1997; Venkitachalam & Willmott, 2015), companies must identify the knowledge that they need. In addition, knowledge management is the basis for implementing business strategies. Therefore, different business strategies require different coordinating knowledge management practices. In other words, knowledge management strategies supplement the implementation of business strategies. Accordingly, the coordination and integration of knowledge management and business strategies are critical (Clarke, 2001; Cook, 1999; Geisler & Wickramasinghe, 2015; Lerro & Jacobone, 2013; Wang, 2015).

The business strategy is one of the most important factors that guide knowledge management. Before implementing knowledge management, corporations must clearly define their goals and strategies for reaching them. These definitions are important because different goals require different strategies and practices. The linking of strategies and knowledge management systems and appropriate supporting measures allows organizations to understand and utilize timely and crucial knowledge and information. This increases overall business performance and thereby maintains competitive advantages.

H7: The intermediary effect of the operator's strategy through knowledge management has a significant impact on innovation performance.

3.2 Strategy and innovation performance

Firms that adopt a technologically aggressive strategy are recommended to deeply open their innovation process in order to foster innovation performance (Aloini et al., 2015). Based on competitive priority (such as quality, cost, delivery, and flexibility), innovation has been recognized as one of the sources of competitive advantage (Lao, Prajogo, & Adebanjo, 2013). Many studies have explored the relationship between business strategy and innovation performance. The findings reveal that business strategy has a positive effect on innovation performance (Lao et al., 2013; Lechner & Gudmundsson, 2014; Prajogo, 2016). Open innovation fully mediates the relationship between technological strategy and innovation performance (Aloini et al., 2015). It may have the greatest impact on innovation performance (Blackburn et al., 2013; Kim et al., 2014; Samiee & Lee, 2014). Miles and colleagues (1978) divide organizations into one of four major categories and propose an adaptive cycle concept to describe these categories. The cycle has three components and 11 dimensions.

H1: The business strategy has a significant impact on the product innovation performance.
H2: The business strategy has a significant impact on the process innovation performance.
H3: The business strategy has a significant impact on R & D innovation performance.

3.3 Knowledge management and innovation performance

Innovation will prove to be a key driver in maintaining transient advantage (Dobni, Klassen, & Nelson, 2015). Firms have increasingly relied on sources of knowledge in their R&D process to develop and profit from innovation (Berchicci, 2013). Demarest (1997) postulates that the most obvious correlation between knowledge management and improved economic performance is in the area of innovations. Previous studies have found that creation of new knowledge within a corporation signals the beginning of innovation. This new knowledge may be process-oriented rather than product-oriented and can mean new ways of working. As the market becomes more fragmented, corporations gain better control of when to enter the market.

Past studies have devoted much attention to empirical examination of the relationship between knowledge management and innovation performance (Cohen & Olsen, 2015). However, innovation requires capital, a key component of which is knowledge. In other words, continuous innovations are supported through knowledge management (Berchicci, 2013; Birasnav, 2014; Cohen & Olsen, 2015; Lai et al., 2014; Noruzy et al., 2013).

H4: Knowledge management capabilities have a significant impact on product innovation performance.

H5: Knowledge management capabilities have a significant impact on process innovation performance.

H6: Knowledge management capabilities have a significant impact on R & D innovation performance.

A review of literature shows that close relationships exist between business strategy, knowledge management, and innovation performance.

4 THE ANALYSIS RESULT

This study used a 5-point Likert scale to measure the level of agreement of the respondents for items in the questionnaire. Several technologies are listed for respondents to choose from as the target technology being introduced. The questionnaire is designed in four parts. The first part collects the data of categories about business strategy. The second compiles the data of technique about knowledge management. The third gathers data about product innovation performance, process innovation performance, and R&D innovation performance.

The target population included members of the Taiwan Medical and Biotechnology Industry Association. Stratified random sampling (by business type) was used to select 300 companies in Taiwan. Random sampling was used to select R&D directors, general managers or vice presidents, or professionals in a higher position, and distributed questionnaires were given to these respondents. After questionnaire design and survey sampling, a total of 229 questionnaires were collected; 201 were valid and 28 were ineffective. The effective response rate was 67%. The descriptive statistics are shown in Table 1.

The regression analysis performed in this study verified that both business strategies and knowledge management capabilities significantly affect innovation.

4.1 Regression analysis

The regression analysis performed in this study verified that both business strategies and knowledge management capabilities significantly affect innovation performance. The medical device industry relies on market and customer skills and financial information to a greater extent than other industries. Information and document management have always been at the core of medical device industry management, and the need for effective knowledge management systems is quite urgent. The goal of knowledge management is the creation and accumulation of knowledge and knowledge creation is of utmost importance (Bloodgood, 2015; Leonard, 1995; Ragab & Arisha, 2013; Ukko et al., 2016). Therefore, effective knowledge

Table 1. Descriptive statistics.

Company categories		Manufacturer	37.1%
Company capital		Under 10 million	23.6%
		50–100 million	14.8%
		Over 1 billion	10.9%
the % of R&D expenditure of the revenue		1–5%	29.7%
The ratio of R&D staff		Under 10%	32.3%
		Items	Cronbach's alpha
Business strategy	Explore	3	0.833
	Analyze	2	0.628
	Defend	2	0.445
Knowledge management	Create	5	0.870
	Share	5	0.896
Innovation performance	Product	6	0.917
	Process	4	0.854
	R&D	9	0.965

creation capabilities can result in significantly positive innovation performance for healthcare businesses (Tables 2 and 3).

4.2 *Mediation analysis*

Path analysis showed that knowledge management is a mediator of the relationships between prospector strategy and product innovation or R&D innovation. The prospector strategy directly and significantly affects both product innovation performance and R&D innovation performance. However, the relationships between prospector strategy and product innovation performance or R&D innovation performance could also be achieved by knowledge management. This implies that choosing the prospector strategy directly affects knowledge management performance, which in turn affects product and R&D innovation performance via a mediating effect.

Individual hierarchical regression analyses of knowledge creation only, knowledge sharing only, and the combination of knowledge creation and knowledge sharing shows that all three models lower the effect of the prospector strategy on product innovation and on R&D innovation. However, knowledge creation has a greater mediating effect than knowledge sharing. Therefore, healthcare businesses should choose the prospector strategy to achieve more beneficial product and R&D innovation performance. This proactive business strategy can create an environment conducive to creation and innovation. Achieving superior knowledge creation will naturally lead to superior product and R&D innovation performance for the business.

Table 2.   Regression analysis.

MODEL		M1-1	M1-2	M1-3	M2-1	M2-2	M3-1	M4-1	M4-2
		Product	Process	R&D	Product	Process	R&D	Create	Share
Knowledge management	Create	0.163*	0.222**	0.179 *					
	Share	0.519***	0.443***	0.525***					
Business strategy	Explore				0.700 ***	0.429***	0.603***	0.383***	0.471 ***
	Analyze				0.045	0.159**	0.141*	0.200***	0.208***
	Defend				0.048	−0.005	-0.013	0.125*	0.040
adjR^2		0.415	0.384	0.440	0.541	0.269	0.459	0.309	0.374
F		81.874	72.182	90.752	90.654	28.924	65.490	35.027	46.388
p		0.000	0.000	0.000	0.000	0.000	0.000	0.000	0.000

*<95%; ** <99%; *** <99.9%

Table 3.   Regression analysis.

MODEL		M3-1	M4-1	M4-2
		R&D	Create	Share
Knowledge management	Create			
	Share			
Business strategy	Explore	0.603***	0.383***	0.471 ***
	Analyze	0.141*	0.200***	0.208***
	Defend	−0.013	0.125*	0.040
adjR2		0.459	0.309	0.374
F		65.490	35.027	46.388
p		0.000	0.000	0.000

*< 95%; ** < 99%; *** < 99.9%

4.3 *Difference analysis*

Companies whose R&D expenditures comprise more than 5% of their total revenue are more likely to choose the prospector strategy and show significantly superior performance in product and R&D innovation. According to data provided by the Industrial Development Bureau, Ministry of Economic Affairs, average R&D expenditures as a percentage of total revenue within Taiwan's medical device industry has increased in recent years, from 4.5% to 4.9% to 5.6%. In comparison, this same metric is 6.8% in the American medical device industry, as reported by the Health Industry Manufacturers' Association. Although Taiwan's medical device industry has increased R&D expenditures, it remains excessively low. Because Taiwan's medical device companies are unable to gain an advantage in the worldwide market for advanced medical devices, they have adopted the defender or analyzer strategies. They have chosen to consolidate the strengths of existing products or to imitate and improve currently available products instead of aggressively developing new markets or products.

Study results show companies with a higher percentage of technical staff demonstrate superior R&D innovation performance. In addition, with sufficient workforce to develop new products, these companies tend to choose the more aggressive prospector strategy. However, technical staff as a percentage of total employees has no effect on knowledge management capabilities. A possible reason for this finding may be that currently, Taiwanese companies that have promoted knowledge management are primarily in industries that are more computerized (e.g., the semiconductor or software industries). However, knowledge management is still in the beginning stages in these industries, and is even less advanced in the emerging medical device industry.

5 CONCLUSION

Taiwan's medical device companies are mostly small and medium-sized enterprises featuring the advantages of flexible manufacturing and product improvement. The industry has seen favorable growth in annual industrial output. However, Taiwan's manufacturers have favored medical device products related to home healthcare, such as electric wheelchairs, electric scooters, blood pressure monitors, blood glucose monitors, thermometers, and contact lenses, and they mostly act as contract manufacturers. They face the disadvantages of not having mastered key technologies or materials, not having established brands, and engendering low demand in Taiwan. Because of these disadvantages, Taiwan's manufacturers have begun to relocate their manufacturing facilities overseas. This currently represents the greatest concern for the development of Taiwan's medical device industry. The global demographic structure is gradually changing and low birthrate and aging societies are a predictable trend. Taiwan's population growth has slowed, and Taiwan will face the challenge of negative population growth in the future. When abundant labor resources are no longer an advantage, new ideas and strategies will be required for the growth of an industry. Therefore, emphasizing R&D innovations and advancements in marketing to enhance the added value of the industry is a positive, viable, and important goal for the present stage.

Past studies of business strategy, knowledge management, and innovation performance have primarily targeted the electronics or information industries. Studies of other industries are lacking in comparison. The government has listed the medical device industry as one of the top 10 emerging industries in Taiwan and as a target of focused development in the 21st century. Thus, the medical device industry was chosen for the present study for an exploration of the correlations between business strategy choices, knowledge management capabilities, and innovation performance. Findings of the study relevant to the research hypotheses are discussed in greater detail in what follows.

(1) Product innovation performance: the prospector and analyzer strategies improve product innovation performance.
(2) Process innovation performance: none of the three strategies significantly improves process innovation performance.
(3) R&D innovation performance: the prospector business strategy effectively improves R&D innovation performance.

In summary, the following suggestions are provided for Taiwan's medical device companies:

(1) Increase investment in R&D expenditures and technical staff.
(2) Adopt the aggressive prospector business strategy and prioritize the development of innovations.
(3) Aggressively develop new products and diversification.
(4) Improve knowledge creation capabilities and construct a corporate knowledge management system using superior knowledge-sharing channels.
(5) Recruit high-tech talent and improve employee education (i.e., train employees in professional areas such as R&D, production, quality control, and international sales).
(6) Adopt a global business strategy.

Simultaneously, these findings should also be applied to R&D expenditures and technical staff. One characteristic of Taiwan's medical device industry is that the product development phase is long. This phase generally lasts from six to 12 years from initial R&D to manufacturing to receiving orders. Therefore, continual investment in R&D expenditures and technical staff is required to increase the added value of products. This establishes a higher barrier to entry and ensures continued high performance. ANOVA results show that higher ratios of R&D expenditures and technical staff produce superior knowledge management and innovation performance results. Thus, we suggest that companies continually invest in R&D and increase the ratio of R&D expenditures as a percentage of total revenue. We also suggest that companies emphasize the recruitment of technical staff and increase the ratio of technical staff. These actions can result in improved knowledge management and innovation performance. Because the prospector strategy is more likely to lead to higher product and R&D innovation performance, companies should consider how to continue developing and innovating products with improved functions. By developing new products and improving product quality, they can progress toward the aggressive prospector strategy and raise the barrier to entry for newcomers. They can also use their existing advantages to enter a new, related niche market and expand their existing competitive advantage. Continued investment in R&D and recruitment of excellent technical staff can ensure continued R&D performance for the business, leading to lasting improvement and growth.

REFERENCES

Akter, S., Wamba, S. F., Gunasekaran, A., Dubey, R., & Childe, S. J. 2016. How to improve firm performance using big data analytics capability and business strategy alignment. *International Journal of Production Economics*, 182, 113–131.

Aloini, D., Pellegrini, L., Lazzarotti, V., & Manzini, R. 2015. Technological strategy, open innovation and innovation performance: Evidences on the basis of a structural-equation-model approach. *Measuring Business Excellence*, 19(3), 22–41.

Andrews, K. R., & Roland, C. 1987. About strategy. We can't solve problems by using the same kind of thinking we used when we created them. 150.

Bea, F. X., & Haas, J. 2016. *Strategisches management*, 8498.

Berchicci, L. 2013. Towards an open R&D system: Internal R&D investment, external knowledge acquisition and innovative performance. *Research Policy*, 42(1), 117–127.

Birasnav, M. 2014. Knowledge management and organizational performance in the service industry: The role of transformational leadership beyond the effects of transactional leadership. *Journal of Business Research*, 67(8), 1622–1629.

Blackburn, R. A., Hart, M., & Wainwright, T. 2013. Small business performance: Business, strategy and owner-manager characteristics. *Journal of Small Business and Enterprise Development*, 20(1), 8–27.

Bloodgood, J. M. 2015. Acquiring external knowledge: How much overlap is best? *Knowledge and Process Management, 22*(3), 148–156.

Bryson, J. M., Edwards, L. H., & Van Slyke, D. M. 2017. Getting strategic about strategic planning research.

Burke, W. W. 2017. *Organization change: Theory and practice.* Thousand Oaks, CA: Sage.

Carlborg, P., Kindström, D., & Kowalkowski, C. 2014. The evolution of service innovation research: A critical review and synthesis. *Service Industries Journal, 34*(5), 373–398.

Chang, W. J., Liao, S. H., & Wu, T. T. 2017. Relationships among organizational culture, knowledge sharing, and innovation capability: A case of the automobile industry in Taiwan. *Knowledge Management Research & Practice,* 1–20.

Clarke, T. 2001. The knowledge economy. *Education+ Training, 43*(4/5), 189–196.

Cohen, J. F., & Olsen, K. 2015. Knowledge management capabilities and firm performance: A test of universalistic, contingency and complementarity perspectives. *Expert Systems with Applications, 42*(3), 1178–1188.

Cook, P. 1999. I heard it through the grapevine: Making knowledge management work by learning to share knowledge, skills and experience. *Industrial and Commercial Training, 31*(3), 101–105.

Davenport, T. H., & Prusak, L. 2001. *Conocimiento en acción como las organizaciones manejan lo que saben.* Upper Saddle River, NJ: Prentice Hall.

Demarest, M. 1997. Understanding knowledge management. *Long range planning, 30*(3), 321374–322384.

Dobni, C. B., Klassen, M., & Nelson, W. T. 2015. Innovation strategy in the US: Top executives offer their views. *Journal of Business Strategy, 36*(1), 3–13.

Drucker, P. F. 1993. *Post-capitalist society.* Abingdon: Routledge.

García-Sánchez, E., García-Morales, V. J., & Bolívar-Ramos, M. T. 2017. The influence of top management support for ICTs on organisational performance through knowledge acquisition, transfer, and utilisation. *Review of Managerial Science, 11*(1), 19–51.

Geisler, E., & Wickramasinghe, N. 2015. *Principles of knowledge management: Theory, practice, and cases.* Abingdon: Routledge.

Hidding, G. J., & Catterall, S. M. 1998. Anatomy of a learning organization: Turning knowledge into capital at Andersen Consulting. *Knowledge and Process Management, 5*(1), 3–13.

Hooge, S., & Le Du, L. 2014, June. Stimulating industrial ecosystems with sociotechnical imaginaries: The case of Renault Innovation Community. In *EURAM* (p. 24).

Huseman, R. C., & Goodman, J. P. 1999. *Leading with knowledge.* Thousand Oaks, CA: Sage.

Industrial Technology Research Institute. 2015. *2015 medical device industry yearbook.* Ministry of Economic Affairs, R.O.C.

Jansen, D. 2017. Networks, social capital, and knowledge production. In *Networked Governance* (pp. 15–42). New York: Springer International Publishing.

Khedhaouria, A., Gurău, C., & Torrès, O. 2015. Creativity, self-efficacy, and small-firm performance: The mediating role of entrepreneurial orientation. *Small Business Economics, 44*(3), 485–504.

Kim, T. H., Lee, J. N., Chun, J. U., & Benbasat, I. 2014. Understanding the effect of knowledge management strategies on knowledge management performance: A contingency perspective. *Information & Management, 51*(4), 398–416.

Lai, Y. L., Hsu, M. S., Lin, F. J., Chen, Y. M., & Lin, Y. H. 2014. The effects of industry cluster knowledge management on innovation performance. *Journal of Business Research, 67*(5), 734–739.

Lao, S. T., Prajogo, D. I., & Adebanjo, D. 2014. The relationships between a firm's strategy, resources and innovation performance: Resources-based view perspective. *Production Planning & Control, 25* (15), 1231–1246.

Lechner, C., & Gudmundsson, S. V. 2014. Entrepreneurial orientation, firm strategy and small firm performance. *International Small Business Journal, 32*(1), 36–60.

Leonard, D. 1995. *Wellsprings of knowledge.*

Lerro, A., & Jacobone, F. A. 2013. Technology districts (TDs) as driver of a knowledge-based development: Defining performance indicators assessing TDs' effectiveness and impact. *International Journal of Knowledge-Based Development 7, 4*(3), 274–296.

Miles, R. E., Snow, C. C., Meyer, A. D., & Coleman, H. J. 1978. Organizational strategy, structure, and process. *Academy of Management Review, 3*(3), 546–562.

Nonaka, I., & Takeuchi, H. 1995. *The knowledge-creating company: How Japanese companies create the dynamics of innovation.* New York: Oxford University Press.

Noruzy, A., Dalfard, V. M., Azhdari, B., Nazari-Shirkouhi, S., & Rezazadeh, A. 2013. Relations between transformational leadership, organizational learning, knowledge management, organizational innovation, and organizational performance: An empirical investigation of manufacturing firms. *International Journal of Advanced Manufacturing Technology,* 1–13.

Pais, L., & Santos, N. R. D. 2015. Knowledge-sharing, cooperation, and personal development. In *The Wiley Blackwell handbook of the psychology of training, development, and performance improvement* (pp. 278–302).

Porter, M. E. 2008. *Competitive strategy: Techniques for analyzing industries and competitors.* New York: Simon & Schuster.

Prajogo, D. I. 2016. The strategic fit between innovation strategies and business environment in delivering business performance. *International Journal of Production Economics, 171,* 241–249.

Quintas, P., Lefere, P., & Jones, G. 1997. Knowledge management: A strategic agenda. *Long range planning, 30*(3), 322385–322391.

Richard, H. C., & Goodman, P. J. 1999. *Leading with knowledge: The nature of competition in the 21st century.*

Samiee, S., & Lee, R. P. 2014. The influence of organic organizational cultures, market responsiveness, and product strategy on firm performance in an emerging market. *Academy of Marketing Science. Journal, 42*(1), 49.

Slater, S. F., Mohr, J. J., & Sengupta, S. 2014. Radical product innovation capability: Literature review, synthesis, and illustrative research propositions. *Journal of Product Innovation Management, 31*(3), 552–566.

Sultani, A. R. 2016. Integrating knowledge management (KM) strategies and processes to enhance organizational creativity and performance. *Journal of Modelling in Management, 11*(1), 154–179.

Ukko, J., Saunila, M., Parjanen, S., Rantala, T., Salminen, J., Pekkola, S., & Mäkimattila, M. 2016. Effectiveness of innovation capability development methods. *Innovation, 18*(4), 513–535.

Utterback, J. 1994. *Mastering the dynamics of innovation: How companies can seize opportunities in the face of technological change.*

Venkitachalam, K., & Willmott, H. 2015. Factors shaping organizational dynamics in strategic knowledge management. *Knowledge Management Research & Practice, 13*(3), 344–359.

Wang, M. 2015. An analysis model for knowledge-based industry development from a case study of China. *International Business and Management, 10*(3), 100–115.

Innovation in Design, Communication and Engineering – Lam et al. (eds)
© 2020 Taylor & Francis Group, London, ISBN 978-0-367-17777-5

The analysis of financial poverty alleviation funds performance in China

Zengjun An
College of Business Administration, Fujian Jiangxia University, China

Qixue Gao
School of Economic Management, Fuzhou University, China

Shih-Ming Ou*
Department of Logistics Management, College of Business Administration, Fujian Jiangxia University, China

ABSTRACT: Basically eliminating poverty is the important strategical target required for comprehensively constructing a moderately prosperous society in China, and the distribution and management of governmental poverty alleviation funds in the poverty alleviation development engineering have played an obvious role in promoting the income increase of poor farmers, increasing the agricultural total value of output and decreasing the poverty rate. Through the literature review on the management of financial funds, as well as carrying out the quantitative analysis to the performance of financial poverty alleviation funds, this paper raises the recommendations and advice that in order to enhance the use efficiency of power alleviation funds. Our finding includes that to continue strengthening the input of poverty alleviation funds, optimize the funds input structure according to the power alleviation performance, improve the management system of financial poverty alleviation funds, as well as fully exert the guiding function of power alleviation funds is necessary.

1 INTRODUCTION

Basically, eliminating poverty is the important sign of comprehensively construction a moderately prosperous society, which is also the necessary requirement to realize the great rejuvenation of the Chinese nation "Chinese Dream". After the continuous striving of several generations, the power alleviation development work in China has made the great achievements, over 600 million people have been alleviated the poverty successfully, and the poverty rate has been decreased from 10.2% in 2000 to 2.8% in 2010. As the "money for saving life" of poor masses and the "booster" for reducing and alleviating the poverty, the national financial poverty alleviation funds have played an important role in improving the production and living conditions in the poor areas, enhancing the life quality and comprehensive qualification of poor population, as well as supporting to develop economy and social industries of poor areas. In order to complete the poverty alleviation task of poor population in 2020 in a timely manner, it is necessary to further reasonably plan the use and management of financial special poverty alleviation funds, and improve the use efficiency of funds. Therefore, it is necessary to carry out the study to the use and development status of current financial poverty alleviation funds so as to provide the advice of reference significance for conducting the poverty alleviation in the next stage, enhancing the funds benefit and improving the anti-poverty measures.

*Corresponding author: benao13@gmail.com

2 LITERATURE REVIEW OF USE AND MANAGEMENT OF POVERTY ALLEVIATION FUNDS

Since the foundation of the People's Republic of China, we have been always focusing on developing the rural production and eliminating poverty. However, the rural poverty alleviation work has started from the reform and opening-up and implemented in a large scale due to the limited financial power of government at the beginning of foundation. And the use of financial poverty alleviation funds was initiated from then and gradually regulated. According to division of rural poverty alleviation work stage of our government, the literature review for the use and management of financial poverty alleviation funds can be divided into five stages.

The first stage: Initial poverty alleviation study stage (1949-1978). Due to the slow development of productivity in the early days of new China, the governmental financial income was low and the broad masses of people were in the common poverty status, the focus of Chinese government anti-poverty was concentrating on promoting the development of collective economy and enhancing the domestic grain output, but without issuing the relevant management regulations on the poverty alleviation funds. Therefore, the scholars have made less achievements in the study of use and management of poverty alleviation funds.

The second stage: Large-scaled development poverty alleviation study stage (1979-1993). During this stage, China has gradually increased the input of poverty alleviation funds, and some problems have been exposed on the use and management of poverty alleviation funds, such as the funds have been issued by several institutes and the use project were scattered [1], the investment selection had blindness and error [2], the poverty alleviation funds had distribution without management, had responsibility without undertaking [3], the poverty alleviation funds were diverted and embezzle [4], the funds spending had one-sidedness [5] etc. Therefore, the Ministry of Finance issued the Interim Procedures for Supporting Underdeveloped Areas to Develop Funds Management in 1983, which clearly regulated that all levels of management institutes of developing funds and relevant departments should reinforce the monitoring and inspection to the use of developing funds; the use departments for developing funds should compile the accounting statement and annual settlement in a timely manner, as well as report to the management institutes of developing funds and financial departments after review and approval.

The third stage: Poverty alleviation assault fortified positions study stage (1994-2000). The key work in this period focused on the six large areas of bad natural conditions, insufficient public service, low population qualification, short resources and poor capital forming capacity. Therefore, the scholars have reinforced the power alleviation funds study for these six large areas. Chen and Hao (1996) raised that China should carry out the financial system reform, implement the standardization in the system of tax distribution and financial transfer payment system according to the problems that one rate for one province, uneven distribution occurred in the use of poverty alleviation funds of Puan Country in Guizhou Province [6]. Wu (1997) believed that the use of poverty alleviation funds in Shanxi Province had the problems that the funds investment was not put into place properly, the funds were paid in a delayed manner, he raised the advice that it was required to strengthen the project management of poverty alleviation loads, optimize the funds delivery system from the regulations, as well as reinforce the inclination of poverty alleviation funds to the poor population etc [7]. Pang and Chen (2000) applied C-D production function to carry out the empirical analysis to the agricultural poverty alleviation funds of "Sanxi" area, as well as raised to put the science and technology into the poverty alleviation, improve the mechanical level of agricultural production so as to enhance benefit of financial poverty alleviation funds [8].

The fourth stage: New times poverty alleviation study stage (2011-now). After accessing to the new times of 21st century for comprehensively constructing a moderately prosperous society, the investment of financial poverty alleviation funds has been stably increased along with the continuously extended coverage and continuously expanded source of poverty alleviation funds. Meanwhile, the use and management of financial poverty alleviation has shown the features of variety and complication. According to the different aspects, the scholars had the following opinions: In terms of the funds investment, the Ministry of Finance Agriculture

275

Department Poverty Alleviation Subject Team believed that it is necessary to balance the fairness and efficiency, reduce the unfair degree of funds investment, and decrease the leakage of poverty alleviation resource [9]. As for the subject of poverty, Chen (2007) thought that it is necessary to develop the subject construction, strengthen the capacity modification of poverty subject, as well as enhance the participation of rural community and poor rural households [10]. As for the poverty alleviation system, Jiang (2008) believed that it is necessary to position the relationship between government and market, introduce the market system in the poverty alleviation work [11]. With the respect of funds performance evaluation, Wu (2011) thought it is necessary to further improve the performance evaluation system, improve the variety degree of evaluation subject, as well as encourage the non-governmental institutes to participate in the evaluation [12].

Generally speaking, the studies in the financial poverty alleviation funds in China are helpful to solve the problems encountered during the use and management of poverty alleviation funds e.g. reasonably use the financial funds, strengthen the social funds collaboration and improve the poverty alleviation development benefit etc., promoting the development and issuance of relevant management regulations of national financial poverty alleviation funds, as well as assisting to smoothly realize the goals of targeted poverty alleviation and targeted poverty overcoming. However, comparing with the study achievement of poverty alleviation in China, the empirical analysis studies on the financial poverty alleviation funds performance are less, most of the studies focus on the theoretical analysis of poverty alleviation and less on analyzing the performance of poverty alleviation funds use through the actual data. Therefore, this paper will mainly focus on the use efficiency of poverty alleviation funds, through analyzing the input of poverty alleviation funds from 1998 to 2010, as well as understanding the source and composition of poverty alleviation funds, it will conclude the political advice that is significant to the use and management of poverty alleviation funds.

3 SOURCE AND COMPOSITION OF FINANCIAL POVERTY ALLEVIATION FUNDS

The financial special poverty alleviation funds are the special funds of national budget arrangement used for supporting each poor area, expediting the economical and social development, improving the basic production and life conditions of poverty alleviation subject, as well as promoting to eliminate the rural poverty [13]. The source of poverty alleviation funds of each poor area mainly includes the financial poverty alleviation development funds arranged by the Central Government, poverty alleviation subsidized loan funds, work-relief funds, grain for green allowance funds etc., as well as each provincial financial accessory poverty alleviation funds, the funds from foreign governments and institutes, and the other social supporting funds.

Central Government funds for poverty alleviation indicates the funds of Central Government financial special arrangement used for poverty alleviation arranged to the country for use in the same year, which includes agricultural construction special allowance funds. The financial poverty alleviation funds are mainly used for the poor countries to develop the production, infrastructure, science promotion and training, as well as supporting the rural education, medical health, cultural industry etc. [14]. The poverty alleviation subsidized loan funds apply the form of compensated use funds. Central Government poverty alleviation subsidized loan is mainly issued by the Agricultural Bank of China to use for supporting to develop the planting and breeding industry for the poor population with low income of poverty alleviation key countries, labor intensive enterprises, agricultural product processing enterprises and market circulation enterprises, as well as partial infrastructure project. The work-relief funds indicate the Central Government financial investment infrastructure engineering, the people who receives relief take part in the engineering construction to obtain the remuneration, which is mainly used for improving the production and living conditions and ecological environment of poor areas [15]. The Central Government special grain for green engineering allowance funds mainly include three parts: the grain subsidy, seeding cost subsidy and

Table 1. National poverty alleviation key country poverty alleviation investment composition and proportion.

Unit: %, 100 million RMB

Index	2006	2007	2008	2009	2010
1. Input proportion of poverty alleviation	100	100	100	100	100
1.1 Central Government poverty alleviation subsidized loans accumulative amount	20.0	22.3	22.8	23.8	19.2
1.2 Central Government poverty alleviation funds	19.4	19.0	21.3	21.8	19.8
1.3 Work-relief funds	13.8	11.2	10.7	8.6	6.7
1.4 Specialized grain for green engineering subsidy	16.6	20.0	14.0	14.1	8.6
1.5 Minimum living standard funds issued by the Central Government	—	—	—	—	15.0
1.6 Provincial arranged poverty alleviation funds	3.9	4.5	5.1	5.1	4.2
1.7 Foreign funds (actual investment amount)	11.1	6.0	3.8	4.7	3.3
1.8 Other funds	15.3	17.1	22.1	21.9	23.3
2. Total amount of poverty alleviation	278.30	316.70	367.70	456.70	606.20

management cost subsidy for the farmer households who have returned grain for green. The provincial arranged poverty alleviation funds indicate all poverty funds covered in the provincial financial budget and specialized arrangement for supporting the key countries and poor villages. Other funds are the funds used for the poverty alleviation development projects except the above-mentioned poverty alleviation funds, e.g. the enterprises establish the enterprises' investment amount in the form of joint venture in the poor countries, the science and technology input of research and development institution to the poor countries and various donations etc.

According to the statistic data reported by 592 poverty alleviation key countries, the funds relating to the poverty alleviation acquired by the poverty alleviation key countries was 60.62 billion yuan in 2010. In which, the Central Government poverty alleviation funds were 1.199 billion yuan, the work-relief funds were 4.04 billion yuan, the Central Government poverty alleviation subsidized loans were accumulated to issue 11.61 billion yuan, the specialized grain for green engineering subsidy issued was 5.21 billion yuan, the minimum living standard funds issued by the Central Government were 9.11 billion yuan, the provincial arranged poverty alleviation funds were 2.54 billion yuan, the foreign funds were 2.01 billion yuan and the other poverty alleviation funds were 14.1 billion yuan. In can be seem from the Table 1 that the Central Government poverty alleviation funds investment takes up 69.1%, and it is the strong force of funds investment of poverty alleviation for key countries. The local government accessory funds and foreign funds take up 4.2% and 3.3% respectively, as well as the other funds e.g. counterpart support, non-governmental organization funds and social public donation etc. take up 23.3%. The increase of other poverty alleviation funds have shown that more and more strength have been added to the team of poverty alleviation development work, which makes the source of poverty alleviation funds to become more various, and the funds investment structure to become ore rationalized.

4 EMPIRICAL ANALYSIS FOR FINANCIAL POVERTY ALLEVIATION FUNDS PERFORMANCE

Due to the different investment of financial poverty alleviation, the contribution rates to the economy of poor countries are also different. In order to further quantitatively analyze the effect of each poverty alleviation fund to the economic growth of poor countries, carry out the empirical analysis to the agricultural total value of output, occurrence rate of farmer poverty and the contribution of rural per capita net income with the different investment input of poor farmers and government in 198-2010, so as to conclude some experience and enlightenment.

4.1 Model setup

The model setup applies C-D production function form, which takes the logarithm to production function, and carries out OLS linear regression model analysis to the contribution rates of each financial fund investment to the economic growth with STATA system software. Meanwhile, mainly categorize the financial poverty alleviation expense into three types including the government poverty alleviation subsidized loan funds, government financial poverty alleviation funds and work-relief funds.

Theory model 1: The multivariate regression analysis between the different input forms of government to the rural poverty alleviation funds and agricultural total value of output, in which y1 is the rural total value of output, x1 is the financial poverty alleviation funds, x2 is the subsidized loan funds, x3 is the work-relief funds.

Agricultural total value of output=F (government poverty alleviation subsidized loan funds, government financial poverty alleviation funds, work-relief funds etc.)

$$\ln(y_1) = \beta_0 + \beta_1 \ln(x_1) + \beta_2 \ln(x_2) + \beta_3 \ln(x_3) + \varepsilon \qquad (1)$$

Theory model 2: The multivariate regression analysis between the different input forms of government to the rural poverty alleviation funds and the occurrence rate of rural poverty, in which, y2 is the occurrence rate of rural poverty, x4 is the financial poverty alleviation funds, x5 is the subsidized loan funds and x6 is the work-relief funds.

Occurrence rate of rural poverty=F (government poverty alleviation subsidized loan funds, government financial poverty alleviation funds, rural per capita net income, work-relief funds etc.)

$$\ln(y_2) = \beta_0 + \beta_1 \ln(x_4) + \beta_2 \ln(x_5) + \beta_3 \ln(x_6) + \varepsilon \qquad (2)$$

Theory mode 3: The multivariate regression analysis between the different investment forms of government to the rural poverty alleviation and the rural per capita net income, in which, y3 is the rural per capita net income, x7 is the per government subsidized loan, x8 is the per government financial poverty alleviation funds, x9 is the per work-relief funds.

Rural per capita net income=F (rural per government subsidized loan funds, rural per government financial poverty alleviation funds, rural per work-relief funds etc.).

$$\ln(y_3) = \beta_0 + \beta_1 \ln(x_7) + \beta_2 \ln(x_8) + \beta_3 \ln(x_9) + \varepsilon \qquad (3)$$

4.2 Result

The data used in this paper is from 1998-2011 Chinese Rural Poverty Monitoring Report and Chinese Statistics Annual. Due to the problem of cross variables are directly easy to have the multi-collinearity, take each variable for the logarithmetic treatment, the variable after change can reduce the possibility of serial correlation, which can better reflect the relationship of variable percentage change. Due to the subsidized loan funds and the work-relief funds are the investment for the specific projects, which are difficult to obtain the apparent effect in the same year, thus postpone these two variables for one year for observation. For example, 1n (financial poverty alleviation funds)(-1) means the subsidized loan funds for the postponed one year. (See Table 2)

It can be seen from Table 2 that the adjustment R2 are all large, which indicates the model fitting effect is good. The coefficient of explanatory variable are positive, which indicates to promote the improvement of agricultural total value of output. In terms of the government

Table 2. Result 1.

Explanatory variable	Explained variable 1n (Agricultural total value of output)		
	Coefficient	Standard deviation	t statistics
Government poverty alleviation expense			
ln (poverty alleviation subsidized loan) (-1)	-0.15807	0.086071	-1.84
ln(financial poverty alleviation funds)	0.633635	0.045331	13.98
ln (work-relief)(-1)	-0.35863	0.279928	-1.28
Constant term	9.455924	1.14804	8.24
Adjusted R^2	0.9541		

poverty alleviation expense, it can be seen from the classification that the poverty alleviation development funds promote the increase of agricultural total value of output, under 95% confidence coefficient every 1% increase of poverty alleviation development funds can significantly promote the agricultural total value of output by 63.36%, and the work-relieve and subsidized loan all can't reach the significance after postponing one year, which also doesn't exert the promoting function to the increase of agricultural total value of output. It may relate to the year-by-year decrease in the number of work-relief and the weaker and weaker color of "relief" of work-relief projects, as well as focusing on the construction of infrastructure. Therefore, it will have a negative effect to the increase of agricultural total value of output. Bo and Li (2013) believed that the work-relief is acquired through the form of project application, but the work-relief projects have become less and the funds input is insufficient due to the shortage of talents and the limited capacity of project application in the poor countries, thus the work-relief investment method has not achieved the due results [16]. Li et al. (2006) believed that the use of poverty alleviation subsidized loans have seriously deviated from the target, the poverty alleviation loan resource flows to the general enterprises rather than the extremely poor villages or the poor-oriented enterprises [17].

It can be seen from Table 3 that the adjustment R2 are all large, which indicates the model fitting effect is good. The coefficient is negative, which indicates the explanatory variable promotes the decrease of poverty rate. In terms of the government poverty alleviation expense, it can be seen from the classification that the poverty alleviation development funds have promoted the decrease the occurrence of poverty rate, under 95% confidence coefficient every 1% increase of poverty alleviation development funds can significantly promote the decrease of poverty rate by 93.33%, the effect is very significant, but the poverty alleviation subsidized loans and work-relief don't play the significant role in reducing the poverty rate. The reason for the increase of subsidized loans fail to reduce the poverty rate may be the feature of bank avoiding the risks, so that which cause the crowding-out effect to the poor population, it makes the subsidized loan funds to generate the targeted deviation and causes the reduction of poverty alleviation effect. The reason for the subsidized loans fail to reach the significant effect may be the "uncompleted" engineering of some infrastructure construction. The safety

Table 3. Result 2.

Explanatory variable	Explained variable 1n (Occurrence rate of poverty)		
	Coefficient	Standard deviation	t statistics
Government poverty alleviation expense			
ln (poverty alleviation subsidized loan) (-1)	0.1805753	0.121017	1.49
lln(financial poverty alleviation funds)	-.9332945	0.075337	-12.39
ln (work-relief)(-1)	0.2698961	0.407735	0.66
Constant term	3.743343	1.816146	2.06
Adjustment R^2	0.9554		

Table 4.   Result 3.

Explanatory variable	Explained variable 1n (rural per capita net income)		
	Coefficient	Standard deviation	t statistics
Government poverty alleviation expense			
ln (per poverty alleviation subsidized loan) (-1)	-.214905	0.054026	-3.98
ln(per financial poverty alleviation funds)	0.753993	0.037898	19.90
ln (per work-relief) (-1)	-.127649	0.168341	-0.76
Constant term	5.940521	0.718698	8.27
Adjustment R^2	0.9872		

drinking engineering in some villages of Xiji Country of Nixia Province failed to implement the planning and design according to the actual design of farmers, which caused some engineering uncompleted that can't exert the poverty alleviation effect of engineering [18].

Referring to Table 3 and 4, we can find that the adjustment R2 are all large, which indicates the model fitting effect is good. The coefficient of explanatory variable is positive, which indicates to promote the increase of rural per capita net income. In terms of the government poverty alleviation, it can be seen from the classification that the per poverty alleviation development funds have promoted the increase of rural per capita net income, under 95% confidence coefficient every 1% increase of per poverty alleviation development funds can significantly promote the increase of rural per capita net income by 75.40%. t statistics absolute value of one-year postponed per poverty alleviation subsidized loans is more than 2, which reaches the significance at statistics. But its coefficient is negative, which doesn't play promoting role in the rural per capita net income, which may relate to the poor investment of subsidized loans and failing to target the poor farmers. Therefore, it has the reverse effect on the rural per capita net income. But the work-relief (t=0.76) doesn't access to the regression model, which reveals the use of such poverty alleviation funds doesn't have significant effect on increasing the net income of poor population, that is, the performance is poor.

5   CONCLUSION AND POLITICAL RECOMMENDATIONS

According to the above-mentioned empirical results, it can be concluded the followings: (1) The government poverty alleviation funds are different, and the investment form performance is different. The government poverty alleviation funds' investment form performance is significant, the subsidized loan poverty alleviation funds and the work-relief funds have no significant effect on other indexes; (2) The financial poverty alleviation funds' investment form performance is the highest. Through the performance generated by the different investment of government financial poverty alleviation funds to the increase of agricultural total value of output, increase of rural per capita net income and the reduction of occurrence rate of poverty, it can be seen that the government financial poverty alleviation funds' investment form performance is the highest, it has very significant effect to the above three indexes, which indicates that the investment of government poverty alleviation funds has positive effect to the poverty alleviation development of poor countries. Basing on the conclusion of this paper, the following policies are raised according to how to improve the efficiency of poverty alleviation funds in China:

First of all, continue to strength the investment of poverty alleviation funds. Since the reform and opening-up, the poverty alleviation in China has made the great achievements, in which the investment of poverty alleviation funds has make great contribution to the economic growth of poor areas and the reduction of poverty. However, until the end of 2014 there were still over 70 million rural poor population in China [19]. Along with the approaching of timeline of comprehensively constructing a moderately prosperous society, China is still facing with the arduous poverty alleviation development task, and the government should continue reinforcing the investment of poverty alleviation funds to the poor areas. It can be seen from the historical experience that only with the strong support and positive guidance of government, as well as the joint investment of all parties, the poor areas can alleviate poverty more effectively and access to a moderately prosperous society, as well as realize the target of comprehensive poverty alleviation of rural poor population in 2020.

Secondly, optimize the investment structure of funds according to the poverty alleviation performance. Optimize the investment structure of government poverty alleviation funds and enhance the use efficiency of government poverty alleviation funds. It can be seen from the analysis that the financial poverty alleviation funds have played significant role in the poverty alleviation development work. The financial poverty alleviation funds not only promote the economic growth of poor countries, create the material wealth and spiritual wealth for the poor population, but also increase the rural per capita net income. Therefore, government should pay full attention and strength the investment of financial poverty alleviation funds. In addition, government should reinforce the supervision and management to the poverty alleviation subsidized loan funds and work-relief funds. It can be seen from Table 1 that the proportion of subsidized loan funds and financial poverty alleviation funds is large, but the poverty alleviation effect is not significant. It is shown by the empirical study that the subsidized loan funds and work-relief funds have not achieved the required effect on the poverty alleviation development work, which has certain difference from the expected target. Therefore, according to the poverty alleviation performance, government should perform reform to the subsidized loan funds, pragmatically implement the poverty alleviation subsidized loan polity. It is able to increase the issuance of micro-credit and combined credit to enable the poor farmers to enjoy the preference of financial subsidized interest. As for the work-relief funds, due to its poverty alleviation performance is poor, government should re-plan and arrange the projects that are more appropriate to the poor population in a scientific manner according to the projects of unreasonable checking of poor masses, meeting the requirement of infrastructure construction of farmers, further increase the rural income and promote the poverty alleviation of poor population.

Thirdly, improve the management supervision system of financial poverty alleviation funds. Optimize the supervision system of poverty alleviation funds, pragmatically implement the financial special poverty alleviation funds management regulation, include the monitoring supervision of financial poverty alleviation funds to the normal financial supervision system. Introduce the social institutions to participate in the supervision and motoring. Establish the regular reporting system for the use status of financial poverty alleviation funds, identify and solve any problems in a timely manner. During the process of distributing and using the poverty alleviation, it is required to realize the special funds for the special purpose, avoid the phenomena of "take advantage of every opportunity to benefit themselves". For example, there is one national-level poverty-stricken county in Dabieshan Area has acquired 240 million yuan poverty alleviation funds (including the financial subsidized loan funds, financial supporting funds and work-relief funds etc.) within ten years, in which over 60% funds was used for balancing the financial budget of rural government, and the remaining funds were not used for the special purposes [20]. As for the above-mentioned issue, the government should establish the strict financial poverty alleviation funds distribution system, reinforce the supervision to the distribution and use of poverty alleviation funds, reduce the distribution turnover of funds, reduce the "loss on the road" of funds, and strive for realizing the "One-step" of distributed financial poverty alleviation funds, as well as improve the use efficiency of funds, reduce the leakage rate of funds, realize the special funds for the special purpose for the financial poverty alleviation.

Fourthly, fully exert the guiding function of poverty alleviation funds. In the poverty alleviation development led by the government, the appearance of poverty-stricken areas have had great changes. Under the strength of continuously increased government poverty alleviation funds subsidy, more and more social power has also participated in the poverty alleviation development work, including the local government poverty alleviation accessory funds, donation funds from the foreign government and institutions, as well as the social supporting funds e.g. Hope Project, Happiness Project etc. Generally speaking, the accessory funds of local government take up about 30% of the national poverty alleviation funds, the applicable foreign funds mainly include the loan without interest and low interest from the international banks e.g. the World Bank etc., donation from the Chinese and overseas Chinese, the donation from the foreign institutions etc. Government should encourage and inspire the activity of rural poverty alleviation, increase the input of self-collected production, fully utilize the poverty alleviation expense of government and non-governmental institutions, as well as guide the various parties of society to widely participate and provide more political and funding support for the development of rural enterprises so as to expend the employment opportunity of poor population, promote the increase of per capita net income. Through the investment of government poverty alleviation funds and adjustment of poverty alleviation investment structure, as well as encouraging the social power to jointly participate, it is able to finally realize the strategical target of eliminating poverty.

REFERENCES

Y. Wang, J. Phys, 2009.*Conf. Ser*. 152 012023.

S.K.M. Jo¨nsson, J. Birgerson, X. Crispin, G. Greczynski, W. Osikowicz, A.W. Denier van der Gon, W. R. Salaneck, M. Fahlman, 2003.*Synyh. Met*. 139 1–10.

Hu Yan, Hidenori Okuzaki, 2009.*Synyh. Met*.159 2225–2228.

J.Y. Kim, J.H. Jung, D.E. Lee, J. Joo, 2002. *Synyh. Met*.126 311–316.

Sungeun Park, Sung Ju Tark, Donghwan Kim, Curr., 2011.Appl. Phys. 11 1299–1301.

S. Timpanaro, M. Kemerink, F.J. Touwslager, M.M. De Kok, S. Schrader, Chem, 2004. *Phys. Lett*. 394 339–343.

Yijie Xia, Kuan Sun, and Jianyong Ouyang,2012. *Adv. Mater*. 24 2436–2440.

S. W. Jian, G. X. Wei, 1983 *Rural Fin. Stu*. 9 1–3.

Y. Q. Wang, 1986 *Agr. Econ*. 12 40–41.

F. L. Du,1987 *Rural Fin. Stu*. 4 43–44.

Z. R. Xiong, X.X. Lin, 1991 *Fujiang Forum* 5 36–37.

Q.H. Yan, W. Jing, D. Li, 1993 *Econ. Issues* 11 35–37.

Y.G. Chen, C.H. Hao, 1996 *Stat. Stu* 6 39–45.

B.Y. Wu, 1997 *Econ. Issues* 4 49–52.

S.L. Pang, B.F. Chen, 2000 *Agr. Tech. Econ*. 2 20–23.

Ministry of Finance Agriculture Department Poverty Alleviation Team, 2004 *Econ. Stu. Ref*. 69.

J. Chen, 2007 *Z N U* 138–147.

A.H. Jiang, 2008 *Central U. Fin. Econ. Acad. J*. 2 13–18.

G.Q. Wu, 2011 *Ministry Fin. Sci. Stu. Institute* 120–121.

http://www.gov.cn/gzdt/ 192011-11/29/content_2006260.htm.

National Bureau of Statistics Rural Social and Economical Survey Division 2010 *China Stat. Press* 344–355.

H.Y. Chen, 2013 *Sci. Press* 44–45.

Z.Z. Bo, C.Y. Li, 2013 *Nationality Forum* 89–93.

X.Y. Li, X.M. Zhang, L.X. Tang, 2006 *Soc. Sci. (Academic Press)* 351–352.

F.W. Li, Y.F. Ma, W. Tian 2016 *Theory Trend* 15–19.

Z.J. An, Q.X. Gao, 2016 *Henan U. Tech. Academic J*. 50–55.

B. Xin, S.L. Yu, 2009 *Shandong Business Sch. Academic J*. 2 68–93.

Innovation in Design, Communication and Engineering – Lam et al. (eds)
© 2020 Taylor & Francis Group, London, ISBN 978-0-367-17777-5

Author index

Smart Science, Design and Technology

The main goal of this series is to publish research papers in the application of "Smart Science, Design & Technology". The ultimate aim is to discover new scientific knowledge relevant to IT-based intelligent mechatronic systems, engineering and design innovations. We would like to invite investigators who are interested in mechatronics and information technology to contribute their original research articles to these books.

Mechatronic and information technology, in their broadest sense, are both academic and practical engineering fields that involve mechanical, electrical and computer engineering through the use of scientific principles and information technology. Technological innovation includes IT-based intelligent mechanical systems, mechanics and systems design, which implant intelligence to machine systems, giving rise to the new areas of machine learning and artificial intelligence.

ISSN: 2640-5504
eISSN: 2640-5512

1. Engineering Innovation and Design: Proceedings of the 7th International Conference on Innovation, Communication and Engineering (ICICE 2018), November 9-14, 2018, Hangzhou, China

   Edited by Artde Donald Kin-Tak Lam, Stephen D. Prior, Siu-Tsen Shen, Sheng-Joue Young & Liang-Wen Ji

   ISBN: 978-0-367-02959-3 (Hbk + multimedia device)
   ISBN: 978-0-429-01977-7 (eBook)
   DOI: https://doi.org/10.1201/9780429019777

2. Smart Science, Design & Technology: Proceedings of the 5th International Conference on Applied System Innovation (ICASI 2019), April 12-18, 2019, Fukuoka, Japan

   Edited by Artde Donald Kin-Tak Lam, Stephen D. Prior, Siu-Tsen Shen, Sheng-Joue Young & Liang-Wen Ji

   ISBN: 978-0-367-17867-3 (Hbk)
   ISBN: 978-0-429-05812-7 (eBook)
   DOI: https://doi.org/10.1201/9780429058127

9 780367 537982